Alliant International University
Los Angeles Campus Library
1000 South Fremont Ave., Unit 5
Alhambra, CA 91803

The Science of the Mind

Series Editor
Tetsuro Matsuzawa
Inuyama, Japan

For further volumes:
http://www.springer.com/series/10149

Shigeru Watanabe · Stan Kuczaj
Editors

Emotions of Animals and Humans

Comparative Perspectives

Editors
Shigeru Watanabe
Department of Psychology
Keio University
2-15-45 Mita, Minato-ku
Tokyo 108-8345
Japan

Stan Kuczaj
Department of Psychology
The University of Southern Mississippi
118 College Dr. #5025
Hattiesburg, MS 39406
USA

ISSN 2192-6646
ISBN 978-4-431-54122-6
DOI 10.1007/978-4-431-54123-3
Springer Tokyo Heidelberg New York Dordrecht London

ISSN 2192-6654 (electronic)
ISBN 978-4-431-54123-3 (eBook)

Library of Congress Control Number: 2012946620

© Springer Japan 2013
This work is subject to copyright. All rights are reserved by the Publisher, whether the whole or part of the material is concerned, specifically the rights of translation, reprinting, reuse of illustrations, recitation, broadcasting, reproduction on microfilms or in any other physical way, and transmission or information storage and retrieval, electronic adaptation, computer software, or by similar or dissimilar methodology now known or hereafter developed. Exempted from this legal reservation are brief excerpts in connection with reviews or scholarly analysis or material supplied specifically for the purpose of being entered and executed on a computer system, for exclusive use by the purchaser of the work. Duplication of this publication or parts thereof is permitted only under the provisions of the Copyright Law of the Publisher's location, in its current version, and permission for use must always be obtained from Springer. Permissions for use may be obtained through RightsLink at the Copyright Clearance Center. Violations are liable to prosecution under the respective Copyright Law.
The use of general descriptive names, registered names, trademarks, service marks, etc. in this publication does not imply, even in the absence of a specific statement, that such names are exempt from the relevant protective laws and regulations and therefore free for general use.
While the advice and information in this book are believed to be true and accurate at the date of publication, neither the authors nor the editors nor the publisher can accept any legal responsibility for any errors or omissions that may be made. The publisher makes no warranty, express or implied, with respect to the material contained herein.

Printed on acid-free paper

Springer is part of Springer Science+Business Media (www.springer.com)

Preface

I occasionally take a course removed from my area of expertise to broaden my perspective. In addition, I have found that researchers from different areas often approach the same topics different ways. For example, I attended a philosophy course taught by Dr. Ogawa. Part of the class focused Nussbaum's book "Upheavals of thought: The intelligence of emotion" (2001). Although emotion is a traditional topic in psychology and neuroscience, areas such as economics and politics also focus on emotion (see popular books like "Economia emotive" by Motteriloni 2006 and translated into Japanese in 2008). Dr. Ogawa's course convinced me that it was important to consider ideas about emotions from as many disciplines as possible. As a result, Stan Kuczaj and I organized an interdisciplinary symposium titled "Emotional animals, sensitive humans" in 2009 at Keio University in Tokyo. The chapters in this book are based on the talks given at that symposium, talks which were both lively and stimulating.

The term "emotion" is derived from the French word "emouvoir, which in turn evolved from the Latin word "emovere". Emovere has two elements, "e" and "movere". The former means "out" and the latter "move". This is relevant because the empirical research of emotion divides emotion into internal emotion (such as feeling) and external emotion (expression of the emotion). Despite its longevity in human language, emotion was introduced into the academic study of mind rather recently. Early treatments of emotion tended to refer to "passion" instead of emotion. For example, Descartes wrote a book *"Les passions de l'ame"*. It is only recently that the study of emotion replaced considerations of passion in philosophy and psychology. Hume who introduced the new word "emotion" but the words were not well defined and there were a lot of confusion in usage of the passion and emotion in seventeenth and eighteenth centuries. Gradually, the passion disappeared at least in field of psychology.

According to Nussbaum, one classic view of emotion is that of the Greek and Roman stoics, where emotion is considered a cognitive/appraisal or evaluation system. James (1884) and Lange (1885) proposed another also well-known view of emotion called the "James–Lange" hypothesis. The emphasized the role of autonomic feedback in producing the experience of emotion. Cannon (1927) criticized this idea, noting that many different emotions seemed to be "caused" by the same visceral change. Damasio (1994) developed a more sophisticated version of the James–Lange hypothesis called the "somatic-marker" hypothesis in which the bodily change elicited by an event leads to a bodily feeling that contributes to decision making of the organism. Now we know that the neural activity in the subcortical system (periaquaduct grey, amygdala, etc.) evoked by the visceral change can be modified by cortical system (ventromedial part of prefrontalis and anterior cingulated cortex etc.). So it is not plausible to assume different emotional feelings evoked by the same visceral change in different contexts. Psychologists and neuroscientists sometimes separate emotion, which is the unconscious physical response to event, and the feeling produced by the emotional response. In this view, the feeling is an explanation of the emotional bodily response. The cognitive/appraisal hypothesis does not separate the two processes. Rolls (1999) proposed a more operational definition of emotion. According to him, emotions are states elicited by rewards or punishers. Such states have several functions, namely elicitation of autonomic and endocrine response to prepare for appropriate response, communicative and social functions of emotional expression and so on.

The Greek and Roman stoics denied emotion in children and animals, a view shared by some contemporary scholars as well. This view was disputed but Nussbaum examined emotion in children and animals. In fact we have tradition of comparative study of emotional expression of animals since Darwin (1872), who proposed three principles of emotional expression. The first is the principle of serviceable association habits, the second the principles of antithesis, and the third the principle of actions due to the constitution of the nervous system independently from the first of the will and independently to a certain extent of habit (or simply the direct action of the nervous system). These principles are applicable to both humans and animals. He gathered information on facial expression of different races and pointed out similarity among them. He claimed a single line of modern human evolution based on these observations. Perhaps it is easier to understand continuum of emotion between humans and animals than the ordinal continuum of intelligence for people. Perhaps it is easier to estimate emotional state of your pets than thought of them. Similar phylogenetic contingency or selection pressure might produce a similar expression in different species. Nussbaum mentioned findings in ethology and developmental psychology and claimed emotion in children and animals. The chapters in this book focus on the nature of emotions and the possibility of emotions in pre-verbal children and non-verbal animals. It may be important to point out that emotion is free from language faculty in animals and some children.

Part I of this book concerns the evolutionary origin of emotions. The first chapter describes emotion and social performance. The second chapter examines hormonal factors in aggressive behavior in fish. The famous parrot Alex, who unfortunately died in 2009, showed not only higher cognitive functions but also colorfulness of his emotional life. The third chapter examines the relation between cognition and emotion in parrots. Another non-primate species that shows higher cognitive function is the dolphin. The fourth chapter concerns the emotional expression of dolphins and its function. Humans play but so do many other animals. We feel pleasure during play, in other words, some part of playing behavior is maintained by self-reinforcement. The fifth chapter clarifies the relationship between play and emotion. The last chapter in Part I is "animal aesthetics". Nussbaum wrote that art influences emotions. Her example was music, but other forms of art have reinforcing properties for animals, suggesting that art can evoke emotion in animals. The seventh chapter examines this possibility.

Part II focuses on human studies. The first chapter (Chap. 8) in Part II provides a broad overview of human emotions. Chapter 9 examines "mentalizing" and emotion in human infants. The last chapter in this section relates emotions to neuroscience.

The chapters in Part III examine the relationship between emotion and other aspects of the human experience. The role of emotion in memory is the subject of Chap. 11. The effect of emotions on human logical judgment is the subject of Chap. 12. The final chapter provides a brief summary of the book as well as recommendations for future research on emotions. Finally, Professor Kuczaj arrives at an overall conclusion.

Tokyo, Japan   Shigeru Watanabe

References

Cannon WB (1927) The James-Lange theory of emotion: a critical examination and an alternatives theory. Am J Psychol 39:106–124
Damasio AR (1994) Descartes' error. Putman, New York
Darwin C (1872) The expression of emotions in man and animals. University of Chicago Press, Chicago
Descartes R (1649) Les passions de l'ame. Gallimard, Paris
James W (1884) What is emotion? Mind 9:188–205
Lange C (1885) The emotions. In: Dunlap E (ed) The emotions. Williams and Wilkins, Baltimore
Motteriloni M (2006) Economia emotive. RCS Libri S.P.A., Milano
Nussbaum MC (2001) Upheavals of thought: the intelligence of emotion. Cambridge University Press, New York
Rolls ET (1999) The brain and emotion. Oxford University Press, Oxford

Contents

Part I Emotion in Animals

1 **Emotions Are at the Core of Individual Social Performance**............. 3
  Kurt Kotrschal

   1.1 Introduction... 4
   1.2 Common Mechanistic Grounds in Vertebrate Emotionality
         and Sociality.. 5
   1.3 Conservative Vertebrate Brains.. 6
   1.4 Coping with Stress: In the Context of Emotionality..................... 12
   1.5 Individuality, Temperament, Personality 13
   1.6 Why Are Humans Drawn into Companionships
         with Animals?... 14
   1.7 Social Universals Allow for Social Relationships
         Between Humans and Their Companion Animals......................... 15
   References.. 16

2 **Hormonal Modulation of Aggression:
  With a Focus on Teleost Studies**.. 23
  Kazutaka Shinozuka

   2.1 Introduction... 24
         2.1.1 Aggression in Fish 24
         2.1.2 Approaches to Investigate Mechanisms
              of Emotional Behavior 27
   2.2 Behavioral Effects of Peptide Hormones on Aggression................. 29
         2.2.1 Teleosts.. 29
         2.2.2 Birds.. 31
         2.2.3 Rodents.. 32
         2.2.4 Interspecific Comparison 33

				Page
	2.3	Action Sites of Peptide Hormones in Teleosts		34
		2.3.1	Peripheral Action	34
		2.3.2	Central Action: Anatomical Studies	35
		2.3.3	Central Action: Functional Studies	36
	2.4	Conclusion		41
	References			42

3 Emotional Birds—Or Advanced Cognitive Processing? ... 49
Irene M. Pepperberg

	3.1	The Four Studies		51
		3.1.1	Object Permanence Experiment	51
		3.1.2	Phonological Awareness	52
		3.1.3	Insightful String Pulling	53
		3.1.4	Numerical Tasks	54
	3.2	Implications of the Data		57
	3.3	Conclusions		60
	References			60

4 Why Do Dolphins Smile? A Comparative Perspective on Dolphin Emotions and Emotional Expressions ... 63
Stan A. Kuczaj II, Lauren E. Highfill, Radhika N. Makecha, and Holli C. Byerly

	4.1	Animal Emotions?		66
	4.2	Methods for Studying Animal Emotions		67
		4.2.1	Physiological Measures	68
		4.2.2	Behavioral Measures	69
		4.2.3	Behavioral and Physiological Measures	70
		4.2.4	Studying Emotions in Wild Populations	71
		4.2.5	Animal Personality and Animal Emotion	71
	4.3	Dolphin Emotions?		73
		4.3.1	Dolphin Vocal Expressions of Emotions	74
		4.3.2	Dolphin Use of Posture to Express Emotions	75
		4.3.3	Touch as a Mode of Dolphin Emotional Expression?	76
		4.3.4	Do Dolphins Grieve?	77
	4.4	Conclusions		78
	References			80

5 Play and Emotion ... 87
Stan A. Kuczaj II and Kristina M. Horback

	5.1	Types of Play		90
	5.2	Emotions and Play		95
	5.3	Play Signals		98
		5.3.1	Play Signals Vs. Playing with Signals	102

	5.4	Benefits of Play	102
	5.5	Conclusions	105
	References	106

6 The Use of Emotion Symbols in Language-Using Apes ... 113
Heidi Lyn and Sue Savage-Rumbaugh

	6.1	Emotion and Language-Using Apes	114
	6.2	Methods	116
		6.2.1 Participants	116
		6.2.2 Data Collection	117
		6.2.3 Coding	118
		6.2.4 Database	119
	6.3	Results	120
		6.3.1 Use and Comprehension of Internal State Words	120
		6.3.2 Pragmatic Force and Co-Construction	123
	6.4	Discussion	124
	References	126

7 Animal Aesthetics from the Perspective of Comparative Cognition ... 129
Shigeru Watanabe

	7.1	From Experimental Aesthetics to Comparative Cognition of Art	130
	7.2	Discriminative and Reinforcing Properties of Auditory Art	133
		7.2.1 Reinforcing Property of Music for Animals	133
		7.2.2 Discriminative Stimulus Properties of Music	137
		7.2.3 Conclusion	142
	7.3	Discriminative and Reinforcing Properties of Visual Art	144
		7.3.1 Reinforcing Property of Visual Art	144
		7.3.2 Discriminative Stimulus Property of Complex Visual Stimuli	145
		7.3.3 Discriminative Property of Painting Style	145
		7.3.4 Discrimination of "Beauty"	146
		7.3.5 Strategy of Discrimination of Aesthetic Stimuli	147
		7.3.6 Conclusion	151
	7.4	Creation of Art	151
		7.4.1 Motor Skills	152
		7.4.2 Does Animal Art Symbolize Something Outside?	152
		7.4.3 Functional Autonomy of Animal Art	154
		7.4.4 Do Animals Enjoy their Product?	156
	7.5	Conclusion	156
	References	157

Part II Emotion in Humans

8 The Unique Human Capacity for Emotional Awareness: Psychological, Neuroanatomical, Comparative and Evolutionary Perspectives 165
Horst Dieter Steklis and Richard D. Lane

- 8.1 Introduction 166
- 8.2 Theory of Levels of Emotional Awareness 167
- 8.3 Normative and Clinical Observations with the Levels of Emotional Awareness Scale 169
- 8.4 A Model of the Neural Substrates of Implicit and Explicit Emotional Processes 171
- 8.5 Reflective Awareness 173
- 8.6 The Comparative Approach: A Caution 174
- 8.7 Emotions and Emotional Awareness in Nonhuman Primates 178
- 8.8 Naturalistic Behavior Observations 178
- 8.9 Experimental Studies 182
- 8.10 Neural Correlates 191
- 8.11 Discussion and Evolutionary Considerations 196
- References 199

9 The Development of Mentalizing and Emotion in Human Children 207
Shoji Itakura, Yusuke Moriguchi, and Tomoyo Morita

- 9.1 Introduction 208
- 9.2 The Development of Mentalizing 209
- 9.3 Developmental Cybernetics 210
  - 9.3.1 Inference of Robot Intention 210
  - 9.3.2 False Belief of a Robot 215
  - 9.3.3 Word Learning from a Robot 216
- 9.4 The Neural Basis of the Development of Mentalizing 218
- 9.5 Conclusion 219
- References 220

10 Emotion, Personality, and the Frontal Lobe 223
Satoshi Umeda

- 10.1 Basic and Advanced Emotions 224
- 10.2 Functional Neuroanatomy of Advanced Emotion 225
- 10.3 Neural Substrates of "Theory of Mind" 230
- 10.4 Personality Change After Damage to the Frontal Lobe 231
- 10.5 Neuropsychological Investigations 232
- 10.6 Theory-of-Mind Performance After Brain Injury 234
- 10.7 Acquired Autism Trait After Medial Prefrontal Damage 235
- References 238

Part III Emotion, Consciousness and Memory

11 Origin and Evolution of Consciousness and Emotion: Has Consciousness Emerged from Episodic Memory? 245
Takashi Maeno

    11.1   Introduction 246
    11.2   What Is Evolution? 246
    11.3   Evolutionary Journey of Control Systems Among Living Organisms 249
    11.4   Acquisition of Episodic Memory Through Evolutionary Processes 253
    11.5   Relationship Between Consciousness and Episodic Memory 255
    11.6   Is the Phenomenon of Consciousness Evolutionarily Necessary? 258
    11.7   Death Belongs to Evolution 263
    References 264

12 The Logic of Memory and the Memory of Logic: Relation with Emotion 265
Philippe Codognet

    12.1   Introduction 266
    12.2   Memory in the Flesh 267
    12.3   Memory in the Genes 271
    12.4   Binary Notation 272
    12.5   From Computing to Biology 273
    12.6   Conclusion 275
    References 276

13 Conclusions: Emotions (and Feelings) Everywhere 277
Stan A. Kuczaj II

    References 280

Index 281

Part I
Emotion in Animals

Chapter 1
Emotions Are at the Core of Individual Social Performance

Kurt Kotrschal

Abstract Affects and their conscious representations, emotions, are the central agents of social organization in humans and non-human animals. These are biopsychological phenomena with morphological and physiological substrates, with evolutionary functions, modulated in ontogeny and conservatively preserved in evolutionary history. Emotions "motivate" social interactions/relationships and emotionality is directly linked with basic physiology, particularly with the stress systems and also, with the most important "anti-stress" complex in mammals, the oxytocin-attachment system, which has also a major role in bonding. Emotional phenotype ("temperament") affects social connectedness and finally, fitness in complex social systems. And there is also a clear link to social cognition, as affects are involved in virtually any decision made by the relevant brain centers in mammals and birds. Finally, the communication of emotions and satisfying each others emotional need

---

K. Kotrschal (✉)
Department of Behavioural Biology, University of Vienna and Konrad
Lorenz Research Station, Fischerau 11, Grünau 4645, Austria
e-mail: kurt.kotrschal@klf.ac.at

is also at the core of human–animal relationships, which may be considered as indirect evidence for common principles of social organization.

1.1 Introduction

Affects and emotions as their conscious representations are the central agents of social organization in humans and non-human animals (Panksepp 1998, 2005). Because it is notoriously hard to know to what degree affects are consciously experienced by humans and non-human animals, the term "emotions" will mainly be used in the following chapter, not implying that all animals are necessarily conscious of all their affects.

Social relationships are always and unavoidably emotional. This was already acknowledged by Darwin in one of his most important books: "The expression of the emotions in man and animals" (1872). For a long time, even mechanistically orientated behavioural biologists tended to avoid the psychological terms and topics of affects and emotions, mainly because of their subjective touch. In fact, in non-human animals, affectivity is only indirectly accessible via measurable behavioural and physiological parameters, because verbal self-reporting is generally confined to humans.

Still, focusing on emotions in non-human animals is not just "subjective wishi-washi". In fact, emotions can be tackled at all four "Tinbergian levels" (Tinbergen 1963). Emotions (1) have ultimate functions in motivating individuals to execute their evolutionary strategies and individual tactics towards fitness optimisation; (2) have a mechanistic (neuronal/hormonal/cognitive) substrate and show a species-specific form of behavioural expression; (3) are individually shaped and modulated during ontogeny between genes, maternal effects and early sozialization; individual emotionality also forms the substrate for ones "temperament" and (4) emotions are conservatively maintained in structure and function over considerable periods of vertebrate evolutionary history.

My contribution will discuss the fact that emotions "motivate" social interactions/relationships and are, in turn, affected by all social interactions. It will be argued that emotionality is directly linked with basic physiology, particularly with the two stress axes (HPA and SAA, see below) and, also, with the most important "anti-stress" complex in mammals, the oxytocin-attachment system (De Vries 2002; De Vries et al 2003; Uvnäs-Moberg 1998). As arousal/activation of the stress axes is mainly about mobilising energy for behaviour, emotions are directly linked with energetics and seem to determine, to a significant degree, individual "social efficiency" (i.e. reaching ones goals in a social setting with minimal/optimal energetic costs). Thereby, emotional phenotype ("temperament") affects social connectedness, "social efficiency" and finally, fitness, in complex social systems. And there is also a clear link to social cognition: the ability to empathize and to properly show and control emotions is at the core of "social competence" (i.e. reaching ones goals in a social setting with minimal social costs, avoiding particularly collateral damage in a long-term valuable relationship). Knowing what others know and being able to predict the behaviour of others based on previous experience (e.g. Bugnyar et al. 2007; Kotrschal et al. 2008) are probably among the major drivers of social intelligence. Hence, the underpinning of all social life is emotionality, but the more complex social life is, the more important will be cognitive mechanisms (related with the prefrontal cortex in mammals and the structurally analogous nidopallium caudolaterale in birds, see below), which not only control the behavioural impulses originating from the emotional systems but also, to some degree, the emotional systems themselves.

Social complexity, basically defined by the existence of networks of long-term and diversified dyadic relationships, are not just found in mammals, but also in birds (Emery 2006), for example in greylag geese (Kotrschal et al. 2006, 2010; Scheiber et al. 2007). Although cognition research in birds is catching up quickly, there is still much less known about the socially relevant basic physiological and brain mechanism in birds as compared to mammals. Hence much of the following relies on results in mammals, but it seems quite plausible that homologous mechanisms are also found in birds. As the same/homologous and functionally equivalent emotional mechanisms are the base for being social at least in the mammals, but probably in a much wider range of vertebrates (Panksepp et al. 1997; Panksepp 2005), we hypothesized that they will also be important in the relationships between owners and their companion animals (Kotrschal et al. 2009). Examples at the end of my contribution will underline this point.

1.2 Common Mechanistic Grounds in Vertebrate Emotionality and Sociality

A surprising degree of similarity in social organization and mechanisms has become apparent in the endothermic vertebrates, not only within the mammals, but also between mammals and birds (Bugnyar and Heinrich 2006; Byrne and Whiten 1988; Emery and Clayton 2004; Kotrschal et al. 2008). This includes principles of behavioural organization, such as the appreciation of the structure and rules of long-term valuable relationships (Kotrschal et al. 2006; Weiß et al. 2008), active social support (i.e. active interference in agonistic interactions) and post conflict management, including "reconciliation" and "consolation" (Aureli and De Waal 2000; De Waal 2000a, b), passive/"emotional" social support (i.e. stress reduction by presence of a social partner; Weiss and Kotrschal 2004; Scheiber et al. 2005), of rhythmicity of interaction patterns in dyadic relationships (Wedl et al. 2010; Hirschenhauser and Frigerio 2005), fission-fusion group organization (Dunbar 2007; Marino 2002), tradition forming (Fritz and Kotrschal 2002), etc.

Similarity among species is also manifest in common features of social cognition in non-human animals. These include individual recognition, episodic memory, perspective taking, being knowledgeable about third-party relationships, theory of mind and being and planning for the future planning (Emery and Clayton 2004;

Kotrschal et al. 2008). The "social brain hypothesis" links social complexity with cognition (Byrne and Whiten 1988; Dunbar 1998; Humphrey 1976). Living in large, socially complex groups in the case of mammals and/or in complex dyads in the case of birds (Emery 2006; Scheiber et al. 2007) seemingly provided selection pressures for the evolution of large and cognitively capable brains.

The elements to be used for assembling complex cognition already seemed to evolve early in vertebrate phylogeny (Güntürkün 2005). The ancestral analogue of oxytocin in fish, isotocin, for example, is produced in the homologous preoptic nucleus in fish and humans. Basic functions of isotocin/oxytocin have also remained unchanged; however, its role in bonding, or "love" (parental or between pair partners; Carter et al. 1995; Carter and Keverne 2002), to use an emotional-subjective term, is only needed in species with elaborate parental care, such as the cichlids among teleost fish, and certainly in the mammals. Or in other words, the mere presence of oxytocin homologues in all vertebrates does not per se indicate that all vertebrates have a highly developed bonding system; but where bonding is found, oxytocin or its homologues will be involved.

1.3 Conservative Vertebrate Brains

As an evolutionary principle, novelty is always based on tinkering with already existing structures. The brain is relatively conservative with regards to evolutionary change (Nieuwenhuys et al. 1998). Welkner (1976): "... it appears that the basic adaptive neural mechanisms had been worked out during early vertebrate evolution." This mainly reflects the preservation of important life-sustaining functions. In the following, a few of these shared elements, brain and other, important in social contexts, will be discussed.

A number of diencephalic and tegmental brain areas govern instinctive socio-sexual behaviour, all in context with emotionality. These areas remained essentially unchanged in structure and function over more than four hundred million years of evolution. ".. birds and teleost (bony) fishes possess a core social behavior network within the basal forebrain and midbrain that is homologous to the social behavior network of mammals. The nodes of this network are reciprocally connected, contain receptors for sex steroid hormones, and are involved in multiple forms of social behavior" (Goodson 2005). Among the main components of this network are the medial amygdala, the lateral septum, the preoptic area, the anterior hypothalamus, the ventromedial hypothalamus and the midbrain tegmentum (Goodson 2005; Newman 1999).

This social behaviour network hosts a number of functional components, for example, the universal brain bonding mechanism (Curley and Keverne 2005): the parvocellular preoptic nucleus produces two main peptides: oxytocin (OT) and arginine-vasopressin (AVP; Goodson and Bass 2001), which only differ by a few peptides between major vertebrate. These peptides act as neuromodulators and hormones, have basic regulatory functions and are centrally involved in the modulation of sex-specific socio-sexual behaviour, with a prevalence of oxytocin in the females and arginine-vasopressin in the males (Goodson 2005; Goodson and Bass 2001; Carter and Keverne 2002).

Particularly in mammals, OT and AVP are involved in tuning brains, attention and emotionality towards bonding with offspring or pair partners. The main mechanisms and functions seem universally present at least in the mammals including humans: The pregnancy hormones estrogen and progesterone up-regulate OT receptors. At birth, OT is released under the influence of cortisol, aiding parturition, milk let-down and facilitating olfactory offspring recognition. Under elevated OT, brain reward systems are activated via the appropriate olfactory input, establishing the mother–offspring bond on side of the mother. In the "small-brained" mammals (e.g. rodents) this mechanism is primarily used for establishing and maintaining mother–offspring affective bonding and for monogamous pair bond in some species, as was first shown in monogamous as compared to polyginous voles (Carter and Keverne 2002). Even in the "large-brained" mammals (e.g. primates), where a mixture of instinctive mechanisms, emotions, and cognition prevail (Panksepp 2005), OT is still involved. Notably, OT is released at female orgasms, at breast feeding or at tactile stimulation during allogrooming (De Vries et al. 2003; Uvnäs-Moberg 1998). In fact, OT seems one of the major mechanistic underpinning of "love" between long-term pair partners (i.e. dyadic attachment, as demonstrated in monogamous voles: Carter et al. 1995).

Birds show surprising parallels to mammals and even primates in their social behaviour (Emery et al. 2007). Still, it is not known whether allopreening in altricial birds, in parallel to allogrooming in mammals, also triggers OT release. Precocial birds, such as greylag geese, do not allopreen at all and still show long-term bonding between parents and offspring, between sisters and between pair partners and other complex social features (Kotrschal et al. 2006, 2010; Scheiber et al. 2007; Weiß et al. 2008). This may either mean that the mesotocin (the bird equivalent of OT) - AVP system is not important for social bonding in these birds, or that their MT-AVP system is mainly activated via the visual input. Still, many birds do allogroom, particularly many species with a long-term pair bond, such as corvids and parrots. It is quite likely that allogrooming will trigger the OT system in a similar way in these birds than in mammals.

Even within-species, bonding and attachment styles (i.e. the basal hormonal/instinctive bonding mechanisms plus the cognitive representation of the bonding partner) may vary substantially between individuals, depending on species-specific characteristics and early socialization (e.g. experience with the primary caregiver) in interaction with basic personality traits; in fact, early caregiving style seems to affect both the setpoints of the stress axes as well as the relevant emotion-cognitive representations of caregiving and relating socially in the offspring (below; Ainsworth 1985; Bowlby 1999; Hinde 1998). Via social influence on brain development (Mayes 2006), the maternal style may be socially transmitted over generations (Meaney et al. 1991; Nelson 2000). Hence, on top of the genetic background, maternal style crucially affects the quality of social bonding/attachment individual offspring will be able to engage later in life. This is mainly true for humans, but may also apply to socially complex non-human animals with dependent offspring, such as monkeys (Hinde and Stevenson-Hinde 1987). This concept was even extended to a peculiar interspecies-relationship with asymmetric dependence: companion dogs and their human partners (Topál et al. 1998).

Another functionally important element at least partially associated with the "social behaviour network" of the brain (above; Goodson 2005) are the phylogenetically conservative affective systems which provide the motivational base for "instinctive" sociosexual and other basic behavioural systems. But also, depending on their phylogeny and social organization, there will be tight interactions of the affective systems, with "higher cognition" (Emery and Clayton 2004) and even "consciousness" (Panksepp 2005). The affective systems are involved in every decision the prefrontal cortex (PC) prepares for conscious judgement. In turn, all decisions made in the PC in a complex, iterative process, will feed back to the emotional systems (Koechlin and Hyafil 2007; Sanfey 2007). These extended and elementary trans-diencephalic, limbic, emotional/affective systems include "seeking" (appetitive, interest), "fear", "rage" (aggressive), "lust", "care", "panic" and "play" (Panksepp 1998). Via common origin, these basic emotional systems seem to be shared by all mammals and probably, birds (Jarvis et al. 2005; Reiner et al. 2004). If this phylogenetic distribution is due to common origin, the basic affective systems may be rooted even deeper in phylogeny and may even be present in reptiles and potentially, fish. Because the "social brain network" (above) is part of these emotional systems, these may be found, in different shades of differentiation and functionality, in all vertebrates (Goodson 2005).

The affective brain systems are directly connected with motor systems for the expressions of emotions (Darwin 1872). Clearly, the behaviour patterns expressing emotions are very different in humans, dogs or geese, but there are common principles: Full individual control over the expressions of emotions is hard to achieve, even in humans. In contrast, adequately socialized individuals are experts in "emotional ("mind") reading", i.e. in deciphering even subtle expression of others´ emotions. This seems a core component of empathy and emotional competence in humans (Eibl-Eibesfeldt 2004) and non-human animals alike.

Mirror neurons may have a role in this basic kind of relating to others. Mirror-neuron-based action systems, are opto-motor reflex systems which were selected for turning certain visual stimulation in a reflex-like way into a motor output which mirrors the visual input. Mirror-neuron systems are probably at the core of relating not only to other individuals´ actions but also to their emotions (Gallese et al. 2004; Rizzolatti and Sinigalia 2007). In fact, these systems may provide an opto-emotional "reflexive" base for the social domain. Most vertebrates live in groups for at least part of their life histories (Krause and Ruxton 2002). For example, group cohesion and "social facilitation" with group members (i.e. the expression of the same behaviour, facilitated via synchronization of their "mood") may be mediated by "emotional contagion", based on mirror neuron systems (Rizzolatti and Craighero 2004; Rizzolatti and Sinigalia 2007). Moreover, these brain systems may be the base for the ability to imitate, for grasping of the intentions of others, or for mediating elementary empathy (De Waal 2008; Gallese and Goldman 1998;

Gallese et al. 2004; Ramachandran 2008). Such emotional communication based on an opto-motor reflex system may also be effective between species. The mutually affective relationship between humans and their companion animals may serve as an example (below).

Reflexive action, as for example in the expression and communication of emotions, not only requires quick and adequate perception mechanisms, but also the relevant behavioural elements to execute appropriate motor output (i.e. mimics, body language). Such motor elements were selected for use in particular contexts, where it is important that individuals will function even without much prior experience (for example sex and reproduction) and where too much learning would rather be counterproductive by being too unreliable or too slow. Hence, all animal species not only come with certain morphological and physiological traits, but also are also equipped with basic behavioural elements and systems: In all animals, a species-specific inventory of stereotyped and highly heritable motor patterns develops early in ontogeny (Eibl-Eibesfeldt 1999; Tinbergen 1951). This system of "reflexes" (Pavlov 1954) or "action patterns" (APs; Lorenz 1978; Tinbergen 1951) may be considered as evolutionary pre-fabricated behavioural elements for later use in (mainly stimulus–response) "standard situations". Von (von Holst 1936) and others (Lorenz and Tinbergen 1939) showed that APs are not simply all-or-nothing responses. Their intensity depends on internal motivation and external stimulus strength (Baerends et al. 1955; Lorenz 1978; Tinbergen 1951). APs are as species-specific and are as heritable as bodily characters. Specific APs may be more or less firmly tied to certain releasing "sign stimuli" (Lorenz 1943; Tinbergen 1951; Marler and Hamilton 1966). For example, when an egg or even an object such as a cube, is lying just outside the nest, the greylag goose nest owner will pull it into the nest with a stereotyped AP. The paradigmatic description of this AP of the greylag goose by Lorenz and Tinbergen (1939) served as a conceptual model for all APs. These are relatively fixed in shape, but not in intensity. The functional orientation (stimulus context) of reflexes/APs, however, may be changed by Pavlovian conditioning. Together with the relevant mechanisms of perception these action patterns are tied together over functional domains in so-called "behavioural systems" to ensure proper individual functioning in core areas, such as reproduction and foraging.

These rules of behavioural organization also apply to the species-specific expressions of emotion via the APs of body language and facial expressions (above; Darwin 1872). These APs are species universals, expressed in a similar way, independent of the cultural background in humans (Eibl-Eibesfeldt 2004). In fact, the reflexive motor expressions of emotions are the central element of social communication in all species. Although the expressions of emotions (i.e. the relevant motor patterns) are largely innate, perceiving and deciphering them in detail seemingly requires a lot of implicit learning during early life history. Just as the young in many species, particularly when they are fully developed at birth/hatching, do not come with a fully inherited set of knowledge about how parents and conspecifics look, but need to learn these things, for example, via "imprinting", social perception needs to be developed via early socialization. Disturbances in this early learning process may lead to impaired social competence and empathy later in life.

This was impressively shown in rhesus monkeys which were socially deprived early in their lives (Hinde 1998). Such massive deprivation, of course not only causes an impaired ability to read the expressions of emotions in others, but leads into sociophobia and sociopathy. This principle of early socializing with the expressions of emotions of others through implicit learning also opens a window of opportunity for between-species socializing, for example, in human–animal companionship. Socializing cats and dogs with humans, for example, from week four of their puppyhood not only creates the base for lifelong trust towards humans (Turner 2000), but may also allow a pup in its sensitive period to learn about the expressions of emotions in humans.

Hence, whereas the motor patterns for the expression of emotions are highly heritable and species-specific, the principles of socializing early in ontogeny (i.e. learning to interpret the social communication of others) rather seem species-general. Species and individuals at birth/hatching usually come with dispositions to attend to certain stimulus combinations, which guide learning ("angeborener Lehrmeister": Lorenz 1943; "instinct to learn": Kamil 1998). Such selective attention usually focuses offspring towards social role models, such as parents or play mates. Thereby, the linkage between emotions and cognition, which is a crucial element for decision making (Sanfey 2007) seems to be fine-tuned by implicit social learning with regards to the social do's and dont's during early sensitive periods of development in most, if not all, social birds and mammals, including humans (Ainsworth et al. 1978; Bowlby 1999; Hinde and Stevenson-Hinde 1987; Scott and Fuller 1965).

In pups and kittens, for example, the period between 3 and 4 weeks and a few months of age is crucial for between-species and within-species socializing (Scott and Fuller 1965; Turner 2000; Turner and Bateson 2000; Turner et al. 1986). Animals which were not appropriately confronted with humans in weeks 3–9 may remain shy or at least ambivalent in their bonding and to develop full trust towards humans (Scott and Fuller 1965; Turner 2000). As is true for primates (Bowlby 1999; Hinde 1998), also carnivore pups with no opportunity to interact with conspecific playmates in their first months will lack "social competence" with conspecific later on, i.e. will resort to aggression significantly more often than individuals socialized adequately. In the social context, this is maladaptive, because individual social efficiency is generally compromised by aggression. Social efficiency is here considered as the competence to reach one´s social goals with minimal effort and with minimal strain on valuable long-term valuable social relationships (Aureli and De Waal 2000) and thereby, with minimal costs for the actor and its social net.

Much of these basic mechanisms are the neurological, physiological and behavioural constituents of bonding, attachment and caregiving, which are among the emotionally most highly loaded and functionally most crucial behavioural domains in social animals. Reasonably complex social systems not only rely on these "old" mechanisms, but, to the contrary, these old, "instinctive" mechanisms also need to be controlled for contextual adequacy by "higher" brain centers. This is certainly true for many birds and mammals, which often reproduce and raise their offspring in the frame of a complex social network. Hence, it needs brain control centres, which are the prefrontal cortex in mammals and the nidopallium caudolaterale in birds. Complex ecological and social environments (Aureli and De Waal 2000; Weiß et al. 2008) demand appropriate mechanisms of decision-making (Kotrschal et al. 2008; Paulus 2007; Sanfey 2007). Individuals need their instinctive substrate for social behaviour, but they also need to control their impulses, to be able to respond conditionally to the opponent´s/partner´s behaviour, to keep track of checks and balances over time (emotionally or cognitively, above), to adjust to categorical (e.g. clan membership) and individual (dyadic) relations with others, to integrate information from different domains into "episodic memory" (Emery and Clayton 2004), or, generally, to form relevant concepts about the environment (Koechlin and Hyafil 2007).

The mammalian prefrontal cortex (PC) does exactly that (Damasio 1999; Güntürkün 2005); it is associated with complex learning (Lissek and Güntürkün 2003), economic choice (Kalenscher et al. 2005), context integration (Lissek and Güntürkün 2005), self control (Kalenscher et al. 2006) and thereby controls impulsive behaviours, allows social judgement, concept formation and categorization, guides emotion-based decision making and conducts mental and sub-conscious trial-and-error simulations before a final decision is reached (Koechlin and Hyafil 2007). The PC reaches its greatest relative size in humans. But if such a control centre did not exist in dogs, they could not be trained to suppress their impulses, for example, their urge to chase a hare or to remain attentive to their master even at the presence of distractors, such as other dogs. Hence, dog training, or any kind of higher vertebrate shaping of social performance, can ultimately be seen as PC empowerment for filtering affective pre-decisions.

Structurally, the PC is part of the laminated Pallium and hence, is only found in mammals (Nieuwenhuys et al. 1998). For more than a century, the bird Telencephalon was thought to consist largely of basal ganglia (Striatum). Consequently, birds were considered instinct-driven stimulus–response machines, hardly capable of making intelligent decisions. However, it is now accepted that the Telencephalon in birds and mammals has comparable proportions of Pallium (Reiner et al. 2004). Based on connectivity and neurochemistry, the "Nidopallium caudolaterale", a dorso-caudal telencephalic area in birds, was identified as the functional equivalent of the mammalian PC (Divac et al. 1994; Güntürkün 2005) prefrontal cortex. This cognitive "rehabilitation" of birds (Jarvis et al. 2005; Reiner et al. 2004) reconciles neuroanatomy with evidence that birds frequently innovate (Lefebvre et al. 1997, 2004), show theory of mind-like abilities (Bugnyar and Heinrich 2006; Bugnyar et al. 2007), can be cognitively complex and innovative (as shown extensively by grey parrot "Alex"; Pepperberg 1999) and may develop social systems of similar complexity than mammals (Kotrschal et al. 2006, 2010; Weiss et al. 2008). In humans, the PC is also the "moral brain". This however, does not mean that morality (basically the social do's and dont's; Broom 2003; De Waal 1996, 2000a, b) is simply "innate". Rather, the PC is the substrate receptive for, and depending on, appropriate inputs during socialization (below), and the specific attentiveness, needed for such kind of implicit social learning, will, in turn, be affectively guided (i.e. role models chosen according to bonding and positive emotions related to them). This principle may prepare other animals with a complex sociality for between-species socializing, including carnivores and primates, parrots (Pepperberg 1999), corvids (Emery 2006; Emery and Clayton 2004; Kotrschal et al. 2008, 2010), or even geese (Weiß et al. 2008).

1.4 Coping with Stress: In the Context of Emotionality

Another common principle among homoeothermic vertebrates is that proper social conduct with partners or offspring comes not just straight from in the genes, but is also subject to individual variation due to maternal effects (pre-partum) and effects of early caregiving/socialization (post-partum). This modulates not only attachment and personality, but in particular stress coping and whether and how social context later in life will affect dealing with stress, or whether and which social contexts themselves will be perceived as stressful. This is highly relevant in the present context, because individual stress modulation is closely related with emotionality. Chronically high stress levels, for example, are often associated with anxiety. In contrast, socially triggered oxytocin dampens glucocorticoid release and promotes trust, relaxation and exploration (Petersson et al. 1999; Uvnäs-Moberg 1998). The two stress axes were conservatively maintained over long periods of evolutionary history (at least 450 million years). These are usually individually modulated to the greatest degree by social context and, vice versa, are themselves important modulators of social behaviour. In the following, we will give a short overview of the basics of these stress mechanisms.

Stress coping is a central element of emotionality and crucial in social performance (Creel et al. 1996; Creel 2005; De Vries et al. 2003; Mayes 2006; McEwen and Wingfield 2003; Sachser et al. 1998; Sapolsky 1992; von Holst 1998). Therefore, it is not surprising that the "social brain network" (above) is also involved and that the degree a species tends to be social or territorial, for example, is contingent with these highly conservative (brain) mechanisms of stress coping (Goodson 2005). The sympathico-adrenergic system (SA), which provides a rapid alarm response (Selye 1951) is associated with a quick release of catecholamines from the adrenals triggered by the sympathetic nerves and with rapid changes in heart rate and blood pressure. In contrast, the hypothalamo–pituitary–adrenal system (HPA) produces a slower, but longer-lasting response (Sapolsky 1992; Sapolsky et al. 2000; von Holst 1998). In the latter, a hypothalamic releasing factor (CRF) enters the anterior pituitary via the portal vessels and triggers the release of adrenocorticotrophic hormone (ACTH), which initiates the synthesis of glucocorticoids (cortisol in most mammals, corticosterone in birds) in the adrenals. These major metabolic hormones also mediate individual decisions whether to allocate energy to growth, behaviour, storage or reproduction (McEwen and Wingfield 2003; Sapolsky 1992; von Holst 1998).

Thereby, the two stress systems directly link social bonding, attachment and social support with the stress hormone-mediated investments into the social domain (Kotrschal 2005). This linkage seems to be to a large degree affective, or emotions at least orchestrate these contexts in parallel. Glucocorticoid release increases arousal, blood glucose and stimulates food uptake. Their frequent, short-term activation is part of coping with any challenge, positive or negative, and even is part of the appetitive interest and of the neurotrophic support of learning (Sapolsky et al. 2000). Chronic elevation of glucocorticoids is pathenogenic, for example mediating type II-diabetes (Vanltallie 2002).

Depending on the nature of social interactions, their stimulatory effects on the two stress axes may vary (Kvetnansky et al. 1995). While many social stimuli effectively activate the SA and HPA axes (De Vries 2002; von Holst 1998; Wascher et al. 2008a, b), passive social support (closeness, or sociopositive interactions with a social partner) dampens stress responses via activating the brain OT system. Thereby, the three crucial steps of the HPA cascade are inhibited: CRF, ACTH and glucocorticoid synthesis (De Vries et al. 2003). Such social support dampens anxiety and aggressiveness, enhances positive emotions, socio-positive interactions, positively re-enforces social bonds and decreases the physiological and energetic costs of social life. These physiological, bio-psychological and behavioural effects of social support are not restricted to primates or mammals, but are evidently also found, in a similar way, in social birds (Emery et al. 2007; Scheiber et al. 2005; Weiß et al. 2008). In fact, also the mechanisms of bonding and social support seem to be part of the basic vertebrate "social behaviour network" in the brain (Goodson 2005).

1.5 Individuality, Temperament, Personality

The two stress axes not only vary between individuals, as related to early social history, but are also main basic factors in the differentiation of individual temperament (i.e. the specific mix of emotionality) and personalities (i.e. relatively stable dispositions to respond to the challenges of life in a particular way). This may also be a main reason why humans and non-human animals seem to be differentiated in their personality characteristics along similar axes. Therefore, predictable emotionalities/temperaments and personalities may provide yet another common principle which enables human to engage in individualized dyadic and social relationships with their companion animals (below).

A decade of comparative and experimental research has revealed a non-random and parallel variation of individual behavioural phenotypes ("personality"), basically rooted in the inter-individual differentiation of emotionality in a variety of vertebrate species and even in invertebrates (Sih et al. 2004a, b). The main axis found in most species is essentially "reactive-proactive" (Koolhaas et al. 1999; "proactive" corresponding with "aggressive or assertative", Huntingford 1976; "shy-bold": Wilson 1998; Wilson et al. 1994; "slow-fast": Drent and Marchetti 1999). Proactive individuals in comparison with reactives tend to approach the challenges of life more actively, they tend to become dominant, are quick, but superficial to explore, are prone to form routines, but are reluctant to change them again and they usually do not excel at solving difficult tasks themselves, but profit from the actions of others by copying or scrounging (Giraldeau and Caraco 2000). These behavioural differences are contingent with differently set physiological systems for coping with stress: When challenged, proactives usually show a strong sympathico-adrenergic response, but only a quickly passing glucocorticoid peak and vice versa in the reactives (Koolhaas et al. 1999).

Individual personality features in animals can either be fairly well and reliably rated by human observers (Gosling 2001; Gosling and John 1999) or be behaviourally tested in a standardized way, for example, via open field, novel object or tonic immobility tests, etc. (Daisley et al. 2005). In human personality testing, verbal approaches prevail, such as attributing a number of features to subjects, as in the "big five" (Costa and McCrae 1999); independent of cultural background (McGrae et al. 1998), this feature theory approach results in five reasonably independent axes: neuroticism, extroversion, openness, agreeableness, conscientiousness. Personality features are genetically heritable, may be greatly modulated by "maternal effects" (i.e. via the direct hormonal interaction of mothers with their developing offspring; Groothuis and Carere 2004) and will also be affected by post-natal parenting and early socialization (Mayes 2006), particularly in species with long-dependent offspring. Early exposure of bird embryos to androgens, for example, generally shifts individuals towards a more "proactive" behaviour style (Daisley et al. 2005; Groothuis et al. 2005).

Per definition, personality attributes are usually consistent over situations and time and also predispose individuals to assume certain social roles (Dingemanse and De Goede 2004; Pfeffer et al. 2002; Sih et al. 2004a, b). In a way, a non-random variation of individual behavioural phenotypes in virtually any group of vertebrates is one of the preconditions for the development of social complexity. If all individuals would be the equal, fission–fusion organization, for example (discussed as one of the driving forces of complex cognition; Dunbar 2007; Marino 2002) would make no functional sense, because sub-groups may primarily form according to competence/personality features. Also, mate choice, particularly in the case of long-term pair bonds, should not only be based on straight indicators of genetic quality of the partner, but also on affective and operational partner compatibility (Dingemanse et al. 2004), including how well partners will provide active and passive social support for each other (Scheiber et al. 2005) and how successful they will act together. In general, a predictable and accountable variation of temperaments and behavioural phenotypes creates a choice of potential partners and allies for different tasks, makes individuals dependable, and dyadic combinations functional. In fact, social support and hence, stress modulation, was suggested as the central mechanism shaping the social system of greylag geese (Kotrschal et al. 2010).

1.6 Why Are Humans Drawn into Companionships with Animals?

Initiation of human bonding with companion animals may be facilitated by the "cuteness" of puppies and chicks, triggering spontaneous stroking (Spindler 1961) and an urge to provide care. Furry or feathery creatures seemingly activate caregiving in humans (Eibl-Eibesfeldt 1999, 2004), via causing emphatic feelings. In contrast, interest in insects, fish or amphibians may be mainly motivated by an exploratory interest in nature (Wilson 1984). Recently, the relevant brain mechanism for the "instinctive" parental response, which may be considered the base of all caregiving, has been identified (Kringelbach et al. 2008). Although it has not yet been demonstrated that this also works between species, for example, via the "Kindchenschema" (Lorenz 1943), such a mechanism very likely exists not only in humans, but probably in most, if not all species which provide elaborate parental care to their offspring. Lorenz´ Kindchenschema concept has stood some scientific scrutiny (Hückstedt 1965; Gardner and Wallach 1965), but more importantly, it has stood the test of time and is often found in commercial applications appealing to our caregiving, by linking products with children, puppies or generally, "cute" creatures. Also, the "evolution" of Mickey Mouse (Gould 1980) and of the Teddy Bear (Hinde and Barden 1985) underline the generality and relevance of the Kindchenschema principle.

Parental behaviours of brood care primarily evolved in species with fully dependant, altricial offspring at birth/hatching. Secondarily, caregiving behaviours, including "kissing" and allogrooming/allopreening in mammals and birds, respectively shifted in function in a range of mammals and birds and are also be employed in adult bonding (Eibl-Eibesfeldt 1970). "Sign stimuli" such as the "Kindchenschema" features seemingly trigger care giving (Eibl-Eibesfeldt 1999) and thereby, activate the oxytocin-related social reward systems (De Vries et al. 2003; Panksepp et al. 1997; Schultz 2000). This is probably sex-specific, to a certain degree, because of the prevalence of the oxytocin system in females, which is linked with estrogen, cortisol and prolactin (above; Curley and Keverne 2005). This fits with generally greater emphatic engagement and social interest in women (Paul 2000), their prevalence in social care professions and in pet keeping.

1.7 Social Universals Allow for Social Relationships Between Humans and Their Companion Animals

Companion animals may satisfy the need of individual humans for a reasonably compassionate partner (Olbrich and Otterstedt 2003; Podberscek et al. 2000; Serpell 1986; Williams and Weinberg 2003) at comparatively low "social costs". In addition, dogs in particular have increasing roles as partners in animal assisted activities and as social lubricants in society (Kotrschal and Ortbauer 2003; Kotrschal et al. 2004; Wilson and Turner 1998). In a way, dogs may be seen as neotenised wolves (Lorenz 1949) and the asymmetric human-companion dog relationship generally seems to fit well a parent–offspring model (Kotrschal et al. 2009). Also, cats and dogs do not argue verbally, do not judge the way a human partner would and are less demanding in many respects. At their core, relationships with companion animals are "essentialised" by emphasizing the emotional level, with only few of those cognitive and cultural components which may complicate relationships between humans. Companion animals may be able to adjust more asymmetrically and uncompromisingly to the needs and peculiarities of their human partners (Wedl et al. 2010) than most conspecific partners would.

Companion animals may be partners for allogrooming and thereby provide social support, dampening stress-related parameters, such as glucocorticoids, heart rate or blood pressure. Some information on the stress-reducing effects of interactions is available for humans (e.g. Friedmann et al. 2000; Uvnäs-Moberg 1998), but surprisingly, hardly for the potential symmetric benefit on side of the animal partner. Stress dampening is not only relevant with respect to long-term health, but there is also a link with the expression of aggressive behaviour, in rats and humans (Haller and Kruk 2006). Chronic glucocorticoid elevation may suppress sexual steroids and aggression in subordinates, but in dominant carnivores, for example, elevated cortisol levels or acute cortisol peaks may precede, facilitate and prime aggressive behaviour (Creel 2005). This may well be relevant with respect to human–animal relations. There is probably a deterministic relationship between individual attachment style (Topál et al. 1998) and personality and attitude of the human involved (Bagley and Gonsman 2005; Dodman et al. 1996). Kotrschal et al. (2009) found contingencies of dog behavior and dyadic performance with owner personality. Hence, owner psychology affected dog behavioral expression, dyadic functioning and the stress loads of the animal companion. However, living with an animal will only be emotionally and physiologically rewarding when there is some social bond and emotional relationship (Olbrich and Otterstedt 2003; Podberscek et al. 2000). In fact, social support may be a particularly important topic in tightly bonded human–animal dyads (Friedmann et al. 2000; Kotrschal et al. 2009; Olbrich and Otterstedt 2003).

This suggests that human–animal dyads may show social, physiological and cognitive structural elements characteristic for dyads in homoeothermic vertebrates in general. This is only possible because of the amazingly common and partially even homologous social toolbox elements discussed above. More than that: the fact that humans can be social with their companion animals and that therefore the elements of the vertebrate social toolbox can also be used between species, also underlines their functional generality, adaptability and robustness. Hence, human–animal dyads, in addition to being interesting in their own right, may have a considerable potential as research models towards the basics of human dyadic relationships.

References

Ainsworth MDS (1985) Patterns of attachment. Clin Psychol 38:27–29
Ainsworth MDS, Blehar MC, Waters E, Wall S (1978) Patterns of attachment: a psychological study of the strange situation. Erlenbaum, New York
Aureli F, de Waal FBN (2000) Natural conflict resolution. University of California Press, Berkeley
Baerends GP, Brower R, Waterbolk HT (1955) Ethological studies on Lebistes reticulatus Peter. I. Analysis of the male courtship pattern. Behaviour 8:249–334
Bagley DK, Gonsman VL (2005) Pet attachment and personality type. Anthrozoös 18:28–42

Bowlby J (1999) Attachment and loss. Basic Books, New York (reprint from 1974)
Broom DM (2003) The evolution of morality and religion. Cambridge University Press, Cambridge
Bugnyar T, Heinrich B (2006) Pilfering ravens, *Corvus corax*, adjust their behaviour to social context and identity of competitors. Anim Cogn 9:369–376
Bugnyar T, Schwab C, Schlögl C, Kotrschal K, Heinrich B (2007) Ravens judge competitors through experience with play caching. Curr Biol 17:1804–1808
Byrne RW, Whiten A (1988) Machiavellian intelligence: social expertise, and the evolution of intellect in monkeys, apes, and humans. Clarendon, Oxford
Carter C, Keverne EB (2002) The neurobiology of social affiliation and pair bonding. In: Pfaff D (ed) Hormones, brains and behavior. Academic, San Diego, pp 299–337
Carter CS, De Vries AC, Getz LL (1995) Physiological substrates of mammalian monogamy: the prairy vole model. Neurosci Biobehav Rev 19:203–214
Costa PT, McCrae RR (1999) A five factor theory of personality. In: Pervine LA, John OP (eds) Handbook of personality: theory and research, 2nd edn. Guilford Press, New York
Creel S (2005) Dominance, aggression and glucocorticoid levels in social carnivores. J Mammal 86:255–264
Creel S, Creel NM, Monfort S (1996) Social stress and dominance. Nature 379:212
Curley JP, Keverne EB (2005) Genes, brains and mammalian social bonds. Trends Ecol Evol 20:561–567
Daisley JN, Bromundt V, Möstl E, Kotrschal K (2005) Enhanced yolk testosterone influences behavioural phenotype independent of sex in Japanese quail (*Coturnix coturnix Japonica*). Horm Behav 47:185–194
Damasio AR (1999) The feeling of what happens. Body and emotion in the making of consciousness. Harcourt Brace, New York
Darwin C (1872) The expression of the emotions in man and animals. Murray, London
De Vries AC (2002) Interaction among social environment, the hypothalamo-pituitary-adrenal axis and behavior. Horm Behav 41:405–413
De Vries AC, Glasper ER, Dentillion CE (2003) Social modulation of stress responses. Physiol Behav 79:399–407
De Waal FBM (1996) Good natured: the origins of right and wrong in humans and other animals. Harvard University Press, Harvard
De Waal FBM (2000a) Primates—a natural heritage of conflict resolution. Science 289:586–590
De Waal FBM (2000b) Chimpanzee politics. Power and sex among apes. JHU-Press, Baltimore
De Waal FBM (2008) Putting the altruism back into altruism: the evolution of empathy. Annu Rev Psychol 59:279–300
Dingemanse NJ, De Goede P (2004) The relation between dominance and exploratory behavior is context-dependent in wild great tits. Behav Ecol 15:1023–1030
Dingemanse NJ, Both C, Drent PJ, Tinbergen JM (2004) Fitness consequences of avian personalities in a fluctuating environment. Proc R Soc Lond B 271:847–852
Divac I, Thibault J, Skageberg G, Palacios JM, Dietl MM (1994) Dopaminergic innervation of the brain in pigeons. The presumed "prefrontal cortex". Acta Neurobiol Exp (Wars) 54:227–234
Dodman NH, Moon R, Zelin N (1996) Influence of owner personality type on expression and treatment outcome of dominance aggression in dogs. J Am Vet Med Assoc 209:1107–1109
Drent PJ, Marchetti C (1999). Individuality, exploration and foraging in hand raised juvenile great tits. In: Adams NJ, Slotow RH (eds) Proceedings of the 22nd international ornithological conference. Bird Life South Africa, Durban/Johannesburg
Dunbar RIM (1998) The social brain hypothesis. Evol Anthropol 6:178–190
Dunbar RIM (2007) Evolution of the social brain. In: Gangestad SW, Simpson JA (eds) The evolution of mind. Guilford Press, New York
Eibl-Eibesfeldt I (1970) Liebe und Haß. Zur Naturgeschichte elementarer Verhaltnsweisen. Serie Piper, München, p 113
Eibl-Eibesfeldt I (1999) Grundriß der vergleichenden Verhaltensforschung. Ethologie, 8th edn. Piper, München

Eibl-Eibesfeldt I (2004) Die biologie des menschlichen verhaltens. Grundriss der humanethologie. Blank, Vierkirchen-Pasenbach

Emery NJ (2006) Cognitive ornithology: the evolution of avian intelligence. Philos Trans R Soc Lond B 361:23–43

Emery NJ, Clayton NS (2004) The mentality of crows: convergent evolution of intelligence in corvids and apes. Science 306:1903–1907

Emery NJ, Seed AM, von Bayern AMP, Clayton NS (2007) Cognitive adaptations of social bonding in birds. Philos Trans R Soc Lond B 362:489–505

Friedmann E, Thomas SA, Eddy TJ (2000) Companion animals and human health: physical and cardiovascular influences. In: Podberscek AL, Paul E, Serpell JA (eds) Companion animals and us: Exploring the relationships between people and pets. Cambridge University Press, Cambridge

Fritz J, Kotrschal K (2002) On avian imitation: cognitive and ethological perspectives. Invited contribution. In: Dauterhahn K, Nehaniv CL (eds) Imitation in animals and artifacts. MIT Press, Cambridge

Gallese V, Goldman A (1998) Mirror neurons and the simulation theory of mind reading. Theor Comput Sci 2:493–501

Gallese V, Keysers C, Rizzolatti G (2004) A unifying view on the basis of social cognition. Theor Comput Sci 12:396–403

Gardner RA, Wallach L (1965) Shapes and figures identified as a baby's head. Percept Mot Skills 20:135–142

Giraldeau L-A, Caraco T (2000) Social foraging theory. Monographs in behavior and ecology. Princeton University Press, Princeton

Goodson JL (2005) The vertebrate social behavior network: evolutionary themes and variations. Horm Behav 48:11–22

Goodson JL, Bass AH (2001) Social behavior functions and related anatomical characteristics of vasotocin/vaspressin systems in vertebrates. Brain Res Rev 35:246–265

Gosling SD (2001) From mice to men: what can we learn about personality from animal research? Psychol Bull 127:45–86

Gosling SD, John OP (1999) Personality dimensions in nonhuman animals: a cross species review. Curr Dir Psychol Sci 8:69–75

Gould SJ (1980) The Panda's thumb. WW Norton and Co, New York

Groothuis TGG, Carere C (2004) Avian personalities: characterization and epignesis. Neurosci Biobehav Rev 29:137–150

Groothuis TGG, Müller W, von Engelhardt N, Carere C, Eising C (2005) Maternal hormones as a tool to adjust offspring phenotype in avian species. Neurosci Biobehav Rev 29:329–352

Güntürkün O (2005) The avian "prefrontal cortex". Curr Opin Neurobiol 15:686–693

Haller J, Kruk MR (2006) Normal and abnormal aggression: human disorders and novel laboratory models. Neurosci Biobehav Rev 30:292–303

Hinde RA (1998) Mother-infant separation and the nature of inter-individual relationships: experiments with rhesus monkeys. In: Bolhuis J, Hogan JA (eds) The development of animal behaviour: a reader. Blackwell, Oxford

Hinde RA, Barden LA (1985) The evolution of the teddy bear. Anim Behav 33:1371–1372

Hinde RA, Stevenson-Hinde J (1987) Interpersonal relationships and child development. Dev Rev 7:1–21

Hirschenhauser K, Frigerio D (2005) Hidden patterns of male sex hormones and behavior vary with life history. In: Anolli L, Duncan S, Magnusson M, Riva G (eds) The hidden structure of interaction: from neurones to culture patterns. IOS Press, Amsterdam

Hückstedt B (1965) Experimentelle untersuchungen zum "kindchenschema". Z Exp Angew Psychol 12:421–450

Humphrey NK (1976) The social function of intellect. In: Bateson P, Hinde R (eds) Growing points in ethology. Cambridge University Press, Cambridge

Huntingford FA (1976) The relationship between antipredator behaviour and aggression among conspecifics in the three-spined stickleback. Anim Behav 24:245–260

Jarvis ED, Güntürkün O, Bruce L, Csillag A, Karten H, Kuenzel W et al (2005) Avian brains and a new understanding of vertebrate brain evolution. Nat Rev Neurosci 6:151–159

Kalenscher T, Widmann S, Diekamp B, Rose J, Güntürkün O, Colombo M (2005) Single units in the pigeon brain integrate reward amount and time-to-reward in an impulsive choice task. Curr Biol 15:594–602

Kalenscher T, Ohmann T, Güntürkün O (2006) The neuroscience of impulsive and self-controlled decisions. Int J Psychophysiol 62:203–211

Kamil AC (1998) On the proper definition of cognitive ethology. In: Balda RP, Pepperberg IM, Kamil AC (eds) Cognitive ethology. Academic, San Diego

Koechlin E, Hyafil A (2007) Anterior prefrontal function and the limits of human decision making. Science 318:594–598

Koolhaas JM, Korte SM, Boer SF, Van Der Vegt BJ, Van Reenen CG, Hopster H et al (1999) Coping styles in animals: current status in behavior and stress physiology. Neurosci Biobehav Rev 23:925–935

Kotrschal K (2005) Why and how vertebrates are social: physiology meets function. Plenary contribution IEC Budapest, August 2005

Kotrschal K, Ortbauer B (2003) Behavioural effects of the presence of a dog in the classroom. Anthrozoös 16:147–159

Kotrschal K, Bromundt V, Föger B (2004) Faktor hund. eine sozio-ökonomische bestandsaufnahme der hundehaltung in Österreich. Czernin, Wien

Kotrschal K, Hemetsberger J, Weiss B (2006) Homosociality in greylag geese. Making the best of a bad situation. In: Vasey P, Sommer V (eds) Homosexual behaviour in animals: an evolutionary perspective. Cambridge University Press, Cambridge

Kotrschal K, Schloegl C, Bugnyar T (2008) Lektionen von rabenvögeln und gänsen. Biologie in unserer Zeit 6:366–374

Kotrschal K, Schöberl I, Bauer B, Thibeaut A-M, Wedl M (2009) Dyadic relationships and operational performance of male and female owners and their male dogs. Behav Processes 81: 383–391

Kotrschal K, Scheiber IBR, Hirschenhauser K (2010) Individual performance in complex social systems: the greylag goose example. In: Kappeler P (ed) Animal behaviour: evolution and mechanisms. Springer, Berlin

Krause J, Ruxton GD (2002) Living in groups. Oxford University Press, Oxford

Kringelbach ML, Lehtonen A, Squire S, Hervey AG, Craske MG, Holliday IE, Green AL et al (2008) A specific and rapid neural signature for parental instinct. PLoS One 3:e1664

Kvetnansky R, Pacak K, Fukuhara K, Viskupic E, Hiremagalur B, Nankova B et al (1995) Sympathoadrenal system in stress. Interaction with the hypothalamic-pituitary-adrenocortical system. Ann N Y Acad Sci 177:131–158

Lefebvre L, Whittle P, Lascaris E, Finkelstein A (1997) Feeding innovations and forebrain size in birds. Anim Behav 53:549–560

Lefebvre L, Reader SM, Sol D (2004) Brains, innovations and evolution in birds and primates. Brain Behav Evol 63:233–246

Lissek S, Güntürkün O (2003) Dissociation of extinction and behavioural disinhibition: the role of NMDA receptors in the pigeon associative forebrain during extinction. J Neurosci 23: 8119–8124

Lissek S, Güntürkün O (2005) Out of context: NMDA receptor antagonism in avian "prefrontal cortex" impairs context processing in a conditional discrimination task. Behav Neurosci 119: 797–805

Lorenz K (1943) Die angeborenen formen möglicher erfahrung. Z Tierpsychol 5:235–409

Lorenz K (1949) So kam der mensch auf den hund. Borotha Schoeler, Wien

Lorenz K (1978) Vergleichende verhaltensforschung. grundlagen der ethologie. Springer, Wien

Lorenz K, Tinbergen N (1939) Taxis und instinkthandlung in der eirollbewegung der graugans. Z Tierpsychol 2:1–29

Marino L (2002) Convergence and complex cognitive abilities in cetaceans and primates. Brain Behav Evol 59:21–32

Marler P, Hamilton WJ (1966) Mechanisms of animal behavior. Wiley, New York
Mayes LC (2006) Arousal regulation, emotional flexibility, medial amygdala function and the impact of early experience. Ann N Y Acad Sci 1094:178–192
McEwen BS, Wingfield JC (2003) The concept of allostasis in biology and biomedicine. Horm Behav 43:2–15
Meaney MJ, Mitchell JB, Aitken DH, Bhatnagar S, Bodnoff SR, Iny LJ, Sarrieau A (1991) The effects of neonatal handling on the development of the adrenocortical response to stress: implications for neuropathology and cognitive deficits later in life. Psychoneuroendocrinology 16:85–103
Nelson RJ (2000) An introduction to behavioural endocrinology. Sinauer, Sunderland
Newman SW (1999) The medial extended amygdala in male reproductive behavior: a node in the mammalian social behavior network. Ann N Y Acad Sci 877:242–257
Nieuwenhuys R, Ten Donkelaar HJ, Nicholson C (1998) The central nervous system of vertebrates, vol I-III. Springer, Berlin
Olbrich E, Otterstedt E (2003) Menschen brauchen tiere: grundlagen und praxis der tiergestützten pädagogik und therapie. Frankh-Kosmos, Stuttgart
Panksepp J (1998) Affective neuroscience. The foundations of human and animal emotions. Oxford University Press, New York
Panksepp J (2005) Affective consciousness: core emotional feelings in animals and humans. Conscious Cogn 14:30–80
Panksepp J, Nelson E, Bekkedal M (1997) Brain systems for the mediation of social separation-distress and social-reward. Evolutionary antecedents and neuropeptide intermediaries. Ann N Y Acad Sci 807:78–100
Paul E (2000) Love of pets and love of people. In: Podberscek AL, Paul E, Serpell JA (eds) Companion animals and us: exploring the relationships between people and pets. Cambridge University Press, Cambridge
Paulus MP (2007) Decision-making dysfunctions in psychiatry-altered homoeostatic processing? Science 318:602–606
Pavlov IP (1954) Sämtliche werke. Akademie, Berlin
Pepperberg IM (1999) The Alex studies. Cognitive and communicative abilities of grey parrots. Harvard University Press, Cambridge
Petersson M, Hulting AL, Uvnäs-Moberg K (1999) Oxytocin causes a sustained decrease in plasma levels of corticosterone in rats. Neurosci Lett 264:41–44
Pfeffer K, Fritz J, Kotrschal K (2002) Hormonal correlates of being an innovative greylag goose. Anim Behav 63:687–695
Podberscek AL, Paul E, Serpell JA (eds) (2000) Companion animals and us: exploring the relationships between people and pets. Cambridge University Press, Cambridge
Ramachandran VS (2008) Reflecting on the mind. Nature 452:814–815
Reiner A, Perkel DJ, Bruce LL, Butler AB, Csillag A, Kuenzel W et al (2004) Revised nomenclature for avian telencephalon and some brain stem nuclei. J Comp Neurol 473:377–414
Rizzolatti G, Craighero L (2004) The mirror-neuron system. Annu Rev Neurosci 27:169–192
Rizzolatti G, Sinigalia C (2007) Mirrors in the brain: how our minds share actions and emotions. Oxford University Press, Oxford
Sachser N, Dürschlag M, Hirzel D (1998) Social relationships and the management of stress. Psychoneuroendocrinology 23:891–904
Sanfey AG (2007) Social decision making: insights from game theory and neuroscience. Science 318:598–602
Sapolsky R (1992) Neuroendocrinology of the stress response. In: Becker S, Breedlove S, Crews D (eds) Behavioral endocrinology. MIT Press, Cambridge
Sapolsky RM, Romero LM, Munck AU (2000) How do glucocorticoids influence stress responses? Integrating permissive, supressive, stimulatory and preparative actions. Endocr Rev 21: 55–89
Scheiber IBR, Weiss BM, Frigerio D, Kotrschal K (2005) Active and passive social support in families of Greylag geese (*Anser anser*). Behaviour 142:1535–1557

Scheiber IBR, Weiß BM, Hirschenhauser K, Wascher CAF, Nedelcu IT, Kotrschal K (2007) Does "relationship intelligence" make big brains in birds? Open Behav Sci J 1:6–8

Schultz W (2000) Multiple reward systems in the brain. Nat Rev Neurosci 1:199–207

Scott JP, Fuller JL (1965) Genetics and the social behavior of the dog. University of Chicago Press, Chicago

Selye H (1951) The general-adaptation-syndrome. Annu Rev Med 2:327–342

Serpell JA (1986) In the company of animals. Basil Blackwell, Oxford

Sih A, Bell AM, Johnson JC, Ziemba RE (2004a) Behavioral syndromes: an integrative overview. Q Rev Biol 79:241–277

Sih A, Bell AM, Johnson JC (2004b) Behavioral syndromes: an ecological and evolutionary overview. Trends Ecol Evol 19:372–378

Spindler P (1961) Studien zur vererbung von verhaltensweisen 3. Verhalten gegenüber jungen Katzen Anthropologischer Anzeiger 25:60–80

Tinbergen N (1951) The study of instincts. Oxford University Press, London

Tinbergen N (1963) On aims and methods of ethology. Z Tierpsychol 20:410–433

Topál J, Miklósi A, Csányi V, Dóka A (1998) Attachment behavior in dogs (Canis familiaris): a new application of Ainsworth's (1969) strange situation test. J Comp Psychol 112:219–229

Turner DC (2000) The human-cat relationship. In: Turner DC, Bateson P (eds) The domestic cat, 2nd edn. Cambridge University Press, Cambridge

Turner DC, Bateson P (eds) (2000) The domestic cat, 2nd edn. Cambridge University Press, Cambridge

Turner DC, Feaver J, Mendl M, Bateson P (1986) Variations in domestic cat behaviour towards humans: a paternal effect. Anim Behav 34:1890–1892

Uvnäs-Moberg K (1998) Oxytocin may mediate the benefits of positive social interaction and emotions. Psychoneuroendocrinol 23:819–835

Vanltallie TB (2002) Stress: a risk factor for serious illness. Metabolism 51:40–45

Von Holst E (1936) Versuche zur theorie der relativen koordination. Pflugers Arch 236:149–158

Von Holst D (1998) The concept of stress and its relevance for animal behavior. Adv Stud Behav 27:1–131

Wascher CAF, Arnold W, Kotrschal K (2008a) Heart rate modulation by social contexts in greylag geese (*Anser anser*). J Comp Psychol 122:100–107

Wascher CAF, Scheiber IBR, Kotrschal K (2008b) Heart rate modulation in bystanding geese watching social and non social events. Proc R Soc B 275:1653–1659

Wedl M, Bauer B, Gracey D, Grabmayer C, Spielauer E, Day J, Kotrschal K (2011) Factors influencing the temporal patterns of dyadic interactions between domestic cats and their owners. Behav Processes 86:58–67

Weiss B, Kotrschal K (2004) Effects of passive social support inn juvenile greylag geese (*Anser anser*): a study from fledging to adulthood. Ethology 110:429–444

Weiß BM, Kotrschal K, Frigerio D, Hemetsberger J, Scheiber IBR (in press). Birds of a feather stay together: extended family bonds, clan structures and social support in greylag geese (Anser anser). In: Ramirez RN (ed) Family relations. Issues and challenges. New York: Nova Science Publishers.

Welkner W (1976) Brain evolution in mammals: a review of concepts, problems and methods. In: Masterton RB, Bitterman ME, Campbell CBC, Hotton N (eds) Evolution of brain and behavior in vertebrates. Lawrence Erlbaum Assoc, Hillsdale

Williams CJ, Weinberg MS (2003) Zoophilia in men: a study of sexual interest in animals. Arch Sex Behav 32:523–535

Wilson EO (1984) Biophilia. Harvard University Press, Cambridge

Wilson DS (1998) Adaptive individual differences within single populations. Phil Trans R Soc Lond. B 353:199–205

Wilson CC, Turner DC (eds) (1998) Companion animals in human health. Sage, London

Wilson DS, Clark AB, Coleman K, Dearstyne T (1994) Shyness and boldness in humans and other animals. Trends Ecol Evol 9(442):446

Chapter 2
Hormonal Modulation of Aggression: With a Focus on Teleost Studies

Kazutaka Shinozuka

Convict cichlid (*Amatitlania nigrofasciata*)

Abstract Aggression is one of the important emotional behaviors displayed by animals while acquiring and/or defending resources. Recent studies have revealed that peptide hormones modulate various social behaviors, including aggression, among vertebrates. Comparison among teleosts, birds, and rodents shows marked species differences in effects of peptides on aggression. A correlation between peptide function and social structure has been suggested to explain these differences; however, the species differences reported in part might also be due to the methodological differences among the studies. In teleosts, the action sites for peptide

K. Shinozuka (✉)
Department of Neurosurgery and Brain Repair, University of South Florida,
Collage of Medicine, 12901 Bruce B. Downs Boulevard, MDC-78, Tampa, FL 33612, USA
e-mail: kshinozuka@gmail.com

hormones in modulating aggression are unclear, because in most behavioral studies these peptide hormones were administered systemically rather than locally. This chapter reviews the modulatory actions of peptide hormones, with particular emphasis on the known action sites for peptide hormones in teleosts. Further studies are necessary to precisely determine localized peptide modulation of aggression in teleosts, as well as to compare its modulatory effects among vertebrates.

2.1 Introduction

2.1.1 Aggression in Fish

Aggression is one of the important emotional behaviors displayed by animals, and has very different patterns of behaviors and functions. Wilson (1975) defined aggression as "a physical act of threat of action by one individual that reduces the freedom or genetic fitness of another," and divided it into the following eight forms: (1) territorial aggression, (2) dominance aggression, (3) sexual aggression, (4) parental aggression, (5) weaning aggression, (6) moralistic aggression, (7) predatory aggression, and (8) antipredatory aggression. In teleosts, the well-known models of aggression are the three-spined stickleback (*Gasterosteus aculeatus*) (Tinbergen 1951) and the Siamese fighting fish (*Betta splendens*) (Thompson 1963). Many species other than these models are also known to show aggressive behavior. This chapter aims to review recent findings on the mechanisms that modulate such aggressive behavior, especially by peptide hormones in teleosts. This section will introduce the behavioral characteristics of territorial aggression and related cognition in teleosts, mainly in cichlids.

The family Cichlidae includes more than 1,500 species, and many cichlid species form a territory. A territory is defined as "an area occupied more or less exclusively by animals or groups of animals by means of repulsion through overt aggression or advertisement" (Wilson 1975). A very wide range of species, including teleosts, form territories for defending a variety of resources such as food, shelter, spawning site, nest site, space for sexual display, and so forth (Table 12.1 in Wilson 1975). Like other territorial freshwater fishes, cichlids also form a territory for shelter and breeding, but rarely form a feeding territory because freshwater environment is not stable enough to supply foods constantly in comparison with marine environment (Barlow 2000). In either case, it is obvious that appropriate modulation of aggression by organisms to defend its territory is very important for increasing its fitness.

Cichlids show some typical behaviors while fighting: (1) lateral display: the fish spread their fins and the branchiostegal membrane, and try to get alongside each other; (2) tail beating: while laterally displaying, the fish beat the tail sideways pushing the water in a direction perpendicular to the median plane of the fish; (3) frontal display: the fish approach each other frontally, and erect the gill covers and the branchiostegal membrane; and (4) mouth fighting: the fish try to grip jaws among themselves, and push and pull, beating heavily (Baerends and Baerends-van

Roon 1950). The sequential assessment model (Enquist and Leimar 1983; Enquist et al. 1990) predicts that fighting begins with the display of aggressive behavior without any physical contact to assess an opponent's fighting ability at a low cost. Applying this model to cichlid fighting, lateral-display and tail-beating enable individuals to obtain information about the opponent's body size and the strength of the water current generated. Thus, these fish can evaluate its opponent by utilizing less energy. If the fish recognizes that the opponent is larger and causes more powerful water current than itself, it will avoid further fights to avoid injury. However, when aggressive behavioral display does not settle the fight, it escalates to fighting with physical contact for a direct evaluation of strength. In convict cichlid (*Amatitlania nigrofasciata*), for example, fighting escalates from lateral display to biting to mouth fighting to circling; the latter involves chasing each other's tail fins. It is known that separation of two convict cichlids by a transparent partition before fighting shortens the time spent in fighting rather than separation by an opaque partition (Keeley and Grant 1993). When the fight escalates to physical contact, the fish that bites its opponent first tends to win the fight (Bronstein and Brain 1991). Collectively, these results indicate that fish actually evaluate each other initially before involving in direct physical contact.

Cichlids can modulate their aggressive behavior to efficiently defend their territory. This is evident in convict cichlids, which have been shown to display aggressive behavior to more than one intruder at a time; however, the degree of aggression may be unevenly distributed between intruders based on the intruder's body size. For example, when confronted with two intruders of similar size, each intruder will be shown the same amount of aggression. However, if one of the intruders is larger, it will receive more aggression (Leiser and Itzkowitz 1999). For territorial individuals, the risk of intrusion by individuals of the neighboring territories is low, because they have their own territory. In such cases, territorial individuals are known to decrease their aggressive behavior toward neighboring territorial individuals. This is called "dear enemy effect" (Fisher 1954). Experimental manipulation demonstrates "dear enemy effect" in convict cichlids. One of two individuals was presented to a subject a day before the experiment as the neighboring individual, and another was presented from beginning of the experiment as the unfamiliar individual. In this condition, the subject shows more aggression toward the unfamiliar individual than toward the neighboring individual, even if the neighboring individual is larger than the unfamiliar individual (Leiser and Itzkowitz 1999). These results suggest that display of aggression toward a number of individuals is not fixed in convict cichlids; their aggressive responses are plastic and are modulated depending on which target takes priority.

The suppression of aggressive behavior toward "the individuals in neighboring territories" suggests that even fish, classically considered as an organism of the lower level, may actually be capable of spatial and individual recognition. The following two reports experimentally demonstrate such recognition in relation to territorial defense in teleosts. Male bicolor damselfish (*Stegastes partitus*) form their territories in a coral reef. They produce characteristic chirp sounds in their territories and show behavioral displays of courtship termed "dips." The dip is described as a vertical or near-vertical dive by the actor toward the substrate from a position, that is, 0.5–1 m above the substrate (Myrberg and Spires 1972). In addition, an individual who hears other's chirp sounds responds by dipping as a competitive courtship display. The experiment, which plays back a number of individuals' sounds, demonstrates that subjects show significantly higher frequency of dips when a sound other than the territorial owner's is played from a nearest neighboring territory. Such territorial owner's sound-dependent suppression of aggressive behavior is also shown when the second nearest territory is used as a playback site (Myrberg 1985). These results suggest that the bicolor damselfish recognizes "who is where" on the basis of sounds. *Astatotilapia burtoni*, which is one of the cichlid fish and an inhabitant of Lake Tanganyika, can learn dominant/subordinate relationships among individuals only by observing pair wise fights among others (Grosenick et al. 2007). The experimental setting is that the subject as a bystander is kept in the center of the experimental tank, and five rival males (A–E) are arranged in isolated compartments around the central bystander compartment. The subject is first trained by observing pair wise fights to realize that A dominates B, B dominates C, C dominates D, and D dominates E. Such scheduled results are possible because of the resident effect (e.g., A dominates B whenever they fight in A's compartment). These relationships imply a dominance hierarchy, that is, $A>B>C>D>E$. After the training, the subject's preference between rivals A and E and between rivals B and D, whom the subject had never seen together, is tested. As a result, subjects spend more time near E (who has always lost fights) than A (who has always won fights). Moreover, subjects also spend more time near D than B, despite the fact that these rivals win and lose the same number of fights during training. These results suggest that subjects recognize the implied dominance hierarchy and infer dominance relationships in novel pairings. Such a preference for D than B is maintained even when they are tested in a novel aquarium, so that the spatial information available during training is unavailable. The authors suggest that *A. burtoni* usually use both individual recognition and spatial recognition for maintaining stable dominant relationships. However, they can maintain such relationships only by using individual recognition, because the spatial information of the established territory in natural environment is regularly disrupted by external factors such as wind, predation, the movements of hippopotamuses, or the natural changes in the shore pools.

As shown in the damselfish, cognitive behavior in teleosts is not limited to the Cichlidae. For example, pairs of the anemonefish (*Amphiprion bicinctus*) show selective aggression against nonpartner individuals even when they are tested outside of their home (Fricke 1973). Moreover, bystanders of the Siamese fighting fish show longer latency to approach a winning individual rather than a losing individual only when they were allowed to witness the fight (Oliveira et al. 1998; for detailed review of fish cognition, also see Bshary et al. 2002; Peake and McGregor 2004). As described first, a territory is formed to provide important resources and increase the fitness of the organism. Existing evidence suggests that many fish have evolved flexible cognition such as evaluation of other fish, individual and spatial recognition, and drawing inferences from dominant relationships to modulate aggressive behavior for efficient acquisition and for maintenance of such important resources.

2.1.2 Approaches to Investigate Mechanisms of Emotional Behavior

As described above, many fish can modulate their behavior in territorial aggression. However, brain mechanisms underlying such behavioral modulation are not yet well known. One approach to studying this is by comparing brain morphology. The comparison of brain morphology among closely related but ecologically different species can reveal a correlation between the local enlargement of brain areas and the microhabitat. The enlarged brain areas might reflect important functions of these areas in adapting to the microhabitat by focal species. For example, African cichlids living in deep, middle, and shallow waters have relatively enlarged eyes, optic tectum/cerebellum, and telencephalon, respectively (Huber et al. 1997). The telencephalon shows the greatest variation compare to other brain areas among the species (van Staaden et al. 1994). Variation in the size of the telencephalon is correlated with spatial complexity of the microhabitat (Huber et al. 1997). Thus, the telencephalon in teleosts is suggested to have an important role for spatial cognition. This observation is consistent with the results of lesion studies that show impairment of spatial memory in goldfish after telencephalic lesion (Portavella and Vargas 2005; Saito and Watanabe 2004, 2006). The same approach can be applied to social environment. For example, the lek-forming polygamous cichlid *Enantiopus melanogenys* has better visual acuity than the monogamous cichlid *Xenotilapia flavipinnis*. In *E. melanogenys*, monitoring the health (e.g., color and vigor) of conspecifics for mating and perceiving minute imperfections in own crater construction for courtship influence reproductive success. These visual demands might lead to higher visual acuity in *E. melanogenys* (Dobberfuhl et al. 2005). On the other hand, monogamous cichlids possess about 20% larger telencephalon compared to the polygamous species. Thus, the telencephalon is suggested to be involved in the recognition of a mate and/or in parental behavior (Pollen et al. 2007). Perciforms, including cichlids, are good candidates for these kinds of interspecies comparisons, because they are sufficiently species-rich, with marked diversity in body size, social organization, and lifestyle. In these fish, a shift from specific perception to cognitive skills may have changed the brain, and such "cognitive brains" might be one of the factors that lead perciforms to marked successful radiation (Kotrschal et al. 1998).

Another approach is the central functions of peptide hormones on social behavior. Arginine vasopressin (AVP) and oxytocin are known as posterior pituitary hormones. Both peptides consist of nine amino acids, namely, Cys-Tyr-Phe-Gln-Asn-Cys-Pro-Arg-Gly for AVP and Cys-Tyr-Ile-Gln-Asn-Cys-Pro-Leu-Gly for oxytocin. These peptides perform peripheral functions. AVP is known as an antidiuretic hormone; it facilitates the reabsorption of water from the collecting duct of the kidney. Oxytocin causes uterus contraction during delivery and milk ejection.

Recent studies also demonstrate a role for these hormones in performing central functions such as regulation of various social behaviors in mammals. In particular, studies in the monogamous prairie vole (*Microtus ochrogaster*) have shown that both AVP and oxytocin play important roles in pair-bond formation (Young and

Wang 2004). Prairie voles normally form a pair after mating. However, local administration of vasopressin 1a (V1a) receptor antagonist into the male ventral pallidum blocks pair-bond formation after mating (Lim et al. 2004). Inversely, forced expression of V1a receptor gene in the male ventral pallidum facilitates pair-bond formation, and cohabitation without mating is enough to form a pair (Pitkow et al. 2001). The meadow vole (*Microtus pennsylvanicus*), which is closely related to the prairie vole, is a promiscuous species. They have fewer V1a receptors in the ventral pallidum than do prairie voles. The forced expression of the V1a receptor gene in the male meadow vole's ventral pallidum alters his partner preference and makes it more like that of monogamous species (Lim et al. 2004). In the female prairie vole, on the other hand, oxytocin receptor antagonist administration into the nucleus accumbens inhibits pair-bond formation (Young et al. 2001), whereas AVP does not have an effect (Insel and Hulihan 1995). These studies are particularly notable, because they demonstrate that a single substance is enough to control complex behavioral phenotypes such as mating behavior. These peptides modulate not only pair-bond formation but also other behaviors. Parental behavior, for example, is also modulated by these peptides. AVP administration into the lateral septum of a paternal prairie vole facilitates paternal behavior such as grooming (Wang et al. 1994). Spontaneous maternal behavior in the prairie vole is inhibited by the administration of oxytocin receptor antagonist into the nucleus accumbens (Olazábal and Young 2006). Oxytocin receptor-knockout female mice show longer latency to retrieve pups (Takayanagi et al. 2005).

In humans also, oxytocin is known to modulate various emotional/cognitive functions. For example, nasal oxytocin administration to adult men increases the amount of investment by a human investor in a "trust game" situation. In this game, a subject acting as an investor is required to decide the amount of monetary units that can be transferred to a human trustee. The trustee receives tripled transferred money units, and can decide how many monetary units are to be given back to the investor. Oxytocin-administered subjects significantly transfer a larger amount of monetary units to the trustee than the vehicle control-administered subjects. This effect disappears when subjects are informed that the amount of back transfer is determined randomly. Thus, oxytocin might cause a selective increase in human trust (Kosfeld et al. 2005; Zak et al. 2005). Other pro-social cognition in humans is also affected by oxytocin. The recognition of human facial identity is improved by post-learning intranasal oxytocin administration (Savaskan et al. 2008). Intranasal oxytocin administration selectively impairs the semantic implicit memory of reproduction-related words (e.g., orgasm and pacifier), but not neutral words (e.g., cake and brake) (Heinrichs et al. 2004). Intravenous administration of oxytocin improves maintenance of effective speech comprehension in autism (Hollander et al. 2007).

Nonmammalian vertebrates possess homologous peptides of AVP and oxytocin, namely, arginine vasotocin (AVT) and isotocin (fish) or mesotocin (amphibians, lizards, and birds). These peptides also consist of nine amino acids: Cys-Tyr-Ile-Gln-Asn-Cys-Pro-Arg-Gly for AVT, Cys-Tyr-Ile-Ser-Asn-Cys-Pro-Ile-Gly for isotocin, and Cys-Tyr-Ile-Gln-Asn-Cys-Pro-Ile-Gly for mesotocin. Comparative studies have revealed that social behaviors such as vocalization and aggression are commonly regulated by peptide hormones in vertebrates (Goodson and Bass 2001). The next section reviews the role of peptide hormones in the modulation of aggressive behavior among teleosts, birds, and rodents, followed by a review of the possible action sites of these hormones in the teleost brain.

2.2 Behavioral Effects of Peptide Hormones on Aggression

2.2.1 Teleosts

Table 2.1 shows a summary of results collated from studies that investigated the effects of peptide hormones on aggressive behavior in teleosts. The Amargosa pupfish (*Cyprinodon nevadensis amargosae*) is known to form loose shoals for reproduction. Whilst most male pupfish do not show aggressive behavior, few males form and defend their territories. The aggressive behavior in these territorial males can be decreased by AVT administration (Lema and Nevitt 2004). The brown ghost knifefish (*Apteronotus leptorhynchus*), which is a weakly electric fish, communicates with each other by electric organ discharge (EOD). It is known that males increase their EOD frequency for courtship with females, while decreasing the frequency for agonistic display against other males. AVT administration decreases this display of aggressive EOD in the brown ghost knifefish (Bastian et al. 2001). On the other hand, the male beaugregory damselfish (*Stegastes leucostictus*) displays increased aggressive behavior during territorial defense after administration of AVT (Santangelo and Bass 2006). These studies indicate that AVT administration, in general, influences aggression in teleosts, but that the direction of the effect is species dependent. Some explanations for these differences are discussed later.

Not only species difference, but also intraspecific difference is another factor, that is, reported to influence the role of AVT in aggression. The male plainfin midshipman (*Porichthys notatus*) is divided into two phenotypes, the territorial type I male and nonterritorial type II male; the latter is a sneaker male and shows female-like appearance. Both male and female fish emit an aggressive sound known as "grunt." The grunt in type I males is modulated by AVT. AVT administration to the preoptic area–anterior hypothalamus (POA–AH) reduces grunting, while V1a receptor antagonist administration to the same region increases grunting. On the other hand, grunting in females is modulated by isotocin. Isotocin administration to the same region decreases grunting, whereas administration of oxytocin receptor antagonist increases it. Interestingly, grunting in type II males is modulated by isotocin, similar to the observation in females, and AVT administration does not have an effect, whereas type II males have an ability for reproduction as a sneaker male (Goodson and Bass 2000a). As summarized in Sects. 2.1 and 2.2, studies in rodents have demonstrated that vasopressin and oxytocin mainly modulate male and female behavior, respectively. In this species, however, it is suggested that the role of peptide hormones on modulation of aggression depends on reproductive tactics, and is independent of gonadal sex.

Table 2.1 A summary of the effects of peptide administration on the aggressive behavior in teleosts

Species	Behavior	Arginine vasotocin	V1aR antagonist	Isotocin	OTR antagonist
Bluehead wrasse (*Thalassoma bifacsiatum*)[a]					
T-TP males	Bite	↔	↓ (towards IP males), ↔ (towards TP males)		
NT-TP males		↑			
Amargosa River pupfish (*Cyprinodon nevadensis amargosae*)[b]					
Territorial males	Bite	↓	↔		
Beaugregory damselfish (*Stegastes leucostictus*)[c]					
Males	Bite	↑	↓	↔	
Rainbow trout (*Onchorhynchus mykiss*)[d]					
Juveniles	Dominance	↓	↑		
Brown ghost knifefish (*Apteronotus leptorhynchus*)[e]					
Males	Electric display	↓			
Plainfin midshipman (*Porichthys notatus*)[f]					
Type I males	Acoustic display	↓	↑	↔	↔
Type II males		↔	↔	↓	↑
Females		↔	↔	↓	↑

T-TP territorial-terminal phase, *NT-TP* non-territorial-terminal phase, *IP* initial phase, *TP* terminal phase, *V1aR* vasopressin 1a receptor, *OTR* oxytocin receptor
Symbols: (↑) facilitation, (↓) inhibition, (↔) no effect
Authors:
[a]Semsar et al. (2001)
[b]Lema and Nevitt (2004)
[c]Santangelo and Bass (2006)
[d]Backström and Winberg (2009)
[e]Bastian et al. (2001)
[f]Goodson and Bass (2000a)

The bluehead wrasse (*Thalassoma bifasciatum*) is also known to show intraspecific differences in AVT function. They change form from an initial phase (IP), which includes females and males with a female-like appearance, to a terminal phase (TP). TP males are divided into nonterritorial TP (NT-TP) and territorial TP (T-TP) males. AVT administration increases aggressive behavior in NT-TP males, but not in T-TP males (Semsar et al. 2001). Removal of T-TP males from the population causes a behavioral change in IP males, and they show more TP-like behavior.

This behavioral change is modulated by AVT because V1a receptor antagonist administration inhibits such a behavioral change. On the other hand, AVT administration to IP individuals in the presence of T-TP males does not cause a behavioral change. Thus, the role of AVT in the bluehead wrasse depends not only on phenotypic differences, but also on social context (Semsar and Godwin 2004).

2.2.2 Birds

Studies using localized administration of AVT have revealed that aggressive behavior in birds is modulated by AVT in the septum. AVT administration induces the white-crowned sparrows (*Zonotrichia leucophrys gambelii*) (Maney et al. 1997) and field sparrows (*Spizella pusilla*) (Goodson 1998a) to sing agonistic songs. However, local administration of AVT decreases overtly aggressive behaviors such as chase and peck in field sparrows and violet-eared waxbills (*Ureaginthus granatina*) (Goodson 1998a, b). In the male zebra finch (*Taeniopygia guttata*), a nonterritorial species, septal administration of AVT increases aggressive behavior toward same-sex conspecifics, whereas V1a receptor antagonist decreases it (Goodson and Adkins-Regan 1999). Similarly, the female zebra finch also increases aggressive behavior in response to septal administration of AVT; however, the V1a receptor antagonist does not have an effect. Thus, intrinsic AVT might be more important in the male zebra finch to modulate its aggressive behavior (Goodson et al. 2004). In contrast to the aggressive behavior, which is modulated by septal injection of AVT, preference for opposite-sex individuals and courtship behaviors are not affected by AVT (Goodson and Adkins-Regan 1999; Goodson et al. 2004).

The medial part of the bed nucleus of the stria terminalis (mBNST) is another area known to be sensitive to social stimuli. In aggregative species, the presentation of a same-sex individual increases Fos expression in AVT neurons located in the mBNST. Conversely, in territorial species, presentation of a breeding mate increases Fos expression, while presentation of a same-sex individual decreases it (Goodson and Wang 2006). Thus, in general, response of AVT neurons in mBNST is sensitive to socially preferable stimuli in focal species (Goodson and Wang 2006). In fact, bathing, which is preferable, but not a social stimulus, does not alter Fos expression (Goodson et al. 2009b). These studies suggest that the role of AVT differs according to the social structure.

It is also known that phenotypic differences within a species also influence AVT function. In the dominant violet-eared waxbill male, administration of V1a receptor antagonist reduces aggressive behavior for acquiring a mate, while aggressive behavior for territorial defense is unchanged. In subordinate individuals, however, the same treatment increases aggressive behavior in territorial defense (Goodson et al. 2009a). The activity of AVT neurons in the nucleus paraventricularis is negatively correlated with frequency of aggressive behavior for territorial defense (Goodson and Evans 2004), so that phenotypic differences in AVT function might be due to the differences in the activity of AVT neurons in the nucleus paraventricularis (Goodson et al. 2009a).

Another peptide hormone, mesotocin, which is homologous to mammalian oxytocin, does not affect aggressive behavior in the zebra finch by intracerebroventricular (ICV) administration (Goodson et al. 2004). Rather, mesotocin might affect the tendency for aggregation because lateral septal administration of an oxytocin receptor antagonist in the female zebra finch causes reduction of preference for familiar/larger flocks (Goodson et al. 2009c).

2.2.3 Rodents

In rodents, in contrast to birds, the hypothalamus rather than the lateral septum is known to play a role in the peptidergic modulation of aggressive behavior. In the golden hamster (*Mesocricetus auratus*), administration of AVP to the anterior hypothalamus or ventrolateral hypothalamus facilitates aggressive behavior in territorial defense (Delville et al. 1996; Ferris et al. 1997). The administration of V1a receptor antagonist to the same regions inhibits aggressive behavior (Ferris and Potegal 1988; Ferris et al. 2006; Cheng and Delville 2009). In support of this, anatomical studies show that the dominant golden hamsters show higher V1a receptor expression in the anterior hypothalamus and ventromedial hypothalamus than subordinates (Cooper et al. 2005). Furthermore, social isolation of golden hamsters increases their aggressive behavior, and this behavioral change is correlated with increasing amounts of V1a receptor in the anterior hypothalamus, lateral hypothalamus, and paraventricular nucleus (Albers et al. 2006). In contrast to AVT administration, oxytocin administration to the anterior hypothalamus reduces aggressive behavior in these hamsters, whereas administration of oxytocin receptor antagonist increases it (Harmon et al. 2002). Thus, oxytocin has an opposing effect compared to AVT on modulation of aggressive behavior in rodents.

The lateral septum is also known to be involved in the vasopressinergic modulation of aggressive behavior in rodents, but results are inconsistent across studies. High expression of V1a receptors in the lateral septum is observed in the highly aggressive California mouse (*Peromyscus californicus*) as compared with the less aggressive white-footed mouse (*Peromyscus leucopus*) (Bester-Meredith et al. 1999; Bester-Meredith and Marler 2001). The administration of a V1a receptor antagonist reduces aggressive behavior for territorial defense in California mice (Bester-Meredith et al. 2005). These studies, therefore, suggest a positive correlation between aggressive behavior and AVP in the lateral septum in rodents. In contrast to these results, in male rats, the amount of AVP neurons in the lateral septum negatively correlates with their aggressive behavior toward same-sex individuals (Everts et al. 1997). Similarly, rats selectively bred to display low anxiety show higher aggressiveness in territorial defense and lower AVP secretion in the lateral septum. This pattern is reversed in rats selectively bred for high anxiety (Beiderbeck et al. 2007). These studies suggest a negative correlation between aggressive behavior and AVP in the lateral septum in rats. Beiderbeck et al. (2007) suggest that AVP in the lateral septum does not modulate aggressive behavior directly because both AVP administration in low-anxiety rats and V1a receptor antagonist administration in high-anxiety rats alter their anxiety-related behavior, but not their aggressive behavior.

Peptide hormones also influence female aggression in rodents. Maternal rats show aggressive behavior toward an intruder. This maternal aggression is decreased by ICV administration of AVP, and increased by the administration of V1a receptor antagonist (Nephew and Bridges 2008; Nephew et al. 2010). The presentation of an intruder to maternal rats increases oxytocin secretion in the paraventricular nucleus and central amygdala. Conversely, local administration of oxytocin in these areas increases aggressive behavior, while administration of an oxytocin receptor antagonist reduces it (Bosch et al. 2005). Thus, maternal aggression is inhibited by AVP and is facilitated by oxytocin. This pattern is the inverse of that observed in male aggressive behavior in rats.

2.2.4 Interspecific Comparison

As described above, effects of peptide hormones on aggressive behavior show marked species differences and phenotypic differences. Thus, one important issue is what factor(s) can predict AVT function in aggression. Based on studies in birds, Goodson (Goodson 1998a, b; Goodson and Adkins-Regan 1999) suggests that the differences in AVT function between species on overt aggression can be explained by their social structure, namely, whether it is territorial or colonial. He found that septal AVT injection in the territorial male field sparrow inhibited overt aggression such as chasing an intruder (Goodson 1998a), whereas the same treatment facilitated chase and peck in the colonial male zebra finch (Goodson and Adkins-Regan 1999). An important result is that septal AVT administration in the territorial violet-eared waxbill (*Uraeginthus granatina*), which is closely related to colonial zebra finch, inhibited overt aggression similar to that observed in the territorial field sparrow (Goodson 1998b). Thus, in these species the difference observed in the AVT function in aggression is correlated with social structure rather than phylogenetic classification. This correlation between social structure and the AVT function is also supported by a biochemical study. The zebra finch, spice finch (*Lonchura punctulata*), and Angolan blue waxbill (*Uraeginthus angolensis*) are all aggregative species that show greater binding of the V1a receptor antagonist in the lateral septum than the melba finch (*Pytilia melba*) and violet-eared waxbill, which are territorial (Goodson et al. 2006).

However, such a correlation is not fully consistent with the findings from rodents. The prairie vole, which is a highly affiliative monogamous species, shows facilitation of aggressive behavior after AVP administration. On the other hand, closely related montane vole (*Microtus montanus*), which is a nonmonogamous asocial species, shows no behavioral change on AVP administration (Young et al. 1997). Moreover, the male golden hamster, which is also a territorial species, shows facilitation of aggression after AVP administration (Ferris and Delville 1994).

In teleosts, a direct comparison of peptide hormone effects on aggression between closely related, but ecologically different species has not been examined. As shown in Table 2.1, in the bluehead wrasse AVT administration facilitates overt aggression only in nonterritorial males (Semsar et al. 2001). On the other hand, it has no effect on the territorial bluehead wrasse (Semsar et al. 2001) or shows inhibition of overt aggression in the territorial male Amargosa pupfish (Lema and Nevitt 2004). A result contrary to that predicted by Goodson's hypothesis is also reported. In the beaugregory damselfish, overt aggression for territorial defense is facilitated by a medium dose of AVT administration. The dose–response curve of AVT administration in the beaugregory damselfish had an inverted-U shape, showing no effect on the behavior at low and high doses of AVT (Santangelo and Bass 2006), leading the authors to emphasize the importance of first acquiring a dose–response curve to test the behavioral effects of AVT. As Santangelo and Bass authors pointed out, the inconsistent results reported in the literature might partly be because of the lack of a dose–response curve to determine the optimum dose for the peptide to be tested for each individual species.

Another important factor is that the aggressive behavior escalates from agonistic display to overt aggression in general. These behaviors might be modulated differentially (Santangelo and Bass 2006). Some fish studies report inhibition of agonistic display after AVT administration (Goodson and Bass 2000a; Bastian et al. 2001). In these cases, however, whether AVT inhibits or facilitates aggression is unclear. In future research, it will be important to test both agonistic display and overt aggression simultaneously to determine whether AVT influences both these behaviors.

In contrast to the studies in birds and rodents, which involved local administration of the peptide hormone, studies on teleosts except those by Goodson and Bass (2000a) and Backström and Winberg (2009) involved systemic administration of the peptide hormones. Thus, it is difficult to determine the action site for AVT in modulating aggression in teleosts based on the existing literature. One striking difference between teleosts and land vertebrates is that teleosts have AVT/isotocin neurons only in the hypothalamic areas, whereas land vertebrates also have these neurons in other brain areas such as the amygdala and BNST. Therefore, teleosts might use different mechanisms of actions in the hormonal control of aggression compared to land vertebrates. The following section reviews the known action sites of peptide hormones for various behaviors in teleosts.

2.3 Action Sites of Peptide Hormones in Teleosts

2.3.1 Peripheral Action

The distribution of AVT receptors in the whole body is known for the white sucker (*Catostomus commersoni*) (Mahlmann et al. 1994) and flounder (*Platichthys flesus*) (Warne 2001). In the white sucker, the AVT receptor gene is expressed in the pituitary, liver, gill, lateral line, and swim bladder, but not in the brain, aorta, heart, kidney, and reproductive glands (Mahlmann et al. 1994). In the flounder, the AVT receptor gene is expressed in the brain, gill, kidney, and liver (Warne 2001). The reason why the white sucker brain lacks AVT receptor expression is unclear. It might be due to low detection sensitivity of the assay used by Mahlmann et al. (1994) because many species are known to have vasotocinergic projections to the entire brain, described later. AVT receptors in the gill and kidney might be related to osmoregulation. In other regions, however, the function of AVT is unclear (Warne 2001).

Administering mammalian AVP or V1 receptor agonist to the flounder aorta causes elevation in the blood pressure preceding temporal reduction in the dorsal aorta (Warne and Balment 1997a). On the other hand, in the ventral aorta, blood pressure elevation without preceding temporal reduction occurs immediately after AVP or V1 receptor agonist administration (Warne and Balment 1997b). Such changes in blood pressure might be due to constriction of blood vessels in the gill by the action of AVT receptor. Thus, AVT might play a part in local distribution of blood flow rather than systemic modulation of blood pressure (Warne and Balment 1997a, b). In addition, teleosts might also have AVT receptor subtypes similar to those in mammals, because administration of the V2 receptor agonist causes preceding blood pressure reduction, but no elevation after the reduction (Warne and Balment 1997a).

The isotocin receptor gene is expressed in the brain, spleen, lateral line, muscle, gill, intestine, liver, heart, kidney, and ovary in the white sucker (Hausmann et al. 1995). However, the peripheral functions of isotocin are not as clear as that of AVT. Isotocin administration causes blood pressure reduction in the dorsal aorta in the flounder (Warne and Balment 1997b), but not in certain other species (Bennett and Rankin 1986; Kulczykowska 1998; Le Mevel et al. 1993). It is unclear whether such differences in the action of isotocin are due to species differences, and this requires further investigation.

2.3.2 Central Action: Anatomical Studies

Anatomical studies in many species report that AVT and isotocin neurons are distributed in the preoptic nucleus: the goldfish (*Carassius auratus*) (Goodson and Wang 2006; Reaves and Hayward 1980), European plaice (*Pleuronectes platessa*) (Goossens et al. 1977), platy (*Xiphophorus maculatus*) (Schreibman and Halpern 1980), European sea bass (*Dicentrarchus labrax*) (Moons et al. 1988, 1989), white sucker (Gill et al. 1977), Atlantic salmon (*Salmo salar*) (Holmqvist and Ekström 1991), and *Astatotilapia burutoni* (Greenwood et al. 2008).

The preoptic nucleus in teleosts can be subdivided into the parvocellular, magnocellular, and gigantocellular regions (Braford and Northcutt 1983). The number and size of AVT neurons in these regions are known to correlate with individual phenotypes. In general, the number and size of AVT neurons decrease in the parvocellular region, and increase in the gigantocellular region in the territorial and reproductive males, as exemplified by the zebrafish (*Danio rerio*) (Larson et al. 2006), *A. burutoni* (Greenwood et al. 2008), plainfin midshipman (Foran and Bass 1998), peacock blenny (*Salaria pavo*) (Grober et al. 2002), rock-pool blenny (*Parablennius parvicornis*) (Miranda et al. 2003), and masu salmon (*Oncorhynchus masou*) (Ota et al. 1999). On the basis of these studies, it is thought that the AVT neurons in the parvocellular region influence neural circuits related to avoidance or affiliation centrally or through the hypothalamus–pituitary–adrenal axis. Moreover, AVT neurons in the gigantocellular region are thought to influence neural circuits related to courtship or aggressive behavior centrally or through the hypothalamus–pituitary–gonadal axis (Greenwood et al. 2008).

Vasotocinergic/isotocinergic projections to the brain have been investigated in several species: goldfish (Thompson and Walton 2009), rainbow trout (*Salmo gairdneri*) (Van den Dungen et al. 1982), plainfin midshipman (Goodson and Bass 2000b; Goodson et al. 2003), green molly (*Poecilia latipinna*) (Batten et al. 1990), Hawaiian sergeant fish (*Abudefduf abdominalis*) (Maruska 2009), halfspotted goby (*Asterropteryx semipunctata*) (Maruska et al. 2007), multiband butterflyfish, and millet butterflyfish (*Chaetodon multicinctus, Chaetodon milliaris*) (Dewan et al. 2008). However, the distribution of AVT receptors has only been investigated in the sea bass (Moons et al. 1989). As seen in Table 2.2, peptidergic projections and its receptors are broadly distributed throughout the brain.

2.3.3 Central Action: Functional Studies

2.3.3.1 Telencephalon

Vasotocinergic projections to the telencephalon are found in all the species investigated (Table 2.2). In particular, projections to the ventral telencephalon might be more common among species than projections to either the dorsal telencephalon or olfactory bulb. The telencephalon in teleosts undergoes a developmental process known as "eversion," in which the dorsal part of the neural tube extends laterally, whereas the neural tube of land vertebrates develops medially. Thus, homologous regions of the dorsal telencephalon to the mammalian brain remain inconclusive. On the other hand, the ventral telencephalon, which originates from the subpallium, can be considered as homologous to the mammalian striatum and septum (Yamamoto 2009; Yamamoto et al. 2007). In land vertebrates, as described earlier, peptide hormones have been shown to modulate social behavior by acting locally in the subpallial areas.

In teleosts, the telencephalon and preoptic nucleus have been known to be involved in reproductive behaviors. Damage to the telencephalon impairs their nest-building and spawning activities (Overmier and Gross 1974; Davis et al. 1976; Kassel et al. 1976; Schwagmeyer et al. 1977; Kyle et al. 1982; Kyle and Peter 1982; Koyama et al. 1984). Inversely, electronic stimulation of the telencephalon and

Table 2.2 A summary of the distributions of peptidergic projections and AVT receptors in teleost brains

Superoder	Species	AVT/IT	Forebrain		Midbrain	Hindbrain
			Telencephalon	Diencephalon		
Protacan-thopterygii	Rainbow trout (*Salmo gairdneri*)[a]	AVT	Olfactory bulb, Dm, saccus dorsalis of the telencephalon	Preoptic nuclei, nucleus habenularis, nucleus preglomerulosus pars medialis, horizontal commissure	Tectum mesencephali, saccus vasculosus, nucleus recessus lobus lateralis, nucleus recessus posterior, nucreus recessus lateralis	Insertion of cranial nerves III and X, spinal cord
Ostariophysi	Goldfish (*Carassius auratus*)[b]	AVT	Olfactory bulb, Vv, Vl, Vs, Vd, Dm, Dl	Nucleus preopticus, anterior tuberal hypothalamus, preopticohypophysial tract, habenula, posterior tuberculum, nucelus lateralis of the ventral hypothalamus, around the preglomerular complex	Dorsal tegmental nucleus, dorsomedial torus semicircularis, secondary visceral nucleus, optic tectum	Dorsal motor vagus, nucleus of Cajal, area postrema
Paracan-thopterygii	Plainfin midshipman (*Porichthys notatus*)[c,d]	AVT	Vs (dorsal telencephalon)	Preoptic nuclei, ventral tuberal hypothalamus, auditory thalamus, posterior commissure, dorsal and ventral nuclei of the periventricular hypothalamus	Periaqueductal, dorsal tegmental region, isthmal nuclei, paralemniscal regions	Area postrema, dorsolateral to the sonicmotor nuclei
		IT	Vd, Vv, Vp, Vs, (dorsal telencephalon)	Preoptic nuclei, ventral and anterior tuberal hypothalamus, central posterior thalamic nucleus, adjacent to the lateral preglomerular nucleus, preoptico-hypophysial tract, between the medial preglomerular nucleas and periventricular hypothalamus, around the lateral and medial preglomerular nucleus and corpus glomerulosum, pretectum, periventricular nucleus of the posterior tuber, lateral hypothalamus	Optic tectum, paratoral tegmentum, periaqueductal gray, lateral lemniscus, ventral tegmentum, reticular formation, isthmal nucleus, isthmal paraventricular cell group, paralemnis-cal tegmentum, (torus semicircularis)	Adjacent to the sensory nucleus of tigeminal nerve, neuropil region of the ventral medulla, ventral medullary nucleus, ventrolateral to the sonic motor nuclei, area postrema

(continued)

Table 2.2 (continued)

Superoder	Species	AVT/IT	Forebrain		Midbrain	Hindbrain
			Telencephalon	Diencephalon		
Acanthopterygii	Green molly (*Poecilia latipinna*)[e]	AVT and IT	Vs, (dorsal telencephalon)	Preoptic nuclei, nucleus posterior tuberis, nucleus preglomerulosus pars centralis	Nucleus interpeduncularis, isthmal region, nucleus recessus lateralis, nucleus recessus posterioris	Trigeminal nucleus, ventral reticular formation, nucleus solitarius, dorsal vagal motor nucleus
	Damselfish (*Abudefduf abdominalis*)[f]	AVT	Ventral telencephalon, (dorsal telencephalon, olfactory bulb)	Preoptic nuclei, preopticohypophysial tract, (thalamic nuclei, hypothalamus)	Tegmentum, torus semicircularis, (optic tectum)	Facial lobe, sensory vagal lobe, secondary gustatory tract, nuclei of the reticular formation, octavolateralis nuclei, vagal motor nucleus, area postrema, (valvula cerebelli, corpus cerebelli)
	Halfspotted goby (*Asterropteryx semipunctata*)[g]	AVT	Ventral telencephalon, (dorsal telencephalon)	Preoptic nuclei, preopticohypophysial tract, (thalamic nuclei, hypothalamus)	Tegmentum, torus semicircularis, (optic tectum)	Ventral medulla, reticular formation, glossopharyngeal nucleus, vagal motor nucleus, area postrema, (octavolateralis nuclei, valvula cerebelli, corpus cerebelli)

Multiband butterflyfish (*Chaetodon multicinctus*)[b]	AVT	Ventral telencephalon, (dorsal telencephalon)	Preoptic area	Torus semicircularis, tegmentum, optic tectum	Ventral medulla, reticular formation, octavolateralis nuclei, vagal motor nucleus, area postrema
Millet butterflyfish (*Chaetodon milliaris*)[h]					
Sea bass (*Dicentrarchus labrax*)[i]	AVT receptor	Dc, Vv, Vl	Nucleus preopticus, tuberal hypothalamus, around the posterior recess	Optic tectum	Non-cellular layer of thecorpus cerebelli

AVT arginine vasotocin, *IT* isotocin, *Dm, Dl, Dc* medial, lateral, and central zones of area dorsalis telencephali, *Vv, Vd, Vl, Vp, Vs* ventral, dorsal, lateral, postcommissural, and supracommissural nuclei of area ventralis telencephali

Authors:
[a]Van den Dungen et al. (1982)
[b]Thompson and Walton (2009)
[c]Goodson and Bass (2000b)
[d]Goodson et al. (2003)
[e]Batten et al. (1990)
[f]Maruska (2009)
[g]Maruska et al. (2007)
[h]Dewan et al. (2008)
[i]Moons et al. (1989)

preoptic nucleus facilitates nest building, courtship, ejaculation, and aggressive behavior (Demski and Knigge 1971; Demski et al. 1975; Satou et al. 1984). The anatomical and behavioral evidence described above strongly suggest that the peptide hormones modulate such reproduction-related behaviors by acting at the telencephalon. However, no study to date has directly examined such a relationship. This challenging topic requires future research.

Brain areas other than the telencephalon might also play an important role in the modulation of reproduction-related behaviors under the influence of peptide hormones. Intraperitoneal (IP) injections of crude fish pituitary extracts cause spawning reflex responses in hypophysectomized killifish (Pickford 1952). This spawning reflex response is effectively caused by AVT injections (Pickford and Strecker 1977). ICV administration of AVT has longer latency in mediating this spawning reflex (Pickford et al. 1980). More critically, telencephalic ablation does not disrupt the spawning reflex (Knight and Knight 1996) suggesting that areas other than the telencephalon might be the action site of AVT in this case.

2.3.3.2 Diencephalon/Midbrain

The plainfin midshipman, a sound-producing fish, has brain nuclei devoted to sound production, which receive projections from the AVT and isotocin neurons (Bass et al. 1994; Goodson and Bass 2000b, 2002; Goodson et al. 2003). The local administration of AVT to the anterior hypothalamus, which is their diencephalic sound-producing center, decreases duration of sound production in the territorial male. Conversely, administration of the V1a receptor antagonist increases the duration of sound production. In a similar manner, local administration of isotocin and oxytocin receptor antagonist to the anterior hypothalamus in females and nonterritorial sneaker males shows similar effects to that of the territorial male. Thus, peptidergic modulation of sound production in the anterior hypothalamus depends on phenotypic differences, but not on the gonadal sex (Goodson and Bass 2000a).

Similarly, local administration of AVT to the paralemniscal midbrain tegmentum, which is the midbrain sound-producing nucleus, increases latency, while decreasing the number of sound production in the territorial male. Midbrain AVT might not be used for normal sound production, but might be used for conditional suppression because administration of V1a receptor antagonist in this region does not have any effect on sound production (Goodson and Bass 2000b).

2.3.3.3 Hindbrain

The goldfish is known to have a tendency to approach visually presented conspecifics. This approaching behavior is decreased by ICV administration of AVT, and increased by isotocin administration (Thompson and Walton 2004). AVT neurons in goldfish are distributed in the preoptic nucleus similar to other teleosts. Its projection to the lateral zone of the dorsal telencephalon might be independently evolved in goldfish because this projection is not found in other teleosts (Table 2.2) (Thompson and Walton 2009). However, the dorsal telencephalon might not be involved in AVT modulation of approaching behavior because lesions in the dorsal telencephalon do not alter this behavior (Shinozuka and Watanabe 2004). Instead, peptide hormones might act on the hindbrain because ICV injection to the fourth ventricle has a stronger effect than an injection to the third ventricle (Thompson et al. 2008). Moreover, seasonal changes in the goldfish's tendency to approach conspecifics correlate with the expression levels of AVT receptors in the hindbrain (Walton et al. 2010).

The dorsal motor vagus, a nucleus in the hindbrain, possesses substance P neurons that receive projections from AVT neurons; thus, substance P release from this nucleus to the periphery might be modulated by AVT neurons (Thompson et al. 2008). In fact, IP injection of substance P antagonist blocks the effect of ICV injection of AVT on the approaching behavior in goldfish (Thompson et al. 2008). Thus, AVT centrally modulates peripheral release of substance P; in turn, the approaching behavior might be altered due to feedback from the peripheral changes caused by substance P release (Thompson et al. 2008).

As already described, a wide variety of action sites for AVT and isotocin in teleosts are known. In future research, it will be important to identify the action site(s) for peptide hormones with respect to specific behavior paradigms, to clearly identify the action site responsible for modulating a particular behavior.

2.4 Conclusion

In this chapter the functions of peptide hormones on aggression are reviewed. Peptide hormones appear to broadly affect aggression among vertebrates. Behavioral studies show that the function of peptide hormones on aggression markedly differs among species. Bird studies and, in part, some studies on teleosts and rodents suggest a correlation between AVT function and social structure, but there are some inconsistencies in the results reported. Two important aspects for future research on peptide modulation of aggressive behavior are to acquire a dose–response curve for each peptide in each species, and to investigate agonistic display and overt aggression simultaneously in the species being investigated. In teleosts, only a few studies have been conducted with local administration of peptide hormones, whereas anatomical studies have indicated that there are many action sites for these peptide hormones in both the brain and the periphery in teleosts. Therefore, studies are required to identify the action site(s) for peptide hormones associated with specific behaviors to compare the mechanism of their action among vertebrates. Future studies incorporating these issues raised will contribute toward understanding and comparing peptide hormone functions in aggression among species, as well as to understand how their functions have evolved.

References

Albers EH, Dean A, Karom MC, Smith D, Huhman KL (2006) Role of V1a vasopressin receptors in the control of aggression in Syrian hamsters. Brain Res 1073:425–430

Backström T, Winberg S (2009) Arginine-vasotocin influence on aggressive behavior and dominance in rainbow trout. Physiol Behav 96:470–475

Baerends GP, Baerends-van Roon JM (1950) An introduction to the study of the ethology of the cichlid fishes. Behaviour Supplement 1:1–243

Barlow GW (2000) The cichlid fishes: nature's grand experiment in evolution. Perseus Publishing, Cambridge

Bass AH, Marchaterre MA, Baker R (1994) Vocal-acoustic pathways in a teleost fish. J Neurosci 14:4025–4039

Bastian J, Schniederjan S, Nguyenkim J (2001) Arginine vasotocin modulates a sexually dimorphic communication behavior in the weakly electric fish *Apteronotus leptorhynchus*. J Exp Biol 204:1909–1923

Batten TFC, Cambre ML, Moons L, Vandesande F (1990) Comparative distribution of neuropeptide-immunoreactive systems in the brain of the green molly, *Poecilia latipinna*. J Comp Neurol 302:893–919

Beiderbeck DI, Neumann ID, Veenema AH (2007) Differences in intermale aggression are accompanied by opposite vasopressin release patterns within the septum in rats bred for low and high anxiety. Eur J Neurosci 26:3597–3605

Bennett MB, Rankin JC (1986) The effects of neurohypophysial hormones on the vascular resistance of the isolated perfused gill of the European eel, *Anguilla anguilla* L. Gen Comp Endocrinol 64:60–66

Bester-Meredith JK, Marler CA (2001) Vasopressin and aggression in cross-fostered California mice (*Peromyscus californicus*) and white-footed mice (*Peromyscus leucopus*). Horm Behav 40:51–64

Bester-Meredith JK, Young LJ, Marler CA (1999) Species differences in paternal behavior and aggression in *Peromyscus* and their associations with vasopressin immunoreactivity and receptors. Horm Behav 36:25–38

Bester-Meredith JK, Martin PA, Marler CA (2005) Manipulations of vasopressin alter aggression differently across testing conditions in monogamous and non-monogamous *Peromyscus* mice. Aggress Behav 31:189–199

Bosch OJ, Meddle SL, Beiderbeck DI, Douglas AJ, Neumann ID (2005) Brain oxytocin correlates with maternal aggression: link to anxiety. J Neurosci 25:6807–6815

Braford MR Jr, Northcutt RG (1983) Organization of the diencephalon and pretectum of the ray-finned fishes. In: Davis RE, Northcutt RG (eds) Fish neurobiology, vol 2. University of Michigan Press, Ann Arbor, pp 117–163

Bronstein PM, Brain PF (1991) Successful prediction of dominance in convict cichlids, *Cichlasoma nigrofasciatum*. Bull Psychonom Soc 29:455–456

Bshary R, Wickler W, Fricke H (2002) Fish cognition: a primate's eye view. Anim Cogn 5:1–13

Cheng SY, Delville Y (2009) Vasopressin facilitates play fighting in juvenile golden hamsters. Physiol Behav 98:242–246

Cooper MA, Karom M, Huhman KL, Elliott Albers H (2005) Repeated agonistic encounters in hamsters modulate AVP V1a receptor binding. Horm Behav 48:545–551

Davis RE, Kassel J, Schwagmeyer P (1976) Telencephalic lesions and behavior in the teleost, *Macropodus opercularis*: reproduction, startle reaction, and operant behavior in the male. Behav Biol 18:165–177

Delville Y, Mansour KM, Ferris CF (1996) Testosterone facilitates aggression by modulating vasopressin receptors in the hypothalamus. Physiol Behav 60:25–29

Demski LS, Knigge KM (1971) The telencephalon and hypothalamus of the bluegill (*Lepomis macrochirus*): evoked feeding, aggressive and reproductive behavior with representative frontal sections. J Comp Neurol 143:1–16

Demski LS, Bauer DH, Gerald JW (1975) Sperm release evoked by electrical stimulation of the fish brain: a functional-anatomical study. J Exp Zool 191:215–231

Dewan AK, Maruska KP, Tricas TC (2008) Arginine vasotocin neuronal phenotypes among congeneric territorial and shoaling reef butterflyfishes: species, sex and reproductive season comparisons. J Neuroendocrinol 20:1382–1394

Dobberfuhl AP, Ullmann JFP, Shumway CA (2005) Visual acuity, environmental complexity, and social organization in African cichlid fishes. Behav Neurosci 119:1648

Enquist M, Leimar O (1983) Evolution of fighting behaviour: decision rules and assessment of relative strength. J Theor Biol 102:387–410

Enquist M, Leimar O, Ljungberg T, Mallner Y, Segerdahl N (1990) A test of the sequential assessment game: fighting in the cichlid fish *Nannacara anomala*. Anim Behav 40:1–14

Everts HGJ, De Ruiter AJH, Koolhaas JM (1997) Differential lateral septal vasopressin in wild-type rats: correlation with aggression. Horm Behav 31:136–144

Ferris CF, Delville Y (1994) Vasopressin and serotonin interactions in the control of agonistic behavior. Psychoneuroendocrinology 19:593–601

Ferris CF, Potegal M (1988) Vasopressin receptor blockade in the anterior hypothalamus suppresses aggression in hamsters. Physiol Behav 44:235–239

Ferris CF, Melloni RH Jr, Koppel G, Perry KW, Fuller RW, Delville Y (1997) Vasopressin/serotonin interactions in the anterior hypothalamus control aggressive behavior in golden hamsters. J Neurosci 17:4331

Ferris CF, Lu S, Messenger T, Guillon CD, Heindel N, Miller M et al (2006) Orally active vasopressin V1a receptor antagonist, SRX251, selectively blocks aggressive behavior. Pharmacol Biochem Behav 83:169–174

Fisher J (1954) Evolution and bird sociality. In: Huxley J, Hardy AC, Ford FB (eds) Evolution as a process. G. Allen & Unwin, London, pp 71–83

Foran CM, Bass AH (1998) Preoptic AVT immunoreactive neurons of a teleost fish with alternative reproductive tactics. Gen Comp Endocrinol 111:271–282

Fricke HW (1973) Individual partner recognition in fish: field studies on *Amphiprion bicinctus*. Naturwissenschaften 60:204–205

Gill VE, Burford GD, Lederis K, Zimmerman EA (1977) An immunocytochemical investigation for arginine vasotocin and neurophysin in the pituitary gland and the caudal neurosecretory system of *Catostomus commersoni*. Gen Comp Endocrinol 32:505–511

Goodson JL (1998a) Territorial aggression and dawn song are modulated by septal vasotocin and vasoactive intestinal polypeptide in male field sparrows (*Spizella pusilla*). Horm Behav 34:67–77

Goodson JL (1998b) Vasotocin and vasoactive intestinal polypeptide modulate aggression in a territorial songbird, the violet-eared waxbill (Estrildidae: *Uraeginthus granatina*). Gen Comp Endocrinol 111:233–244

Goodson JL, Adkins-Regan E (1999) Effect of intraseptal vasotocin and vasoactive intestinal polypeptide infusions on courtship song and aggression in the male zebra finch (*Taeniopygia guttata*). J Neuroendocrinol 11:19–25

Goodson JL, Bass AH (2000a) Forebrain peptides modulate sexually polymorphic vocal circuitry. Nature 403:769–772

Goodson JL, Bass AH (2000b) Vasotocin innervation and modulation of vocal-acoustic circuitry in the teleost *Porichthys notatus*. J Comp Neurol 422:363–379

Goodson JL, Bass AH (2001) Social behavior functions and related anatomical characteristics of vasotocin/vasopressin systems in vertebrates. Brain Res Rev 35:246–265

Goodson JL, Bass AH (2002) Vocal-acoustic circuitry and descending vocal pathways in teleost fish: convergence with terrestrial vertebrates reveals conserved traits. J Comp Neurol 448:298–322

Goodson JL, Evans AK (2004) Neural responses to territorial challenge and nonsocial stress in male song sparrows: segregation, integration, and modulation by a vasopressin V_1 antagonist. Horm Behav 46:371–381

Goodson JL, Wang Y (2006) Valence-sensitive neurons exhibit divergent functional profiles in gregarious and asocial species. Proc Natl Acad Sci USA 103:17013–17017

Goodson JL, Evans AK, Bass AH (2003) Putative isotocin distributions in sonic fish: relation to vasotocin and vocal-acoustic circuitry. J Comp Neurol 462:1–14

Goodson JL, Lindberg L, Johnson P (2004) Effects of central vasotocin and mesotocin manipulations on social behavior in male and female zebra finches. Horm Behav 45:136–143

Goodson JL, Evans AK, Wang Y (2006) Neuropeptide binding reflects convergent and divergent evolution in species-typical group sizes. Horm Behav 50:223–236

Goodson JL, Kabelik D, Schrock SE (2009a) Dynamic neuromodulation of aggression by vasotocin: influence of social context and social phenotype in territorial songbirds. Biol Lett 5:554

Goodson JL, Rinaldi J, Kelly AM (2009b) Vasotocin neurons in the bed nucleus of the stria terminalis preferentially process social information and exhibit properties that dichotomize courting and non-courting phenotypes. Horm Behav 55:197–202

Goodson JL, Schrock SE, Klatt JD, Kabelik D, Kingsbury MA (2009c) Mesotocin and nonapeptide receptors promote estrildid flocking behavior. Science 325:862–866

Goossens N, Dierickx K, Vandesande F (1977) Immunocytochemical localization of vasotocin and isotocin in the preopticohypophysial neurosecretory system of teleosts. Gen Comp Endocrinol 32:371–375

Greenwood AK, Wark AR, Fernald RD, Hofmann HA (2008) Expression of arginine vasotocin in distinct preoptic regions is associated with dominant and subordinate behaviour in an African cichlid fish. Proc R Soc Biol Sci 275:2393–2402

Grober MS, George AA, Watkins KK, Carneiro LA, Oliveira RF (2002) Forebrain AVT and courtship in a fish with male alternative reproductive tactics. Brain Res Bull 57:423–425

Grosenick L, Clement TS, Fernald RD (2007) Fish can infer social rank by observation alone. Nature 445:429–432

Harmon AC, Huhman KL, Moore TO, Albers HE (2002) Oxytocin inhibits aggression in female Syrian hamsters. J Neuroendocrinol 14:963–969

Hausmann H, Meyerhof W, Zwiers H, Lederis K, Richter D (1995) Teleost isotocin receptor: structure, functional expression, mRNA distribution and phylogeny. FEBS Lett 370:227–230

Heinrichs M, Meinlschmidt G, Wippich W, Ehlert U, Hellhammer DH (2004) Selective amnesic effects of oxytocin on human memory. Physiol Behav 83:31–38

Hollander E, Bartz J, Chaplin W, Phillips A, Sumner J, Soorya L et al (2007) Oxytocin increases retention of social cognition in autism. Biol Psychiatry 61:498–503

Holmqvist BI, Ekström P (1991) Galanin like immunoreactivity in the brain of teleosts: distribution and relation to substance P, vasotocin, and isotocin in the Atlantic salmon (*Salmo salar*). J Comp Neurol 306:361–381

Huber R, Van Staaden MJ, Kaufman LS, Liem KF (1997) Microhabitat use, trophic patterns, and the evolution of brain structure in African cichlids. Brain Behav Evol 50:167–182

Insel TR, Hulihan TJ (1995) A gender-specific mechanism for pair bonding: oxytocin and partner preference formation in monogamous voles. Behav Neurosci 109:782–789

Kassel J, Davis RE, Schwagmeyer P (1976) Telencephalic lesions and behavior in the teleost, *Macropodus opercularis*: further analysis of reproductive and operant behavior in the male. Behav Biol 18:179–188

Keeley ER, Grant JWA (1993) Visual information, resource value, and sequential assessment in convict cichlid (*Cichlasoma nigrofasciatum*) contests. Behav Ecol 4:345–349

Knight WR, Knight JN (1996) Telencephalon removal does not disrupt the vasotocin induced spawning reflex in killifish, *Fundulus heteroclitus*. J Exp Zool 276:296–300

Kosfeld M, Heinrichs M, Zak PJ, Fischbacher U, Fehr E (2005) Oxytocin increases trust in humans. Nature 435:673–676

Kotrschal K, Van Staaden MJ, Huber R (1998) Fish brains: evolution and environmental relationships. Rev Fish Biol Fish 8:373–408

Koyama Y, Satou M, Oka Y, Ueda K (1984) Involvement of the telencephalic hemispheres and the preoptic area in sexual behavior of the male goldfish, *Carassius auratus*: a brain-lesion study. Behav Neural Biol 40:70–86

Kulczykowska E (1998) Effects of arginine vasotocin, isotocin and melatonin on blood pressure in the conscious Atlantic cod (*Gadus morhua*): hormonal interactions? Exp Physiol 83:809–820

Kyle AL, Peter RE (1982) Effects of forebrain lesions on spawning behaviour in the male goldfish. Physiol Behav 28:1103–1109

Kyle AL, Stacey NE, Peter RE (1982) Ventral telencephalic lesions: effects on bisexual behavior, activity, and olfaction in the male goldfish. Behav Neural Biol 36:229–241

Larson ET, O'Malley DM, Melloni RH Jr (2006) Aggression and vasotocin are associated with dominant-subordinate relationships in zebrafish. Behav Brain Res 167:94–102

Le Mevel JC, Pamantung TF, Mabin D, Vaudry H (1993) Effects of central and peripheral administration of arginine vasotocin and related neuropeptides on blood pressure and heart rate in the conscious trout. Brain Res 610:82–89

Leiser JK, Itzkowitz M (1999) The benefits of dear enemy recognition in three-contender convict cichlid (*Cichlasoma nigrofasciatum*) contests. Behaviour 136:983–1003

Lema SC, Nevitt GA (2004) Exogenous vasotocin alters aggression during agonistic exchanges in male Amargosa River pupfish (*Cyprinodon nevadensis amargosae*). Horm Behav 46:628–637

Lim MM, Wang Z, Olazábal DE, Ren X, Terwilliger EF, Young LJ (2004) Enhanced partner preference in a promiscuous species by manipulating the expression of a single gene. Nature 429:754–757

Mahlmann S, Meyerhof W, Hausmann H, Heierhorst J, Schönrock C, Zwiers H et al (1994) Structure, function, and phylogeny of [Arg8] vasotocin receptors from teleost fish and toad. Proc Natl Acad Sci USA 91:1342–1345

Maney DL, Goode CT, Wingfield JC (1997) Intraventricular infusion of arginine vasotocin induces singing in a female songbird. J Neuroendocrinol 9:487–491

Maruska KP (2009) Sex and temporal variations of the vasotocin neuronal system in the damselfish brain. Gen Comp Endocrinol 160:194–204

Maruska KP, Mizobe MH, Tricas TC (2007) Sex and seasonal co-variation of arginine vasotocin (AVT) and gonadotropin-releasing hormone (GnRH) neurons in the brain of the halfspotted goby. Comp Biochem Physiol A Mol Integr Physiol 147:129–144

Miranda JA, Oliveira RF, Carneiro LA, Santos RS, Grober MS (2003) Neurochemical correlates of male polymorphism and alternative reproductive tactics in the Azorean rock-pool blenny. *Parablennius parvicornis*. Gen Comp Endocrinol 132:183–189

Moons L, Cambré M, Marivoet S, Batten TFC, Vanderhaeghen JJ, Ollevier F et al (1988) Peptidergic innervation of the adrenocorticotropic hormone (ACTH)-and growth hormone (GH)-producing cells in the pars distalis of the sea bass (*Dicentrarchus labrax*). Gen Comp Endocrinol 72:171–180

Moons L, Cambré M, Ollevier F, Vandesande F (1989) Immunocytochemical demonstration of close relationships between neuropeptidergic nerve fibers and hormone-producing cell types in the adenohypophysis of the sea bass (*Dicentrarchus labrax*). Gen Comp Endocrinol 73:270–283

Myrberg AA (1985) Acoustically mediated individual recognition by a coral reef fish (*Pomacentrus partitus*). Anim Behav 33:411–416

Myrberg AA, Spires JY (1972) Sound discrimination by the bicolor damselfish, *Eupomacentrus partitus*. J Exp Biol 57:727–735

Nephew BC, Bridges RS (2008) Central actions of arginine vasopressin and a V1a receptor antagonist on maternal aggression, maternal behavior, and grooming in lactating rats. Pharmacol Biochem Behav 91:77–83

Nephew BC, Byrnes EM, Bridges RS (2010) Vasopressin mediates enhanced offspring protection in multiparous rats. Neuropharmacology 58:102–106

Olazábal DE, Young LJ (2006) Oxytocin receptors in the nucleus accumbens facilitate "spontaneous" maternal behavior in adult female prairie voles. Neuroscience 141:559–568

Oliveira RF, McGregor PK, Latruffe C (1998) Know thine enemy: fighting fish gather information from observing conspecific interactions. Proc R Soc Lond B Biol Sci 265:1405–1409

Ota Y, Ando H, Ueda H, Urano A (1999) Differences in seasonal expression of neurohypophysial hormone genes in ordinary and precocious male masu salmon. Gen Comp Endocrinol 116:40–48

Overmier JB, Gross D (1974) Effects of telencephalic ablation upon nest-building and avoidance behaviors in East African mouthbreeding fish, *Tilapia mossambica*. Behav Biol 12:211–222

Peake TM, McGregor PK (2004) Information and aggression in fishes. Learn Behav 32:114–121

Pickford GE (1952) Induction of a spawning reflex in hypophysectomized killifish. Nature 170:807–808

Pickford GE, Strecker EL (1977) The spawning reflex response of the killifish, *Fundulus heteroclitus*: isotocin is relatively inactive in comparison with arginine vasotocin. Gen Comp Endocrinol 32:132–137

Pickford GE, Knight WR, Knight JN (1980) Where is the spawning reflex receptor for neurohypophysial peptides in the killifish, *Fundulus heteroclitus*? Revue canadienne de biologie/éditée par l'Université de Montréal 39:97–105

Pitkow LJ, Sharer CA, Ren X, Insel TR, Terwilliger EF, Young LJ (2001) Facilitation of affiliation and pair-bond formation by vasopressin receptor gene transfer into the ventral forebrain of a monogamous vole. J Neurosci 21:7392–7396

Pollen AA, Dobberfuhl AP, Scace J, Igulu MM, Renn SCP, Shumway CA et al (2007) Environmental complexity and social organization sculpt the brain in Lake Tanganyikan cichlid fish. Brain Behav Evol 70:21–39

Portavella M, Vargas JP (2005) Emotional and spatial learning in goldfish is dependent on different telencephalic pallial systems. Eur J Neurosci 21:2800–2806

Reaves TA Jr, Hayward JN (1980) Functional and morphological studies of peptide-containing neuroendocrine cells in goldfish hypothalamus. J Comp Neurol 193:777–788

Saito K, Watanabe S (2004) Spatial learning deficits after the development of dorsomedial telencephalon lesions in goldfish. Neuroreport 15:2695–2699

Saito K, Watanabe S (2006) Deficits in acquisition of spatial learning after dorsomedial telencephalon lesions in goldfish. Behav Brain Res 172:187–194

Santangelo N, Bass AH (2006) New insights into neuropeptide modulation of aggression: field studies of arginine vasotocin in a territorial tropical damselfish. Proc R Soc Biol Sci 273:3085–3092

Satou M, Oka Y, Kusunoki M, Matsushima T, Kato M, Fujita I et al (1984) Telencephalic and preoptic areas integrate sexual behavior in hime salmon (landlocked red salmon, *Oncorhynchus nerka*): results of electrical brain stimulation experiments. Physiol Behav 33:441–447

Savaskan E, Ehrhardt R, Schulz A, Walter M, Schächinger H (2008) Post-learning intranasal oxytocin modulates human memory for facial identity. Psychoneuroendocrinology 33:368–374

Schreibman MP, Halpern LR (1980) The demonstration of neurophysin and arginine vasotocin by immunocytochemical methods in the brain and pituitary gland of the platyfish, *Xiphophorus maculatus*. Gen Comp Endocrinol 40:1–7

Schwagmeyer P, Davis RE, Kassel J (1977) Telencephalic lesions and behavior in the teleost *Macropodus opercularis* (L.): effects of telencephalon and olfactory bulb ablation on spawning and foamnest building. Behav Biol 20:463–470

Semsar K, Godwin J (2004) Multiple mechanisms of phenotype development in the bluehead wrasse. Horm Behav 45:345–353

Semsar K, Kandel FLM, Godwin J (2001) Manipulations of the AVT system shift social status and related courtship and aggressive behavior in the bluehead wrasse. Horm Behav 40:21–31

Shinozuka K, Watanabe S (2004) Effects of telencephalic ablation on shoaling behavior in goldfish. Physiol Behav 81:141–148

Takayanagi Y, Yoshida M, Bielsky IF, Ross HE, Kawamata M, Onaka T et al (2005) Pervasive social deficits, but normal parturition, in oxytocin receptor-deficient mice. Proc Natl Acad Sci USA 102:16096–16101

Thompson TI (1963) Visual reinforcement in Siamese fighting fish. Science 141:55–57

Thompson RR, Walton JC (2004) Peptide effects on social behavior: effects of vasotocin and isotocin on social approach behavior in male goldfish (*Carassius auratus*). Behav Neurosci 118:620–626

Thompson RR, Walton JC (2009) Vasotocin immunoreactivity in goldfish brains: characterizing primitive circuits associated with social regulation. Brain Behav Evol 73:153–164

Thompson RR, Walton JC, Bhalla R, George KC, Beth EH (2008) A primitive social circuit: vasotocin-substance P interactions modulate social behavior through a peripheral feedback mechanism in goldfish. Eur J Neurosci 27:2285–2293

Tinbergen N (1951) The study of instinct. Oxford University Press, Oxford

Van den Dungen HM, Buijs RM, Pool CW, Terlou M (1982) The distribution of vasotocin and isotocin in the brain of the rainbow trout. J Comp Neurol 212:146–157

van Staaden MJ, Huber R, Kaufman LS, Liem KF (1994) Brain evolution in cichlids of the African Great Lakes: brain and body size, general patterns, and evolutionary trends. Zoology 98:165–178

Walton JC, Waxman B, Hoffbuhr K, Kennedy M, Beth E, Scangos J et al (2010) Behavioral effects of hindbrain vasotocin in goldfish are seasonally variable but not sexually dimorphic. Neuropharmacology 58:126–134

Wang Z, Ferris CF, De Vries GJ (1994) Role of septal vasopressin innervation in paternal behavior in prairie voles (*Microtus ochrogaster*). Proc Natl Acad Sci USA 91:400–404

Warne JM (2001) Cloning and characterization of an arginine vasotocin receptor from the euryhaline flounder *Platichthys flesus*. Gen Comp Endocrinol 122:312–319

Warne JM, Balment RJ (1997a) Changes in plasma arginine vasotocin (AVT) concentration and dorsal aortic blood pressure following AVT injection in the teleost *Platichthys flesus*. Gen Comp Endocrinol 105:358–364

Warne JM, Balment RJ (1997b) Vascular actions of neurohypophysial peptides in the flounder. Fish Physiol Biochem 17:313–318

Wilson EO (1975) Sociobiology: the new synthesis. Harvard University Press, Cambridge

Yamamoto N (2009) Studies on the teleost brain morphology in search of the origin of cognition. Jpn Psychol Res 51:154–167

Yamamoto N, Ishikawa Y, Yoshimoto M, Xue HG, Bahaxar N, Sawai N et al (2007) A new interpretation on the homology of the teleostean telencephalon based on hodology and a new eversion model. Brain Behav Evol 69:96–104

Young LJ, Wang Z (2004) The neurobiology of pair bonding. Nat Neurosci 7:1048–1054

Young LJ, Winslow JT, Nilsen R, Insel TR (1997) Species differences in V a receptor gene expression in monogamous and nonmonogamous voles: behavioral consequences. Behav Neurosci 111:599–605

Young LJ, Lim MM, Gingrich B, Insel TR (2001) Cellular mechanisms of social attachment. Horm Behav 40:133–138

Zak PJ, Kurzban R, Matzner WT (2005) Oxytocin is associated with human trustworthiness. Horm Behav 48:522–527

Chapter 3
Emotional Birds—Or Advanced Cognitive Processing?

Irene M. Pepperberg

Abstract Grey parrots (*Psittacus erithacus*) have been shown to exhibit many complex cognitive and communicative abilities in a laboratory setting. The parrots' successes likely rely on two factors: (a) an underlying neurological architecture that supports complex information processing, and (b) training involving social interaction and contextually-applicable rewards that enables them to express their capacities in ways measurable by human researchers (Pepperberg, *The Alex Studies*, 1999). Sometimes their behavior could not, however, be predicted from either their biology or their training, but rather involved what, in humans, would be called an emotional

---

I.M. Pepperberg (✉)
Department of Psychology, Harvard University, William James Hall,
33 Kirkland Street, Cambridge, MA 02138, USA
e-mail: impepper@media.mit.edu; impepper@wjh.harvard.edu

response to a situation. This paper describes four such situations, involving object permanence, phonation, insightful string-pulling, and numerical concepts. Each of these situations also required considerable cognitive processing; interestingly, such processing was seemingly stimulated by the concomitant emotional state. I thus suggest the possible existence in Grey parrots of a connection between emotional responses and cognitive processing such that their interaction synergistically supports successful outcomes, possibly related to how affect influences mental processing in humans.

Grey parrots (*Psittacus erithacus*) have been shown to exhibit many complex cognitive and communicative abilities in a laboratory setting (e.g., Pepperberg 1999, 2004, 2006a, b, 2007; Pepperberg and Gordon 2005; Pepperberg and Shive 2001; Pepperberg and Wilcox 2000; Pepperberg et al. 1997, 2008; Pepperberg and Carey accepted pending revision). Alex, the most studied of these birds, learned, for example, to identify over 50 different objects using sounds of English speech; he labeled seven colors, five shapes; he could label quantities up to eight, even when presented as subsets within heterogeneous collections (e.g., could quantify the number of blue keys in a mixture of red and blue keys and red and blue blocks), could add small sets up to six, and deduced the ordinality of his numbers; he inferred the equivalence among Arabic numerals, vocal English number labels, and sets of objects. He demonstrated full object permanence, concepts of category, same/different, relative size, and absence (i.e., vocally reported on what attribute was same or different between two objects or which was bigger or smaller; he stated "none" if nothing was same or different, if they were of equal size, or if a designated quantity was missing from a collection). His comprehension and production skills were equivalent, and he understood conjunctive conditions (i.e., could provide information about the specific instance of one category of an item uniquely defined by the conjunction of two other categories; for example, "What object is color-A *and* shape-B?"). He used "I want X" and "Wanna go Y" appropriately, where X and Y were, respectively, object or location labels and these requests were intentional: If trainers responded incorrectly (e.g., substituting alternative items), he either refused the item, tossed it back at the trainer, or coupled refusals with a repetition of the initial request. He questioned trainers in order to learn labels for novel objects, shapes, or colors. He combined labels to request, refuse, categorize or quantify over 100 different items. Other Grey parrots (e.g., Griffin) have also acquired some of these abilities. The parrots' successes likely rely on two factors: (a) an underlying neurological architecture that supports complex information processing (e.g., Jarvis et al. 2005), and (b) training involving social interaction and contextually-applicable rewards that enables them to express their capacities in ways measurable by human researchers (Pepperberg 1999).

Sometimes their behavior could not, however, be predicted from either their biology or their training, but rather involved what, in humans, would be called an emotional response to a situation. I describe four such situations, involving object permanence, phonation, insightful string-pulling, and numerical concepts.

Each of these situations also required considerable cognitive processing; interestingly, such processing was seemingly stimulated by the concomitant emotional state. I thus suggest the possible existence in Grey parrots of a connection between emotional responses and cognitive processing such that their interaction synergistically supports successful outcomes, possibly related to how affect influences mental processing in humans (e.g., Ciompi 1991; Pessoa 2008; Salzman and Fusi 2010).

3.1 The Four Studies

A brief description of the various tasks that the parrots experienced demonstrates how each task required a certain level of cognitive processing for its solution and how the experimenters often manipulated the task in some manner to challenge the bird, generally (but not always) with respect to the reward system. Of note is not only that the birds demonstrated particular, often unexpected responses to these challenges, but that their responses were also emotional in nature, and that the emotionality seemingly potentiated their accessing, and our being able to evaluate, the cognitive ability demonstrated.

3.1.1 Object Permanence Experiment

As part of a study on object permanence, we replicated all the standard tasks given young children, in which objects to be recovered are hidden in a myriad of different ways, varying from simple covers to complicated "shell games". We also, however, included a task that required the parrots to demonstrate knowledge of exactly *what* was hidden (Pepperberg et al. 1997). The subjects, Alex and Griffin, were tricked into believing that "X" (a particularly desirable treat such as a whole almond) was hidden, but after responding appropriately by detecting the position of the treat, found "Y" (something less desirable, such as parrot chow). The subjects reacted with what in humans would be described as surprise or anger, engaging in rigorous beak-banging and tossing of the covers used to hide the items. It was, of course, these *emotional* responses that enabled the experimenters to know that the birds had experienced a violation of expectations, a sort of cognitive dissonance, as well as a recognition of the immutability of items. That is, the parrots' responses showed that they knew that they had solved the task correctly, knew that objects did not spontaneously mutate into other objects, and demonstrated that they hadn't just expected to find *something* of interest but rather a very specific item, which was not indeed present. Had they, for example, simply stared at the substituted item, we would have found it more difficult to argue that they were aware of the substitution.

3.1.2 Phonological Awareness

Alex had been trained to label human phonemes (Pepperberg 2007), that is, to associate colored plastic and wooden alphabet letters B, CH, I, K, N, OR, S, SH, T with their corresponding appropriate phonological sounds (e.g., I with /i/). The various letters, each of a different color, with the colors changed on each trial, were spread out on a tray; in the task, Alex would be asked, for example, "What sound is blue?" or "What color is /s/?" His reward was initially the opportunity to chew on the letters; later, when the novelty of the letters wore off, Alex would ask for other desirable items or actions ("Want nut!", "Want tickle") instead.

In the situation of interest, students and I were at the MIT Media Lab, demonstrating this set of phonological abilities to a number of CEOs of various United States and international corporations, whose schedule allowed only about 5–7 min with us. Alex was correctly responding to questions such as "What color /sh/?" and requesting nuts as his reward. So that the short time available for the demonstration would not be spent watching Alex eat, after each response he was told to wait, that he could have his reward after a few more trials. Such a comment from trainers was unusual, in that he was always rewarded immediately after being told he was correct; only after an error, when he heard "no", were his requests denied. He became more and more agitated, asking for nuts with more emphasis after each correct response: His vocalizations became louder, with more emphasis on the label "nut". Finally, he looked at me and said, "Wanna nut...N-U-T!!", actually stating the individual sounds "nnn", "uhh", and "t" (/n/, /ə/, /t/, respectively). He had never been trained to perform this kind of task, and although N and T had been trained individually, U had not.

Whether or not his behavior was a direct consequence of some emotional state, he had clearly jumped ahead of us scientifically, demonstrating that he could not only sound out the alphabet letters but also segment and sound out English labels, a skill that had not been taught. (And, of course, a humorous interpretation, essentially implied, would be something like, "Hey, stupid human, must I spell it out for you?"). Whatever the motivation for his response, it was unlikely to have appeared had he not been thwarted (i.e., had his expectations not been violated), had he not been somewhat angered and in search of some way to obtain his desired goal.

Alex's segmentation abilities and capacity for phonological awareness were not tested further in that manner, but his subsequent pattern of acquisition of two novel labels, *spool* and *seven*, demonstrated these abilities (Pepperberg 2007). I describe their acquisition simply to demonstrate that the N–U–T example may have been the first sign of his capacity for related advanced skills. In contrast to his early acquisition pattern (for a given label, first the vowels, then the less difficult consonants—ones that didn't involve lips—then finally the difficult vowels; see Pepperberg 2007), Alex now began using combinations of existing phonemes and labels to identify, respectively, a wooden bobbin and the Arabic numeral 7: /s/ (unvoiced, trained in conjunction with the alphabet letter, S) with, for the former, *wool*, to form "s" (pause) "wool" ("s-wool"; /s-pause-wUl/); and with, for the latter, *one*, to form "s" (pause) "wun" ("s-wun", /s-pause-/wən/"). The pauses initially provided space for the absent (and difficult sans lips) respective /p/ and /v/ (preserving the number of syllables or prosodic rhythm of the targeted vocalization). Alex retained these forms for months, although usually only days or weeks of training enabled learning of a new label having existent phonemes (Pepperberg 1999). Eventually he spontaneously produced a perfectly formed "spool" (/spul/) and, just before his death, "seben" (/sEbIn/). Although his combinatory rule system was relatively limited, the data show intriguing parallels between Alex's and young children's early label acquisition (Pepperberg 2007).

3.1.3 Insightful String Pulling

This task involved the ability, without training or prior experience, to obtain food suspended by a string by reaching down, pulling up a loop of string onto the perch, stepping on the loop to secure it, and repeating the sequence several times (Heinrich 1995). Not all species nor all individuals within a species succeed on this task (Heinrich 1995; Osthaus et al. 2005; Thorpe 1963; Vince 1961; Werdenich and Huber 2006), suggesting it requires a higher-order cognitive ability that is prevalent neither among species (e.g., ravens succeed more often than crows, dogs mostly failed) nor within a given species (e.g., not all ravens that were tested succeeded). Clearly, this ability might be affected by prior physical manipulative experience, but might it also be tempered by other types of training, such as for a bird that has learned to demand access to various objects vocally? Specifically, if a bird can request the item from a human, rather than work to obtain it on his own, might the bird fail this test of insight? Such was what I set out to explore (Pepperberg 2004).

The task involved four Grey parrots: Alex, Arthur, Kyaaro, and Griffin. Alex, and also Griffin, had had considerable instruction in English speech and had learned to request items via "Want X". Arthur and Kyaaro were just beginning such training and had acquired only a few labels for objects; they did not yet know "Want X".

The task followed procedures described by Heinrich (1995), with an adaptation for the laboratory setting, in that birds were all tested individually on "T" stands, from which a desired object was hung on a long length of chain or cord. No bird had had any training with the apparatus prior to testing. All birds were vocally encouraged to "pick up" the suspended object.

The birds differed considerably in their responses (Pepperberg 2004). Arthur and Kyaaro, the birds with little vocal training in human communication, immediately attempted and succeeded on the task, and repeated their successful behavior several times in several trials in two or three sessions. They succeeded on simple variations of the task as well. When given two objects of different desirability (a nut versus chalk) suspended from two strings, Arthur chose the more desirable one (the nut) first. In contrast, Alex and Griffin, the birds with considerable vocal communication abilities, never made any attempts to retrieve the suspended nut. Instead, they consistently looked at the trainer and stated "Want nut" in multiple trials in two sessions. Even subsequent exposure to Arthur's successful retrieval failed to alter their behavior. As in the phonological "nut" example, Alex's volume and intensity increased with each failure of the trainer to comply with his request. Interestingly, after several trials in which trainers failed to comply, Alex and Griffin stopped making requests, as though they had learned that no one would attend to their demands (i.e., a form of learned helplessness).

Alex and Griffin seemed to be engaging in deliberate communication as a problem-solving strategy. They appeared to act intentionally; they faced the trainers and not the nuts during their requests. They were not treating humans as a physical object to be used (e.g., as a stepping-stone to reach something desired, as observed by Gomez 1990 in his work with apes). Their requests were the same as at other times when they were frustrated by seeing treats just out of their reach. Interestingly, they did not request the nut from Arthur after he succeeded, probably because none of the parrots ever share their treats.

Why Alex and Griffin failed to make any physical attempts at the task is unclear. One possibility is that development of their vocal centers might have inhibited some level of physical development. Griffin, for example, reached a plateau and failed to proceed on object permanence tasks while he was developing flight, as though two separate systems could not simultaneously develop at peak performance. Dominance (at least amongst the birds), however, was not an issue because all birds were tested individually, and Alex and Griffin were not close enough to steal the other birds' nuts. Arguably, Alex's and Griffin's demands that the trainers do the task might be taken as evidence that they consider themselves dominant to the humans in the laboratory.

A possible interpretation, for the purpose of the current discussion, however, is that a bird that responds with vocal demands to a physical task might be considered to have demonstrated an alternative higher-order intelligence, in that it knew how to manipulate another individual to access its wants—and had made its desires evident in what could be seen as an emotional manner. Alex and Griffin acted the way a human might, by stating, "It's not in MY contract to do physical chores."

3.1.4 Numerical Tasks

During a particular task on numerical competence, Alex acted in a way that involved very advanced cognitive processing plus some level of emotionality, although not quite as emotionally strong as in the situations described above. His behavior makes sense only in the context of his significant history with respect to number, and an understanding of how his data compared with results from number studies on other species.

Koehler (1950) was among the first to examine number competence in Grey parrots. He devised a clever procedure wherein birds were presented with boxes whose covers each had different numbers of paint splotches; the splotches were random with respect to size and form, so that the different numbers could not relate to mass, density, brightness, contour, etc. Novel splotches were used on each trial. The boxes were arranged in random order with respect to number, with the order also changing on each trial. Birds of various species were then shown pieces of clay of various sizes and shapes, and required, for reward, to open the box that had the same number of paint splotches. Grey parrots and jackdaws succeeded on quantities up to eight; pigeons succeeded up to about four, and domestic fowl only up to two. Koehler called this "thinking in unnamed numbers" or "non-numerical counting" because the subjects might have been simply performing a one-to-one correspondence between the pieces of clay and paint splotches without actually counting in the human sense (see Fuson 1988). Interestingly, if humans are faced with the task of labeling quantities of objects very quickly, without being given time to count, they are about as accurate as the pigeons (e.g., can label quantity exactly only up to about four; Revkin et al. 2008, unless the sets are in particular patterns, such as on dice or dominoes; the behavior is known as subitizing). Koehler's colleagues (e.g., Lögler 1959) demonstrated that the successful birds could transfer this behavior from simultaneous visual presentations to sequential visual presentations (flashes of light) and sequential auditory presentations (notes on a flute).

Other researchers studied number labeling with apes. Premack (1976) found that his star pupil Sarah performed more accurately on match-to-sample studies with number than on labeling of numerical sets via plastic symbols. (NB: Boysen 1993; Matsuzawa et al. 1991, later found that apes could use Arabic numerals to quantify sets up to about eight, but that such abilities required considerable training.) Interestingly, researchers studying humans (Geschwind 1979) found that match-to-sample and labeling involved different brain areas, further suggesting different mechanisms for the behavior demonstrated by Koehler and that of actual counting.

How might a bird that could label objects fare on such a task? Alex had already shown that he could label objects, colors, and shapes—particularly triangles and squares, with the labels "3-corner" and "4-corner". Could he transfer some understanding of the numbers of corners (which he might or might not have at that time), to that of exact quantities? The answer was that he could, and that he eventually learned to label sets of up to eight (Pepperberg 1987; Pepperberg and Carey accepted pending revision), including heterogeneous sets (of, for example, red and blue balls and blocks; Pepperberg 1994). Subsequent work (Pepperberg 2006a) demonstrated that he was actually counting, and not using some form of estimation, approximation, subitizing, or other nonexact forms of enumeration.

Nevertheless, Alex still might not truly have understood what his numbers represented. For example, young children who can seemingly exactly label small sets of X items, fail when asked to "Give me X marbles" (Wynn 1990). Thus, Alex was given a set of comprehension trials (Pepperberg and Gordon 2005), in which he was, without training, asked "What color/object is [number-x]?" for collections of various simultaneously presented quantities (e.g., subsets of 4–5–6 blocks of three different colors; subsets of 2–4–6 keys, corks, sticks of one color).

In this task, his overall accuracy was greater than 80%, and was unaffected by array quantity, mass or contour. His results demonstrated numerical comprehension competence comparable to that of chimpanzees and very young children. Alex did, however, engage in intriguing types of behavior en route to his success.

After responding without error in his first eight trials (at least one for each targeted number), Alex balked during tests for about two weeks [e.g., would stare at the ceiling, reply repeatedly with an object label not on the tray, ask to go to his cage, for water or various foods (we always responded to food and water requests)]. He might produce labels for the four colors *not* on the tray; that is, avoid any response that could be construed as task-related (*NB*: a statistically unlikely behavior; Pepperberg and Lynn 2000). He might toss all items off the tray. Using possibly more enticing items (e.g., jelly beans) in trials didn't help. Unlike behavior in other number trials, here he neither maintained his gaze on the tray nor made eye contact with the questioner. Eventually, he returned to testing, though he might still occasionally repeat such actions. When he had acted similarly before (usually during training, sometimes during testing, Pepperberg 1992; Pepperberg and Lynn 2000), we discounted the trials, and did so again. Had his responses been errors tied to the trials, they would have been counted. (*NB*: The 16 discarded trials were later repeated with different items and colors, so all collections were tested eventually).

Why did Alex act this way? Not because trials were difficult; he already had succeeded on trials of equal or greater difficulty (e.g., distinguishing among four, five, six items or one, two, three items; Pepperberg and Gordon 2005), and he also refused to work on simple trials (e.g., distinguishing among one, three, six items). In the past, Alex's inattentiveness generally had two causes. First, he often responded negatively if rewards lacked novelty. We did use novel *combinations* of items in each comprehension trial, but each item, by necessity, was familiar because Alex had to be able to label all attributes, so he could not be cued as to the type of query (e.g., "What color?" vs. "What matter?"). Too, he had already completed 32 trials in a simpler version of the task (Pepperberg and Gordon 2005). In previous studies (see Pepperberg 1999), when familiar objects were repeatedly used as rewards, Alex often ceased to work in ways noted above. In contrast, his attention span often increased and accuracy improved on transfer trials, which by definition incorporated novel (i.e., conceivably more interesting) items (Pepperberg 1999). Second, Alex often seemed to find reward in humans' reactions when he persisted in providing incorrect answers (most often, carefully avoiding a correct response, Pepperberg and Lynn 2000). That is, he by this time knew the response that would garner a reward and end the session, but chose instead to inhibit that response to elicit a more engaging caretaker response.

Of even greater interest, however was that on one trial, soon after this period of noncompliance, Alex was asked "What color three?" to a set of two, three, and six objects. He replied "five"; the questioner asked him twice more and each time he replied "five". The questioner finally said "OK, Alex, tell me, what color 5?" He immediately responded "none". He had been taught to respond "none" if no category (color, shape, or material) was same or different when he was queried about the similarity or difference of two objects (Pepperberg 1988), and had spontaneously transferred this response to the query "What color bigger?" concerning two objects of identical size in a study of relative size (Pepperberg and Brezinsky 1991), but had never been taught the concept of absence of quantity nor to respond to absence of an exemplar—that is, to use "none" as a zero-like symbol—something the Western world did not achieve until the 1600s (Pepperberg and Gordon 2005). In the present study, however, he not only provided the correct response, *but also set up the question himself*. That is, he had figured out how to manipulate the trainer into asking the question he seemingly wished to answer. His action, occurring soon after a period of noncompliance, might have resulted from lack of interest in the given task and as a possible attempt to make the procedure more challenging. Thus, again, some level of emotionality (conceivably boredom), coupled with complex cognitive processing, led to data that would not likely otherwise have been obtained. (NB: To ensure that his response of "none" was meaningful and not an odd happenstance, the question was repeated randomly throughout other trials with respect to each possible number. On these "none" trials, Alex's accuracy was 5/6, 83.3%; $p<0.01$, binomial test, chance of ¼ the three relevant color labels plus "none". His one error, interestingly, was to label a color not on the tray.)

3.2 Implications of the Data

The data discussed above raise several questions concerning Alex's capacities. The first is whether—particularly in the number study—Alex demonstrated some level of "theory of mind" (TOM)? That is, by figuring out how to manipulate the experimenter into asking him a particular question, did he exhibit an understanding of the thought processes of that experimenter? The question is not one with a simple answer, and the data merely suggest that birds such as Alex would be appropriate subjects for more traditional tests of TOM.

The second question, and the main point of presenting these data, however, is to ask whether the results would have been obtained had there not been a confluence of advanced cognitive processing and some level of emotional behavior? Interestingly, humans generally classify emotional responses, particularly in nonhumans, as *lacking* in higher order cognition, but Alex and Griffin demonstrated more advanced behavior within emotional responses. Is there a way to investigate how emotions might drive, or at least influence, avian cognition?

The key might come from examining the work of Ciompi (1991), Damasio (1999), Humphrey (2000), and more recent studies by researchers like Pessoa (2008) and Barrett and Barr (2009), who examine how cognitive processing might initially develop from emotional responses in humans. In papers by other researchers (reviewed in Salzman and Fusi 2010), data are presented to demonstrate that such interactions are maintained throughout life—by the bidirectional interconnections between the amygdala, which plays a role in many emotional processes, and the prefrontal cortex (PFC), with its areas responsible for cognitive processing. Might similar connections explain my Grey parrots' behavior?

According to researchers studying developmental issues, the first types of categorization and generalization—basic cognitive processes—may occur when an individual initially categorizes and generalizes emotions with respect to environmental events and then inter-subject interactions. Ciompi (1991) reminds us that even simple conditioned reflexes are based on positive/negative reinforcement—that is, positive versus negative affect, and also of the work of many early researchers, who have shown that almost every action involves emotional elements that can become attached to cognitive content. Damasio (1999), for example, argues that "core" consciousness (the basic form that involves total awareness of the present, but not the future or past) emerges when we interact with an object (including other beings), and is a feeling that accompanies the making of a mental image—even one that is retained for less than a minute. The mental image allows for categorization of emotions and events with respect to their emotional content; then, in intact individuals, eventually leads to categorization of involved objects and actions. Note that a child, for example, initially does not label an emotion, but talks about objects about which it cares and expresses the emotion by displaying positive or negative affect (Bloom 2000). Barrett and Barr (2009) describe neural pathways that might be responsible for how these categorizations lead to a representation of objects and actions that can be manipulated to allow for advanced learning—that is, manipulation of the representations in various complex ways. Thus, at some basic level, categorization that begins in the *emotional* domain, before the emergence of language, seems to lead to the ability to categorize other aspects of the world, which then leads to understanding and use of representation. Development seems to proceed along a rational path, but has its roots in emotionality. Could tapping into that emotional basis continue to assist in cognitive processing? Ciompi (1991) reviews earlier material that suggests how memory and information processing are activated or blocked by motivational states; more recent studies seem to support this view.

Researchers such as Blanchette and Richards (2010) argue that emotional arousal may signal that a situation is important and personally significant, and thus may encourage participants to mobilize cognitive resources to think through the problem carefully (again, note Ciompi 1991). Some data suggest that negative affect enhances the performance of tasks that require a systematic, analytic approach. Although much of such data involve emotions of sadness rather than anger or frustration (e.g., Forgas 2007), Blanchette and Richards (2010) argue that emotions that are related to uncertainty are what signal that something is amiss and that careful information processing is required, leading the individual to engage in more elaborative processing. Such appeared to be the case for Alex in the N–U–T instance and for Griffin and Alex during the object permanence situation: Although their emotional states could have been interpreted as "anger", they were responding to violations of expectations of reward, which are instances of uncertainty. Alex's and Griffin's response to the string-pulling task might also be interpreted in this manner: They had never before been asked to perform physical tasks, and thus switched into their more common mode of behavior—asking the human to perform the task.

Another study (e.g., Barth and Funke 2010), on a task comparing "nice" (e.g., supportive) versus "nasty" (e.g., critical or unrewarding) environments also demonstrated that the latter engendered greater information retrieval, better analytic processing, possibly different motivational states, and somewhat improved performance. Thus the effects of emotional valence appear to be important predictors of behavioral outcomes, and particularly for those involving cognitive processing. Whether Alex's state of "boredom" in the number comprehension task—a possible consequence of a lack of stimulation—could be seen as a "nasty" environment is unclear, but he obviously viewed being asked to repeat the task numerous times as unpleasant, or at least unrewarding. His manipulation of the trainer has at least a passing commonality with his manipulations of humans in the string-pulling task—albeit in a much more sophisticated manner (i.e., by getting the trainer to ask a different question that he wished to answer).

Returning to the work of Barrett and Barr (2009), we can possibly follow the neural connections involved in these functions. As reviewed in their study, in humans the orbitofrontal cortex (OFC, part of the PFC) integrates sensory input from the world and the body, plays a role in representing reward and threat, and in processing olfactory, auditory, and visual information. Their study emphasizes how such integration prepares the body to respond to a *particular* stimulus; however, because OFC is linked to memory (e.g., Frey and Petrides 2000), possibly emotional reactions might actually enable innovative connections to be made between existent knowledge and a novel situation. Pessoa (2008) suggests that numerous possible anatomical connections exist between affective and cognitive brain areas in humans; Salzman and Fusi (2010: 175) argue that "the neural circuits mediating cognitive, emotional, physiological, and behavioral responses may not truly be separable and instead of inextricably linked." Pessoa (2008) and Salzman and Fusi (2010) also suggest, based on their review of experiments, that some of these interconnections extend to nonhuman primates, particularly with respect to determining outcome expectancies of actions and executive control. Although direct correlates between human and avian cortical areas are difficult to determine, parrots do have cortical areas that work in ways similar to those of humans (Jarvis et al. 2005), and parallels in behavior are thus possible. For example, Rose and Colombo (2005) argue that parts of the avian nidopallium caudolaterale (NCL) are analogous to the mammalian PFC; the NCL receives input from, and sends output to, areas now considered part of the avian amygdala (e.g., Jarvis et al. 2005; see also Kröner and Güntürkün 1999). Given that neural categorization occurs when a neural ensemble provides the same output from different inputs, maybe it is an emotional response that connects the processing of, and memory for, related concepts, and allows for neural organization for complex cognitive processing to deal in an innovative manner with unusual and novel situations? Such an idea is, again, foreshadowed by Ciompi (1991) in a review of his and other studies. Maybe the confluence of a specific neural architecture, contextually-applicable and socially interactive training, low-level elements of theory of mind, *and* aspects of emotion and affect is necessary for advanced cognitive processing to occur (Fig. 3.1)?

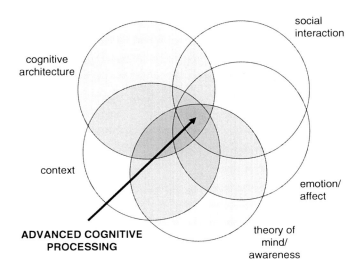

Fig. 3.1 Intersection of elements necessary to produce advanced learning-cognitive processing

3.3 Conclusions

In sum, although we still have only a rudimentary understanding of the functional interactions of the emotional and cognitive parts of the primate brain—and still less for avian species—several research studies demonstrate the existence of the relevant anatomical connections (e.g., review in Salzman and Fusi 2010; note Kröner and Güntürkün 1999). Furthermore, some studies (Barrett and Barr 2009) suggest how, in humans, certain heightened emotional states can provide the basis for more detailed cognitive processing, and other studies (e.g., Barth and Funke 2010; Blanchette and Richards 2010) demonstrate how certain types of affect might improve such processing. Quite possibly, other emotional states may alter—and possibly engender—cognitive processing as well. Thus, the parrot data described above suggest that the question may be not whether my Greys' behavior was emotional *or* cognitive, but how these two aspects of behavior might have interacted to maximize their problem-solving abilities. Furthermore, finding such connections in a nonprimate, nonmammalian species would suggest that another area of convergent evolution exists between birds and primates.

References

Barrett LF, Barr M (2009) See it with feeling: affective predictions during object perception. Philos Trans R Soc B 364:1325–1334

Barth CM, Funke J (2010) Negative affective environments improve complex problem solving performance. Cogn Emot 24:1259–1268

Blanchette I, Richards A (2010) The influence of affect on higher level cognition: a review of research on interpretation, judgement, decision making and reasoning. Cogn Emot 24:561–595

Bloom L (2000) The intentionality model: how to learn a word, any word. In: Golinkoff RM, Hirsh-Pasek K, Bloom L, Smith LB, Woodward AL, Akhtar N, Tomasello M, Hollich G (eds) Becoming a word learner: a debate on lexical acquisition. Oxford University Press, New York, pp 124–135

Boysen ST (1993) Counting in chimpanzees: nonhuman principles and emergent properties of number. In: Boysen ST, Capaldi EJ (eds) The development of numerical competence: animal and human models. Erlbaum, Hillsdale, pp 39–59

Ciompi L (1991) Affects as central organizing and integrating factors: a new psychosocial/biological model of the psyche. Br J Psychiatry 159:97–105

Damasio A (1999) The feeling of what happens. Harcourt, San Diego

Forgas JP (2007) When sad is better than happy: negative affect can improve the quality and effectiveness of persuasive messages and social influence strategies. J Exp Soc Psychol 43: 513–528

Frey S, Petrides M (2000) Orbitofrontal cortex: a key prefrontal region for encoding information. Proc Natl Acad Sci USA 97:8723–8727

Fuson KC (1988) Children's counting and concepts of number. Springer, New York

Geschwind N (1979) Specializations of the human brain. Sci Am 241:180–199

Gomez JC (1990) The emergence of intentional communication as a problem-solving strategy in the gorilla. In: Parker ST, Gibson KR (eds) "Language" and intelligence in monkeys and apes: comparative developmental perspectives. Cambridge University Press, New York, pp 333–355

Heinrich B (1995) An experimental investigation of insight in Common Ravens (*Corvus corax*). Auk 112:994–1003

Humphrey N (2000) The privatization of sensation. In: Heyes C, Huber L (eds) The evolution of cognition. MIT Press, Cambridge, pp 241–252

Jarvis JD, Güntürkün O, Bruce L, Csillag A, Karten H, Kuenzel W et al (2005) Avian brains and a new understanding of vertebrate evolution. Nat Rev Neurosci 6:151–159

Koehler O (1950) The ability of birds to "count". Bull Anim Behav 9:41–45

Kröner S, Güntürkün O (1999) Afferent and efferent connections of the caudolateral neostriatum in the pigeon (*Columba livia*): a retro- and anterograde pathway tracing study. J Comp Neurol 407:228–260

Lögler P (1959) Versuche zur Frage des "Zähl"-Vermögens an einem Graupapagei und Vergleichsversuche an Menschen. Z Tierpsychol 16:179–217

Matsuzawa T, Itakura S, Tomonaga M (1991) Use of numbers by a chimpanzee: a further study. In: Ehara A, Kimura T, Takenaka O, Iwamoto M (eds) Primatology today. Elsevier, Amsterdam, pp 317–320

Osthaus B, Lea SEG, Slater AM (2005) Dogs (*Canis lupus familiaris*) fail to show understanding of means-end connections in a string-pulling task. Anim Cogn 8:37–47

Pepperberg IM (1987) Evidence for conceptual quantitative abilities in the African Grey parrot: labeling of cardinal sets. Ethology 75:37–61

Pepperberg IM (1988) Acquisition of the concept of absence by an African Grey parrot: learning with respect to questions of same/different. J Exp Anal Behav 50:553–564

Pepperberg IM (1992) Proficient performance of a conjunctive, recursive task by an African Grey parrot (*Psittacus erithacus*). J Comp Psychol 106:295–305

Pepperberg IM (1994) Numerical competence in an African Grey parrot. J Comp Psychol 108:36–44

Pepperberg IM (1999) The Alex studies. Harvard University Press, Cambridge

Pepperberg IM (2004) "Insightful" string-pulling in Grey parrots (*Psittacus erithacus*) is affected by vocal competence. Anim Cogn 7:263–266

Pepperberg IM (2006a) Grey parrot (*Psittacus erithacus*) numerical abilities: addition and further experiments on a zero-like concept. J Comp Psychol 120:1–11

Pepperberg IM (2006b) Ordinality and inferential abilities of a Grey parrot (*Psittacus erithacus*). J Comp Psychol 120:205–216

Pepperberg IM (2007) Grey parrots do not always 'parrot': phonological awareness and the creation of new labels from existing vocalizations. Lang Sci 29:1–13

Pepperberg IM, Brezinsky MV (1991) Relational learning by an African Grey parrot (*Psittacus erithacus*): discriminations based on relative size. J Comp Psychol 105:286–294

Pepperberg IM, Carey S (accepted pending revision) Grey parrot number acquisition: the inference of cardinal value from ordinal position on the numeral list Cognition

Pepperberg IM, Gordon JD (2005) Numerical comprehension by a Grey parrot (*Psittacus erithacus*), including a zero-like concept. J Comp Psychol 119:197–209

Pepperberg IM, Lynn SK (2000) Perceptual consciousness in Grey parrots. Am Zool 40:393–401

Pepperberg IM, Shive HA (2001) Simultaneous development of vocal and physical object combinations by a Grey parrot (*Psittacus erithacus*): bottle caps, lids, and labels. J Comp Psychol 115:376–384

Pepperberg IM, Wilcox SE (2000) Evidence for a form of mutual exclusivity during label acquisition by Grey parrots (*Psittacus erithacus*)? J Comp Psychol 114:219–231

Pepperberg IM, Willner MR, Gravitz LB (1997) Development of Piagetian object permanence in a Grey parrot (*Psittacus erithacus*). J Comp Psychol 111:63–75

Pepperberg IM, Vicinay J, Cavanagh P (2008) The Müller-Lyer illusion is processed by a Grey parrot (*Psittacus erithacus*). Perception 37:765–781

Pessoa L (2008) On the relationship between emotion and cognition. Nat Rev Neurosci 9:148–158. doi:10.1038/nrn2317

Premack D (1976) Intelligence in ape and man. Erlbaum, Hillsdale

Revkin SK, Piazza M, Izard V, Cohen L, Dehaene S (2008) Does subitizing reflect numerical estimations? Psychol Sci 19:607–614

Rose J, Colombo M (2005) Neural correlates of executive control in the avian brain. PLoS Biol 3:e190. doi:10.1371/journal.pbio.0030190

Salzman CD, Fusi S (2010) Emotion, cognition, and mental state representation in amygdala and prefrontal cortex. Annu Rev Neurosci 33:173–202

Thorpe WH (1963) Learning and instinct in animals. Methuen, London

Vince M (1961) String pulling in birds. III. The successful response in greenfinches and canaries. Behaviour 17:103–129

Werdenich D, Huber L (2006) A case of quick problem-solving in birds: string pulling in keas, *Nestor notabilis*. Anim Behav 71:855–863

Wynn K (1990) Children's understanding of counting. Cognition 36:155–193

Chapter 4
Why Do Dolphins Smile? A Comparative Perspective on Dolphin Emotions and Emotional Expressions

Stan A. Kuczaj II, Lauren E. Highfill, Radhika N. Makecha, and Holli C. Byerly

S.A. Kuczaj II (✉)
Department of Psychology, The University of Southern Mississippi,
118 College Dr. #5025, Hattiesburg, MS 39406, USA
e-mail: s.kuczaj@usm.edu

L.E. Highfill
Eckerd College, 4200 54th Avenue South, St. Petersburg, FL 33711, USA

R.N. Makecha
The College of the Bahamas, Oakes Field, P.O. Box N-4912, Nassau, The Bahamas

H.C. Byerly
Dolphins Plus, 31 Corinne Pl, Key Largo, FL 33037, USA

Abstract Although emotions have proven difficult to define, they nonetheless influence the lives of humans and non-human animals. Processing emotions is particularly important for members of species with dynamic social lives. For such animals, the ability to recognize and respond to the emotional states of others facilitates successful social interactions. For example, dolphins live in dynamic fission/fusion societies, and it seems likely that dolphins are sensitive to the emotional states of other dolphins. In this chapter, we consider the evidence for emotional states and emotional recognition in dolphins from a comparative perspective. This includes a discussion of methodologies used to study emotions in animals as well as a consideration of the overlap between animal personality research and animal emotion research. We conclude that there are many challenges facing those who study animal emotions, but also believe that a better understanding of animal emotions will increase our understanding of animal behavior and our ability to improve the welfare of wild animals in captivity and domestic animals.

In this chapter, we focus on the possibilities that dolphins experience emotions and that dolphins may be sensitive to the emotional states of other dolphins. However, we wish to frame our considerations within a comparative perspective, and so will couch our discussion and recommendations within the context of both historical and current literature on emotions and emotional expression.

Despite considerable recent advances in neuroscience and cognitive psychology, a number of phenomena remain difficult to define, let alone study. Examples of such phenomena range from play to consciousness, and most certainly include emotions. Darwin's book, *The Expression of the Emotions in Man and Animals* (Darwin 1872), has been the catalyst for the psychological study of emotions, including comparative studies, but contained no explicit definition of emotion. Neither did Bain's *The Emotion and the Will*, published a number of years before Darwin's book (Bain 1859). The lack of operational definitions of emotion is not restricted to historical treatments of the topic. Cacioppo and Gardner's (1999) chapter on emotion in the *Annual Review of Psychology* also lacked a precise definition of the phenomena. The reason that definitions of emotion are rare in the literature is straightforward. "Emotion", like "play" and "consciousness", is a difficult concept to define. Consequently, scholars ranging from Darwin to Caciopppo and Gardner have thoughtfully pondered emotions without actually defining the concept itself.

This does not mean that there have been no attempts to define emotion. However, such attempts have yielded conflicting results. Cornelius (1996) summarized the state of affairs as follows: "Emotions...have been defined in many ways by different psychologists. The answer to the question 'what is an emotion?' depends on whom you ask and when you ask him or her" (p. 9). Cornelius pointed out that individual definitions of emotion depend on the individual's theoretical orientation, and identified four theoretical traditions of research on emotion in psychology that influence definitions and research interpretations: (1) Darwinian (following Darwin 1872)—emotions have adaptive functions, as does the ability to interpret the emotions of others. (2) Jamesian (following James 1884)—the experience of emotions results from the experience of bodily changes; (3) cognitive (following Arnold 1960)—emotions result from an individual's interpretation and assessment of events; and (4) social constructivist (following Averill 1980)—emotions are cultural constructions that can only be understood via a social level of analysis.

To illustrate the range of definitions of emotions that exist in the literature, we provide a non-exhaustive sampling of definitions that have been posited by various authors: according to Young (1943), "an emotion is an acute disturbance or upset of the individual which is revealed in behavior and in conscious experience, as well as through wide-spread changes in physiology, and which is initiated by factors within a psychological situation" (p. 405). The notion that emotions are intense manifestations was echoed by Panksepp (1998)—"When powerful waves of affect overwhelm our sense of ourselves in the world, we say that we are experiencing an *emotion*. When similar feelings are more tidal—weak but persistent—we say we are experiencing a *mood*" (p. 47). But not everyone agrees that changes in feelings are sufficient to produce an emotion. For Kagan (2007), "a detected change in feeling is a necessary, but not sufficient, feature of an emotion" (p. 21). Kagan summarized modern views of emotions as a set of ideas that revolve around four "imperfectly related phenomena": (1) a neurological change, (2) a consciously detected change in feeling, (3) an interpretation and/or labeling of the feeling, and (4) some type of preparedness for a behavioral response. Kagan proposed that "feeling" be used to describe the perceived changes in (2) and "emotion" be reserved for feelings that have been processed and interpreted (3).

Bekoff (2007) used the terms "emotion" and "feeling" in the exact opposite way of Kagan. For Bekoff, emotions are reactions to external stimuli whereas feelings are interpretations of these reactions. Bekoff also suggested that even though an external event may "trigger one emotion," an organism may decide it "feels" differently once it reflects on the experience. Thus, Bekoff believed that we interpret our emotions, and that these interpretations result in different feelings (or moods), while Kagan argued that we experience feelings and that interpretations of these feelings result in emotional experiences.

Given the wide range of definitions of emotion, it is not surprising that there is also disagreement about the number of emotions and the best manner in which to categorize and label emotions. Izard (1992) identified ten "basic" emotions: anger, contempt, disgust, distress, fear, guilt, interest, joy, shame, and surprise. Kagan (2007) argued that the notion of "basic" emotions ignores the richness and complexity of the human emotional experience: "Let us agree to a moratorium on the use of single words, such as *fear, anger, joy,* and *sad,* and write instead about emotional processes with full sentences rather than ambiguous, naked concepts that burden readers with the task of deciding who, why, and especially what (Kagan 2007, p. 216).

Clearly, the scientific study of emotions is a bit of a mess. There is no agreed-upon definition of "emotion", and the terms "emotion," "feeling," and "mood" are used in different ways by different authors. Despite the fact that emotions are difficult to define, we nonetheless recognize emotions in others. However, the extent to which animals can recognize others' emotional states has long been a matter of dispute. Bell (1840) argued that only humans are capable of emotional expression. Darwin (1872) disagreed, but did not argue that emotional expressions had actually evolved to communicate emotions. He instead focused on the notion that many emotional expressions had evolved because they were remnants of important functional behaviors. More importantly from a comparative perspective, Darwin suggested that there might be significant biological similarities in animal and human emotional expressions, a proposal that suggests some sort of evolutionary continuity in emotional expression.

Darwin's notion gains support from the significance of emotions for social beings. The ability to interpret our own emotions and those of others is thought to be critical for social interactions, particularly in contexts in which interactions are dynamic (Bekoff 2007; Reeve 2009; Ridley 1998). This is true for both human and non-human societies (although it is not clear that emotions play any role in the lives of social insects). Emotions seem to be important, then, despite our inability to define them in a universally acceptable manner.

4.1 Animal Emotions?

Humans have long been interested in the possibility that animals experience emotions. Debate about animal emotions remained largely philosophical until Darwin (1872) provided the impetus for the systematic and scientific study of animal emotions. In his pioneering work, Darwin demonstrated the need for a comparative study of emotions by proposing that the emotional differences that exist between at least some species are differences of degree rather than differences of kind. However, the idea that the emotional lives of animals and humans might overlap in meaningful ways was far from universally accepted.

Part of the reason for this was the lack of interest in the mental lives of animals by advocates of Behaviorism and Ethology. Behaviorism dominated American psychology in the early half of the twentieth century (Skinner 1938; Watson 1914), while the ethological approach provided the zeitgeist for the study of animal behavior in Europe (Lorenz 1949; Tinbergen 1951). Although these two traditions differed in many respects, they both viewed emotions and cognition as non-observable and therefore inappropriate for scientific study. The result was a dearth of studies on animal emotions (and animal cognition) during much of the twentieth century despite the early efforts of pioneering comparative psychologists such as Romanes (1881), Tolman (1932), and Washburn (1908).

This began to change, albeit slowly, when researchers became interested in animal welfare and began studying states such as fear, distress and frustration, states that we associate with emotions in humans (e.g. Dawkins 1977, 2000; Dantzer and Mormède 1981; de Waal 2008; Moberg et al. 1980; Sandem et al. 2002). Although much of this work focused on negative emotions such as fear, the importance of positive emotions was not ignored. For example, Dawkins (2000) argued that an animal which engages in successful reinforcement learning must have some sense of what "feels good" versus what "feels worse," and that emotionless stimulus–response mechanisms would severely restrict the behavioral flexibility that occurs in many species (see Kuczaj and Makecha 2008, for a consideration of the significance of behavioral flexibility in communication systems).

Scientists studying animals in the wild also found it increasingly difficult to explain behavior without considering animals' internal states (Bekoff 2000; Dawkins 2000; Fraser 2009; Talan 2006). For example, Goodall (1990) described what seemed like depression in one of the chimpanzees she was studying. After another chimp's death, the depressed chimp became increasingly listless and apathetic, stopped eating and eventually died.

The increasing concern for animal welfare and the belief that understanding affective states was necessary to successfully interpret natural behavior resulted in more comparisons of human and animal emotions (Bekoff 2000; Marler and Evans 1997; Owren and Bachorowski 2007; Panksepp 1998; Parr and Gothard 2007; Paul et al. 2005; Preuschoff and van Hooff 1997; Suomi 1997). For example, Parr, Waller, and Fugate (2005) reported that facial emotional expressions are quite similar across primate species and that the facial musculature of chimpanzees (*Pan troglodytes*) is very similar to that of humans, suggesting that the facial behavior of closely related species may be evolutionarily continuous. Additionally, investigations of brain structure similarities between humans and animals suggested that some of the physiological mechanisms of emotion were evolutionarily linked. For example, Panksepp (2003, 2007) argued that certain neural systems involved in human emotions, such as the limbic system, may be homologous across species.

The recognition that animals experience emotions coincided with increased interest in animals' natural communication systems. In fact, numerous scholars suggested that one of the primary functions of non-human primate communication was the expression of emotional states (Bastian 1965; Lancaster 1965; Luria 1982; Myers 1976; Rowell and Hinde 1962; Siebert and Parr 2003; Young 1943). However, Bachorowski and Owren (2003) suggested that the primary function of signaling is not to express the signaler's emotion, but instead to affect the listener's emotional state (and thus influence the listener's behavior). According to Bachorowski and Owren, the listener's attribution of the caller's emotional state is a secondary outcome that results from the listener's own affective response, its past experience, and the context in which the signaling occurred.

4.2 Methods for Studying Animal Emotions

Arguably the easiest way to study emotions in humans involves verbal self-report. We can simply ask someone, "How did that situation make you feel?" Unfortunately, we cannot do the same with animals. How then, should we approach the study of animal emotions? Here, we discuss various methods and their relevance to studying dolphin emotion.

The majority of animal emotion research has focused on negative emotions (Dawkins 2006), most likely because negative emotions are more easily observed and measured (e.g. fear, anxiety). Although emotions themselves are not directly observable, it is possible to measure their correlates or consequences, such as increased heart rate and facial expressions (Kirkden and Pajor 2006). Over the past few decades, researchers have used a wide variety of methods to examine emotional responses in animals. These methods can be broadly divided into two categories: physiological measures and behavioral measures (Dawkins 2006).

4.2.1 Physiological Measures

Physiological measures used to assess animal emotion examine the relationship between physiological functioning (e.g., brain activity, heart rate, respiration rate, cortisol levels) and potentially emotion-inducing situations. For example, the heart rates of two female rhesus monkeys increased when approached by a dominant animal, whereas the heart rates did not increase if the same females were approached by a subordinate animal or a relative (Aureli et al. 1999). Moreover, the heart rates of the females decelerated faster if they received allo-grooming after the approach by a dominant individual. These findings suggest that it is possible to examine emotional reactions to social interactions using physiological measures. It would be interesting to use such measures to compare emotional responses to threatening and reconciliation behaviors in dolphins and other species.

Due to the many challenges of working with aquatic mammals, the study of dolphin physiology has been somewhat sparse. However, new technologies and training techniques have expanded this area of research. For example, Miksis et al. (2001) examined the heart rate responses of a captive dolphin to various auditory stimuli. They found that the dolphin's heart rate increased when she heard conspecifics' vocalizations, a response that may mirror the physiological startle response seen in some terrestrial animals. Measuring cardiac responses in dolphins could be explored in a variety of situations, which in turn could provide important insights into their emotional experiences.

For many years, brain lesions and electrical stimulation have been used to explore the underlying neural processes of emotion (LeDoux 1996). These research methods have provided much information about the areas of the brain (e.g. the amygdala, rhinal cortex, thalamus) that are involved in emotional processing and expression (Pinel 2009). However, these techniques cannot be conducted without harming the subjects and so are inappropriate for the study of dolphin emotion.

Fortunately, there are less invasive ways to study the physiological underpinnings of emotion. One technique places subjects in an emotion-inducing situation and examines the changes in the living brain using brain-imaging techniques. For example, using a PET scan Takamatsu et al. (2003) were able to monitor activity in the brains of rhesus monkeys after injecting the subjects with a pharmacological stressor. Brain imaging techniques are possible to use with dolphins. MRIs and PET scans have been used to examine sleep patterns in bottlenose dolphins (e.g., Ridgway et al. 2006). However, to date, brain imaging has not been used to study emotional responses in dolphins. As technology continues to advance, this line of research should be explored.

In their review article, Paul et al. (2005) described some limitations concerning the use of physiological measures to study animal emotions. For example, the sampling methods used to collect physiological measures (e.g. blood draws, attaching monitors) may cause confounding emotional reactions in the subjects. Furthermore, they explained how different emotional states sometimes result in the same physiological response (Paul et al. 2005). For example, an animal's heart rate can rise in response to both a rewarding and punishing stimuli. These limitations must be considered when using physiological measures to study animal emotions, and certainly suggest that physiological measures alone will be inadequate.

4.2.2 Behavioral Measures

Emotional reactions can also be measured through observations of overt behaviors. Measurements can be made of spontaneous behaviors, such as approaching or avoiding an object (Paul et al. 2005), or of learned behaviors, such as learning to choose pain-relieving foods (Danbury et al. 2000). Most of the research examining emotional behavioral responses has focused on behaviors such as facial expressions, postures, and vocalizations. For example, chimpanzees have faces quite similar to humans and are able to produce a variety of facial expressions such as a dramatic, full closed grin when confronted with an unexpected and frightening stimulus (Goodall 1986). However, measuring facial expressions is problematic in its own right. Although the use of facial coding systems (Parr and Gothard 2007) is quite useful in measuring facial expressions, these expressions are often accompanied by other facial activity. This extraneous activity may interfere with an accurate assessment of emotional states (Larsen and Fredrickson 1999). The use of facial expressions to study emotions is especially difficult with dolphins, given that they do not produce many facial muscle movements. However, we have observed rough-tooth dolphins with an open-mouth expression engaging in curious-type behaviors, so the possibility of using at least some dolphin facial expressions to interpret dolphin emotion remains a possibility. Nonetheless, the use of facial expressions as indicators of emotional state will be quite limited for those of us interested in dolphin emotions.

The relation of changes in body posture to emotion has been examined in a variety of species. For example, sheep change their ear posture more frequently in negative situations, such as being separated from group members, than in positive situations, such as feeding on fresh hay (Reefmann et al. 2009). However, caution should be used when interpreting changes in body posture and their relation to emotional state. It is important to determine the reliability of postural measures and to examine postural behaviors that are consistently displayed in a variety of situations that seem to have the same affective valence (Paul et al. 2005). Dolphins are able to move their bodies in many different ways, so the relationship between changes in body posture and emotion warrants further exploration in dolphins.

Finally, vocalizations have been considered to represent a direct expression of emotion in animals. For example, one study examined piglet vocalizations before, during, and after castration (Puppe et al. 2005). The piglets vocalized more frequently during the surgical procedure, implying the highest level of pain during this portion of the procedure. While much of the vocalization research has focused on vocal expressions during negative experiences, more recently, vocal expressions have been linked to positive emotions as well. For example, rats produce a short chirping ultrasonic vocalization when engaging in positive behaviors such as sex, play or receiving tickling (Panksepp and Burgdorf 2003). Interestingly, these chirping vocalizations are inhibited during negative situations. However, Paul et al. (2005) point out that both vocalizations and facial expressions most likely evolved to have communicative functions, and therefore may be influenced by an audience. Furthermore, it should be noted that vocalizations may not necessarily represent a direct association to an emotional state, but instead represent the need to communicate an emotional state (Paul et al. 2005) or to influence the listener's emotional state (Bachorowski and Owren 2003). Despite these limitations, dolphins are great candidates for this sort of research because they produce a wide range of vocalizations in a variety of contexts (e.g. Hernandez et al. 2010; Herzing 1996; McCowan and Reiss 1995).

4.2.3 Behavioral and Physiological Measures

It has been suggested that emotion researchers should employ physiological measures along with behavioral observations to attain the most accurate interpretation of animal emotion (Paul et al. 2005). One area of research that combines behavioral and physiological measures is the lateralization of emotional expression. For example, the facial expressions of monkeys and humans begin on the left side of the face and are more pronounced on that side as well (Hauser 1993). This implies that the right hemisphere in both human and non-human primates dominates facial emotional expressions. Hemispheric lateralization of emotion in dogs based on the direction of tail wagging was investigated by Quaranta, Siniscalchi, and Vallortigara (2007). They showed dogs pictures of their owner, a human stranger, a cat or a large, unfamiliar, and dominant-type dog. The dogs' tails wagged consistently to the right when viewing the humans (both familiar and unfamiliar) and the cat. However, the dogs' tails consistently wagged to the left when viewing the intimidating dog or when they were left alone. If viewing humans and cats yields positive emotions in dogs, then it appears that the left hemisphere of the brain is involved in positive emotions in dogs. Moreover, if viewing an intimidating dog or being left alone results in negative emotions, then the right hemisphere of the canine brain seems to process and/or produce negative emotions in dogs.

Left-hemispheric dominance has been implicated in dolphin vision (e.g. Kilian et al. 2000; Delfour and Marten 2006; Thieltges et al. 2011), and dolphins appear to sleep one hemisphere at a time (Mukhametov et al. 1997) all of which suggests that dolphins may demonstrate lateralization in body-posturing related to emotional expression.

4.2.4 Studying Emotions in Wild Populations

Most research on animal emotions has examined animals in controlled or semi-controlled situations, but there have also been observations of emotional expression in wild populations of animals (e.g., Goodall 1986 as cited above). One striking example is the grieving expressions exhibited by elephants toward a dying or deceased matriarch (Douglas-Hamilton et al. 2006). Elephants appear interested in sick, dying or dead elephants regardless of any genetic relationship, suggesting a generalized response to conspecifics in distress. Dolphins have been reported to support sick or dead dolphins (e.g., Fertl and Schiro 1994; Harzen and Santos 1992). Further research needs to be conducted to determine the emotional significance of this behavior (see later discussion on the possibility of dolphin grief).

4.2.5 Animal Personality and Animal Emotion

There are limitations to the methods currently used to study animal emotion, and it is unlikely that any single technique will suffice. We believe that the closely related field of animal personality may provide some useful approaches for the study of animal emotion. As with the study of emotions in animals, the study of animal personality was avoided for many years due to its anthropomorphic nature (Gosling 2001). Furthermore, animal personality and animal emotionality both rely on some sort of assessment of internal states—assessments that must be made without the benefits of human language. However, the study of animal personality has blossomed over the past few decades, while the study of animal emotionality has not. The possibility that individual animals exhibit distinct personality traits has been explored in many species, ranging from butterflies to chimpanzees (see Gosling 2001 for a review), while the study of emotions has been limited to relatively few species.

The study of animal personality and animal emotionality are similar in a number of ways. Perhaps the most important similarity is that much of the terminology used to describe animal personality overlaps with words one would use to describe emotions in animals (e.g. fearful, jealous, anxious; see Plutchik 2001, for an excellent discussion of emotional terms). Another similarity reflects the concern that both areas have with behavioral consistency. By definition, animal personality focuses on the characteristics of an individual that are stable over time and across contexts. For example, a "curious" dolphin consistently explores novel objects or novel environments, whereas a "timid" dolphin does not (Highfill and

Kuczaj 2007). An individual's emotional responses towards certain situations may often be stable as well, especially if these responses reflect the animal's personality. Thus, an individual dolphin might always display fear when it is approached by an unfamiliar animal or encounters a novel object. Or a dolphin might always be jealous when one of its playmates, its mother, or even a human pays attention to another dolphin. One of the goals of animal personality and animal emotion research is to better predict behaviors based upon past behaviors, a goal that is more easily achieved if animals possess stable personality and emotional characteristics. Another similarity between animal personality and animal emotion concerns their relationship to evolutionary fitness. There is a growing consensus that personality differences play important roles in natural selection (Buss 2009; Wolf et al. 2007), but the precise manner in which this occurs is far from clear in humans or other species.

Given the similarities between research on animal emotions and animal personality, it is worthwhile to examine techniques used in animal personality research to assess their effectiveness for animal emotion research. Currently, there are two methods used to study animal personality: coding and rating. The coding method involves researchers coding an animal's behavior within either a novel or familiar situation (Gosling 2001). For example, Svartberg and Forkman (2002) investigated personality traits in dogs by coding each subjects' reactions within certain scenarios. Some of these tasks included reactions to strangers and reactions to "fleeing" prey-like objects. A factor analysis identified the presence of five traits: playfulness, curiosity/fearfulness, chase-proneness, sociability, and aggressiveness. The coding method is analogous to the methods described above which examined the physiological and behavioral expressions of emotion. In these methods, the animal's emotion had to be categorized (coded) by the researcher(s).

Hebb (1946) was one of the first to advocate a version of the coding method to study emotion in humans and animals. He proposed that investigators begin with a list of their "intuitive" notions of which emotions might exist for a given species, next create a set of operational definitions of the emotions that were hypothesized to exist for the target species, and then test the hypotheses via careful observation of behaviors in natural and/or contrived situations. His work with chimpanzees led him to conclude that "classifying emotional behavior is based on a complex set of clues…the stimulus, the subject's experience with this and related stimuli, the response, various aspects of the subject's other behaviors, and behavioral characteristics of the species (pp. 96–97)". Hebb's recognition that many factors contribute to an accurate assessment of an animal's emotional state is but one of the insights provided in this pioneering work. Others include Hebb's finding that an investigator's ideas about an animal's personality influences his decisions about its emotional state in a particular context and his conclusion that our ability to recognize emotion in man and animal is not fundamentally different.

It is essential that the coding method be used with more species and across more emotional contexts in order to increase our understanding of animal emotions, but we also believe that researchers studying animal emotions should take advantage of the rating method. This method requires observers to make judgments about an individual animal's characteristics based on each observer's overall experiences with the animal (Vazire and Gosling 2004), oftentimes using a list of behavioral adjectives or descriptions to rate each animal. Using this technique, Highfill and Kuczaj (2007) demonstrated that dolphins display individual personality traits that are stable over time and across situations, and are reliably rated by human trainers. Within the personality literature, the rating method has been established as a reliable measurement of personality when high levels of agreement occur between the raters (e.g. Dutton et al. 1997; Gosling 1998; Highfill and Kuczaj 2007). Significant inter-rater reliability implies that judges who are well acquainted with the subjects can reliably use certain traits to describe animals. This successful approach has not been applied as often when studying animal emotions (Paul et al. 2005). We believe that humans who are well acquainted with an animal would be able to reliably judge its emotional tendencies. Although we encourage more use of the rating method in the field of animal emotions, it is important to note a limitation of this method. A rater's unique experience with an animal may influence her ratings (Hebb 1946), and so it is essential to know as much as possible about each rater's experience with the target animals in order to correctly interpret their ratings (Highfill et al. 2010). For example, a veterinarian who interacts with a dolphin during uncomfortable situations (e.g. blood draws, ultrasounds) may only witness the animal's negative emotions. However, a dolphin trainer may observe a wider variety of both positive and negative emotions. The importance of considering the nature of the rater's interactions with the animals when using the rating method for studying both animal personality and emotionality cannot be overstated.

4.3 Dolphin Emotions?

Although little is known about dolphin emotions, dolphins are prime candidates for the study of emotion. Dolphins live in dynamic social groups and it is likely that emotions play important roles in their social interactions. Dolphins also have complex communication systems and make use of acoustic signals, body postures, and touch to convey information, including that concerning emotions (Dudzinski et al. 2009; Kuczaj and Kirkpatrick 1993; McCowan and Reiss 2001; Paulos et al. 2008). In addition, the neocortex and paleocortical structures in the dolphin brain are consistent with the notion that dolphins experience emotions and are sensitive to the emotions of others (Jerison 1986). But what is the actual evidence for emotions in dolphins?

Dolphin emotional displays are expressed through a variety of modalities, including sound, postures, and touch. However, the behaviors in such displays are often ambiguous (Samuels 1996; Würsing 2000). For example, leaping behavior in dolphins could be a form of play, some form of social display, or simply the result

Fig. 4.1 Dolphins may communicate emotions through aerial behaviors

of being chased. Similarly, open mouth displays could indicate aggression, fear, sexual arousal, or playful intent. To further complicate matters, dolphins may use multiple modalities simultaneously or sequentially to express emotions. For example, Overstrom (1983) found that dolphins' expressions during aggressive displays consisted of body postures, jaw claps (loud popping sounds produced by a fast closing motion of the jaws), and increasingly intense burst pulse vocalizations. Although these aggressive displays are certainly multi-modal, it is not always clear what emotion is expressed in one of these displays. Such displays could reflect annoyance, anger, frustration, or even fear. This is true for behaviors in general, not just aggressive behaviors. Consequently, it is necessary to understand the motivation that underlies behavior if one wishes to hazard reasonable guesses about the emotion(s) that the behavior expresses (Fig. 4.1).

4.3.1 Dolphin Vocal Expressions of Emotions

Although dolphin vocalizations are still poorly understood, it is clear that they produce many types of vocalizations, including clicks, whistles, burst pulses, squawks, and moans (Au 1993; Tyack 2000). Lilly and Miller (1961) believed that vocal exchanges between dolphins might communicate emotional information, and Lilly (1963) suggested that the intensity of dolphin emotions was correlated with the intensity of the burst pulses they produce. Subsequent work also suggests that dolphin vocalizations may convey information about the signaler's emotional state (Caldwell and Caldwell 1967, 1971; Au 1993; Dudzinski 1996). Caldwell, Haugen, and Caldwell (1962) reported a "loud, sharp, crack" produced by two captive female bottlenose dolphins after being frightened by a plastic model of a dolphin that was placed in their pool. This sound was also reported in three juvenile dolphins, one male and two females, after a researcher shone a bright light in their eyes at night. The jaw pop described earlier is another distinct sound that is often associated with aggression or play. Because of the manner in which it is produced, the jaw clap may provide visual and tactile information as well as acoustic information, all of which may be important for other dolphins' interpretation of the jaw clapping dolphin's emotional state.

The reciprocal whistling between mother–calf pairs during separations likely convey emotional information that indicates stress, distress, or alarm (Tyack 1986; Dudzinski 1996; Cook et al. 2004; Frohoff 2004; Smolker et al. 1993). Given that some situations are more precarious than others, dolphin mothers likely use a variety of signals to communicate emotional state and the significance of their calls to their calves.

If dolphin vocalizations do communicate emotion, it is possible that some of these acoustic emotional signals are universal and thus occur in similar contexts in different populations of dolphins. Evidence for a possible universal emotional signal in killer whales was provided by Rehn et al.(2010). They found that the same "excitement" call was used by three socially and reproductively isolated groups of killer whales, suggesting that this call may be universal for this species. Herman (2010) suggested that the duration of dolphin vocalizations may be used to communicate emotional states, and it is possible that this relationship is another universal characteristic of dolphin vocalizations. Although the precise relationship between dolphin vocalizations and emotional state remains largely speculative at this time, we believe that future research will unravel the complex relationship between dolphin vocalizations and emotional states, including the extent to which such signals are universal or limited to specific dialects.

4.3.2 *Dolphin Use of Posture to Express Emotions*

Postures may serve communicative functions for dolphins and other cetacean species. For example, the S-posture has been observed in bottlenose dolphins, spinner dolphins, Atlantic spotted dolphins, humpback whales, and beluga whales (Dudzinski 1998; Horback et al. 2010). S-postures occur when the animal produces an "S" shape with its body, and has been suggested to be an expression of anger or annoyance. However, S-postures have also been observed during juvenile play and courtship displays (Caldwell and Caldwell 1977; Byerly et al. 2011). Johnson and Norris (1994) suggested that the S-posture of a spinner dolphin was an imitation of a Grey Reef Shark (*Carcharhinus amblyrhinchos*). Helweg et al.(1992) hypothesized that

Fig. 4.2 Dolphins may use touch to convey emotional states

S-posturing in humpback whales is a signal that is used in male competition for females. S-postures are often used in combination with other expressions, such as arching of the body, an open mouth, greater eye white display, and vocal barks and squawks, all of which Herzing (1996) described to be anger displays in spotted dolphins (Fig. 4.2).

4.3.3 Touch as a Mode of Dolphin Emotional Expression?

It is likely that dolphins use touch to express emotions. Tactile interactions are especially salient during social and play bouts (Dudzinski 1998; Dudzinski et al. 2009; Paulos et al. 2008). For example, dolphins may touch each other when they meet or immediately prior to separating (Paulos et al. 2008). Contact behaviors are often used in conjunction with acoustical and postural signals to communicate behavioral states to other group members (Dudzinski 1998; Paulos et al. 2008), and may be used to communicate emotional states as well. Females use their pectoral fins to facilitate and maintain bonds with their offspring and with other females (Dudzinski 1998; Dudzinski et al. 2009). In contrast, contact by the head is often seen during rough play and aggressive bouts, particularly among males (Kaplan and Connor 2007).

Many of the contact behaviors in dolphins are considered to be affiliative, including petting and rubbing behaviors. Petting occurs when one dolphin moves its pectoral fin along another dolphin's pectoral fin, whereas rubbing occurs when one dolphin moves its body along any part of another dolphin's body (Dudzinski 1998; Dudzinski et al. 2009; Kaplan and Connor 2007). Rubbing and petting behaviors in dolphins are analogous to the grooming behavior seen in primates and may be a form of appeasement used to calm excited or aroused individuals. A specific type of rubbing consists of one dolphin rubbing any part of another dolphin's body with its pectoral fin (Dudzinski et al. 2009; Sakai et al. 2006; Tamaki et al. 2006). Rubbing may serve to strengthen social bonds between group members, as well as to advertise the strength of these bonds to other group members. Tamaki et al. (2006) found an increase of pec-rubbing episodes in captive bottlenose dolphins following aggressive interactions, suggesting that pec-rubbing may serve to mend relationships and reduce hostility. In general, tactile interactions serve to maintain the social cohesion of a group, and the ability of such behaviors to convey emotional information is likely an important part of this process. However, the extent to which such tactile behaviors communicate emotional information remains to be determined (Fig. 4.3).

4.3.4 Do Dolphins Grieve?

The controversy surrounding the issue of whether or not animals grieve is much too complex to be resolved in this chapter. However, the possibility that dolphins experience something akin to this overwhelming emotion deserves consideration. Dolphins, like elephants, appear to be quite interested in dead members of their species (Dudzinski et al. 2003). This recognition could reflect dolphins' awareness of death and some form of grieving as a result. Or it could reflect what appears to be a predisposition for dolphins to assist injured or sick members of their group (Caldwell and Caldwell 1966; Cremer et al. 2005; Kilborn 1994; Lodi 1992; Warren-Smith and Dunn 2006). Maternal interactions with dead calves or with inanimate objects following the loss of a calf have been reported for a variety of cetacean species in both wild and captive contexts (Fertl and Schiro 1994; Harzen and Santos 1992; Mann and Barnett 1999; Palacios and Day 1995; Ritter 2007). In one extreme case, Bearzi (2007) found that a dolphin mother continued to interact with her dead calf even after it began to decompose. The mother pulled loose pieces of decomposing flesh off of the lifeless calf, perhaps a form of grooming on the hapless mother's part. Herzing (2000) reported that spotted dolphin mothers appeared to grieve when their calves died. She described the mothers as listless and sometimes disoriented, behaviors that are potentially detrimental to survival since slower swimming speeds and lack of attention to the environment makes one more vulnerable to predators. Although these displays may be interpreted as grief, it is difficult to determine if these displays actually reflect the emotion of grief or some other form of behavioral response to a non-moving offspring.

Fig. 4.3 A pink belly may signal emotional state to other dolphins

4.4 Conclusions

The study of human emotion is replete with controversy concerning definitions and methodology. Clearly, despite our ability to ask other humans about their emotions and introspect about our own personal emotions, we are a long way from understanding human emotions. We cannot ask animals about their emotional states or their ability to correctly interpret the emotions of others, and so it is not surprising that the study of animal emotion is even longer on speculation and shorter on data than is the literature concerning human emotions. However, continued study of animal emotions is essential for more complete theories of animal behavior and improved abilities to enhance animal welfare.

Our stance is that at least some animals do experience emotions and that members of social species are likely to be sensitive to the emotions of other members of their species. Not everyone will agree with these statements. For example, Dixon (2001) argued that some humans are too ready to agree that animals have emotions. Similar to Kagan's (2007) argument that emotions result from the interpretation of feelings, Dixon proposed that animals' abilities to interpret their states will be critical in decisions about the presence or absence of animal emotions. Regardless of how one views this proposal, it is clear that our ability to assess such interpretations by animals, including whether or not animals can actually make these sorts of self-judgments, is sorely lacking. The notion that emotions only exist following some sort of conscious interpretation has been challenged by Ruys and Stapel (2008), who argue that emotional responding is such an important aspect of the social life of humans that emotional responses can occur without conscious awareness of the reason for the response. We agree with their assessment, and believe that emotional responding in animals need not depend on an animal's conscious processing of some sort of underlying information that yields an emotion. If this is the case, then at least some animal emotions are processed unconsciously and it is possible that some recognition of specific emotions in others also occurs without conscious awareness. Our task, then, is to determine which aspects of an animal's emotional life occur without conscious assessment and which aspects require conscious processing. This is likely to vary across species, and the comparative study of animal emotions should reveal the extent to which different species rely on conscious and unconscious processing of emotional information.

Of course, none of this will be easy. It is insufficient to simply state that an animal is experiencing an emotion such as fear, happiness, or jealousy. Moreover, it is unlikely that there will be simple one-to-one correspondences between individual behaviors and individual emotions. As noted earlier, behaviors can be ambiguous, which often leads to different interpretations by different investigators. However, if animals read others' emotions prior to engaging in social interactions, it would benefit them to make as accurate assessments as possible, which suggests that it is possible to accurately interpret the emotional basis of an animal's behavior. In order to do so, we will need to consider the context in which a behavior occurs, the multi-modal nature of the behavior, and the personality of the animal producing the behavior. As our consideration of dolphin emotions demonstrated, sometimes a simple smile just is not enough.

We suspect that some combination of Kagan's (2007) "rich descriptions" of emotional behaviors and Hebb's (1946) "intuitive" methodology will work best. But regardless of the technique that is used to study animal emotions, it is essential that investigators provide detailed descriptions of behaviors that they count as evidence for particular emotions. This is necessary for comparisons across studies, and might eventually yield a behavioral dictionary of emotions that takes into account context, personality, and species. Given the current difficulty of providing agreed-upon operational definitions for individual emotions (let alone for the concept of emotions per se), detailed descriptions such as these are important to advance the field.

To sum up, the study of animal emotion, like the study of human emotion, is anything but straightforward. But unless we deny a priori that animals experience emotions, it is essential that more work on animal emotions be conducted to determine which species experience emotions and/or are capable of recognizing emotions in others, the specific emotions involved, and the extent to which any of this reflects conscious awareness.

References

Arnold MBB (1960) The nature of emotion. Penguin, Harmondsworth
Au WWL (1993) The sonar of dolphins. Springer, New York
Aureli F, Preston SD, de Waal FBM (1999) Heart rate responses to social interactions in free-moving rhesus macaques (*Macaca mulatta*): a pilot study. J Comp Psychol 113:59–65
Averill J (1980) A constructionist view of emotion. In: Plutchik R, Kellerman H (eds) Emotion: theory, research and experience. Academic, New York, pp 305–339
Bachorowski JA, Owren MJ (2003) The sounds of emotion: the production and perception of affect-related vocal acoustics. In: Ekman P, Campos JJ, Davidson RJ, de Waal FBM (eds) Emotions inside out: 130 years after Darwin's the expression of the emotions in man and animals. Ann N Y Acad Sci, New York
Bain A (1859) The emotions and the will. J. W. Parker and Son, London
Bastian J (1965) Primate signaling systems and human languages. In: Devore I (ed) Primate behavior: field studies of monkeys and apes. Holt, Rinehart, and Winston, New York, pp 585–606
Bearzi G (2007) A mother bottlenose dolphin mourning her dead newborn calf in the Amvrakikos Gulf, Greece. Retrieved from http://www.wdcs-de.org/docs/Bottlenose_Dolphin_mourning_dead_newborn_calf.pdf
Bekoff M (2000) Animal emotions: exploring passionate natures. Bioscience 50:862–870
Bekoff M (2007) The emotional lives of animals. New World Library, Novato
Bell C (1840) The hand, its mechanisms and vital endowments, as evincing design: the Bridgewater Treatises on the power, wisdom, and goodness of God as manifested in the Creation, Treatise IV. Harper and Brothers, New York
Buss D (2009) How can evolutionary psychology successfully explain personality and individual differences? Perspect Psychol Sci 4:359–366
Byerly HC, Richardson JL, Kuczaj SA II (2011) Context of S-Posture display in bottlenose dolphins (in preparation)
Cacioppo JT, Gardner WL (1999) Emotion. Annu Rev Psychol 50:191–214
Caldwell MC, Caldwell DK (1966) Epimeletic (care-giving) behavior in Cetacea. In: Norris KS (ed) Whales, dolphins and porpoises. University of California Press, Berkeley, pp 755–789
Caldwell MC, Caldwell DK (1967) Interspecific transfer of information via pulsed sound in captive odontocete cetaceans. In: Busnel RG (ed) Animal sonar systems: biology and bionics. Laboratoire de Physiologie Acoustique, Jouy-en-Josas, pp 879–936
Caldwell MC, Caldwell DK (1971) Statistical evidence for individual signature whistles in Pacific white sided dolphins, *Lagenorhynchus obliquidens*. Cetology 3:1–9
Caldwell DK, Caldwell MC (1977) Cetaceans. In: Sebeok T (ed) How animals communicate. Indiana University Press, Bloomington, pp 794–808

Caldwell MC, Haugen RM, Caldwell DK (1962) High-energy sound associated with fright in the dolphin. Science 137:907–908

Cook M, Sayigh LS, Blum JE, Wells RS (2004) Signature-whistle production in undisturbed free-ranging bottlenose dolphins (*Tursiops truncatus*). Proc Biol Sci 271:1043–1049

Cornelius RR (1996) The science of emotion: research and tradition in the psychology of emotion. Prentice Hall, Saddle River

Cremer MJ, Hardt FAS, Tonello AJ Jr (2005) Evidence of epimeletic behavior involving a *Pontoporia blainvillei* calf. Biotemas 19:83–86

Danbury TC, Weeks CA, Chambers JP, Waterman-Pearson AE, Kestin SC (2000) Self-selection of the analgesic drug Carprofen by lame broiler chickens. Vet Rec 146:307–311

Dantzer R, Mormède P (1981) Can physiological criteria be used to assess welfare in pigs? In: Sybesma W (ed) Current topics in veterinary medicine and animal science, vol 11. Kluwer, Boston/Hingham, pp 53–73

Darwin C (1872) The expression of the emotions in man and animals. Murray, London

Dawkins MS (1977) Do hens suffer in battery cages? Environmental preferences and welfare. Anim Behav 25:1034–1046

Dawkins MS (2000) Animal minds and animal emotions. Am Zool 40:883–888

Dawkins MS (2006) Through animal eyes: what behavior tells us. Appl Anim Behav Sci 100:4–10

Delfour F, Marten K (2006) Lateralized visual behavior in bottlenose dolphins (*Tursiops truncatus*) performing audio-visual tasks: the right visual field advantage. Behav Processes 71:41–50

Dixon B (2001) Animal emotions. Ethics Environ 6:22–30

Douglas-Hamilton I, Bhalla S, Wittemyer G, Vollrath F (2006) Behavioural reactions of elephants towards a dying and deceased matriarch. Appl Anim Behav Sci 100:87–102

Dudinski KM, Sakai M, Masaki K, Kogi K, Hishii T, Kurimoto M (2003) Behavioural observations of bottlenose dolphins towards two dead conspecifics. Aquat Mamm 29:108–116

Dudzinski KM (1996) Communication and behavior in the Atlantic spotted dolphin (Stenella frontalis): relationships between vocal and behavioral activities. Dissertation, Texas A & M University

Dudzinski KM (1998) Contact behavior and signal exchange in Atlantic spotted dolphins (*Stenella frontalis*). Aquat Mamm 24:129–142

Dudzinski K, Gregg J, Ribic C, Kuczaj S (2009) A comparison of pectoral fin contact between two different wild dolphin populations. Behav Processes 80:182–190

Dutton DM, Clark RA, Dickins DW (1997) Personality in captive chimpanzees: use of a novel rating procedure. Int J Primatol 18:539–552

Fertl D, Schiro A (1994) Carrying of dead calves by free-ranging Texas bottlenose dolphins (*Tursiops truncatus*). Aquat Mamm 20:53–56

Fraser D (2009) Animal behaviour, animal welfare, and the scientific study of affect. Appl Anim Behav Sci 118:108–117

Frohoff TG (2004) Stress in dolphins. In: Bekoff M (ed) Encyclopedia of animal behaviour. Greenwood Press, Westport, pp 1158–1164

Goodall J (1986) The chimpanzees of Gombe. Harvard University Press, Cambridge

Goodall J (1990) Through a window: my thirty years with the chimpanzees of Gombe. Houghton Mifflin, Boston

Gosling SD (1998) Personality dimensions in spotted hyenas (*Crocuta crocuta*). J Comp Psychol 112:107–118

Gosling SD (2001) From mice to men: what can we learn about personality from animal research? Psychol Bull 127:45–86

Harzen S, Santos MED (1992) Three encounters with wild bottlenose dolphins (Tursiops truncatus) carrying dead calves. Aquat Mamm 18:49–55

Hauser MD (1993) Right hemisphere dominance for the production of facial expression in monkeys. Science 261:475–477

Hebb D (1946) Emotion in man and animal: an analysis of the intuitive processes of recognition. Psychol Rev 53:88–106

Helweg DA, Bauer GB, Herman LM (1992) Observations of an S-shaped posture in humpback whales. Aquat Mamm 18:74–78

Herman LM (2010) What laboratory research has told us about dolphin cognition. Int J Comp Psychol 233:310–330

Hernandez EN, Solangi M, Kuczaj SA II (2010) Time and frequency parameters of bottlenose dolphin whistles as predictors of surface behavior in the Mississippi Sound. J Acoust Soc Am 127:3232–3238

Herzing DL (1996) Underwater behavioral observations and associated vocalizations of free-ranging Atlantic spotted dolphins, *Stenella frontalis*, and bottlenose dolphins, *Tursiops truncatus*. Aquat Mamm 22:61–79

Herzing DL (2000) Acoustics and social behavior of wild dolphins: implications for a sound society. In: Au WWL, Popper AN, Fay RR (eds) Hearing in whales and dolphins. Springer, New York, pp 225–272

Highfill LE, Kuczaj SA (2007) Do bottlenose dolphins (*Tursiops truncatus*) have distinct and stable personalities? Aquat Mamm 33:380–389

Highfill L, Hanbury D, Kristiansen R, Kuczaj S, Watson S (2010) Rating versus coding in animal personality research. Zoo Biol 29:509–516

Horback KM, Friedman WR, Johnson CM (2010) The occurrences and context of s-posture display by captive belugas (*Delphinapterus leucas*). Int J Comp Psychol 23:689–700

Izard CE (1992) Basic emotions, relations among emotions, and emotion-cognition relations. Psychol Rev 99:561–565

James W (1884) What is emotion? Mind 9:188–205

Jerison HJ (1986) The perceptual worlds of dolphins. In: Schusterman RJ, Thomas JA, Wood FG (eds) Dolphin cognition and behavior: a comparative approach. Erlbaum, Hillsdale, pp 141–166

Johnson CM, Norris KC (1994) Social behavior. In: Norris KS, Würsig B, Wells RS, Würsig M (eds) The Hawaiian spinner dolphin. University of California Press, Berkeley, pp 14–30

Kagan J (2007) What is emotion?: history, measures, and meaning. Yale University Press, New Haven

Kaplan JD, Connor R (2007) A preliminary examination of sex differences in tactile interactions among juvenile Atlantic spotted dolphins (*Stenella frontalis*). Mar Mamm Sci 23:943–953

Kilborn SS (1994) Object carrying in a captive beluga whale (*Delphinapterus leucas*) as possible surrogate behavior. Mar Mamm Sci 10:496–501

Kilian A, von Fersen L, Güntürkün O (2000) Lateralization of visuospatial processing in the bottlenose dolphin (*Tursiops truncatus*). Behav Brain Res 116:211–215

Kirkden RD, Pajor EA (2006) Using preference, motivation and aversion tests to ask scientific questions about animals' feelings. Appl Anim Behav Sci 100:29–47

Kuczaj SA, Kirkpatrick VM (1993) Similarities and differences in human and animal language research: toward a comparative psychology of language. In: Roitblat HL, Herman LM, Nachtigall PE (eds) Language and communication: comparative perspectives. Lawrence Erlbaum Associates, Hillsdale, pp 45–64

Kuczaj SA, Makecha R (2008) The role of play in the evolution and ontogeny of contextually flexible communication. In: Griebel U, Oller K (eds) Evolution of communicative flexibility: complexity, creativity, and adaptability in human and animal communication. MIT Press, Cambridge, pp 253–278

Lancaster J (1965) Primate behavior and the emergence of human culture. Holt, Rinehart and Winston, New York

Larsen RJ, Fredrickson BL (1999) Measurement issues in emotion research. In: Kahneman D, Diener E, Schwarz N (eds) Well-being: the foundations of hedonic psychology. Russell Sage, New York, pp 40–60

LeDoux J (1996) The emotional brain. The mysterious underpinnings of emotional life. Simon and Schuster, New York

Lilly J (1963) Distress call of the bottlenose dolphin: stimuli and evoked behavioral responses. Science 139:116–118

Lilly J, Miller A (1961) Vocal exchanges between dolphins. Science 134:1873–1876

Lodi L (1992) Epimeletic behavior of free-ranging rough-toothed dolphins, *Steno bredanensis*, from Brazil. Mar Mamm Sci 8:284–287

Lorenz K (1949) King Solomon's ring. Methuen, London

Luria A (1982) Language and cognition. MIT Press, Cambridge

Mann J, Barnett H (1999) Lethal tiger shark (*Galeocerdo cuvieri*) attack on bottlenose dolphin (Tursiops sp.) calf: defense and reactions by the mother. Mar Mamm Sci 15:568–575

Marler P, Evans CS (1997) Communication signals of animals: contributions of emotion and reference. In: Segerstrale U, Molnar P (eds) Nonverbal communication: where nature meets culture. Lawrence Erlbaum Associates, Mahwah, pp 151–170

McCowan B, Reiss D (1995) Maternal aggressive contact vocalizations in captive bottlenose dolphins *(Tursiops truncatus)*: wide-band, low-frequency signals during mother/aunt-infant interactions. Zoo Biol 14:293–309

McCowan B, Reiss D (2001) The fallacy of 'signature whistles' in bottlenose dolphins: a comparative perspective of 'signature information' in animal vocalizations. Anim Behav 62:1151–1162

Miksis JL, Grund MD, Nowacek DP, Solow AR, Connor RC, Tyack RC (2001) Cardiac responses to acoustic playback experiments in the captive bottlenose dolphin *(Tursiops truncatus)*. J Comp Psychol 115:227–232

Moberg GP, Anderson CO, Underwood TR (1980) Ontogeny of the adrenal and behavioral responses of lambs to emotional stress. J Anim Sci 51:138–142

Mukhametov L, Supin A, Polyakova I (1977) Interhemispheric asymmetry of the electroencephalographic sleep patterns in dolphins. Brain Res 134:581–584

Myers RE (1976) Comparative neurology of vocalization and speech: proof of a dichotomy. Ann N Y Acad Sci 280:745–757

Overstrom NA (1983) Association between burst-pulse sounds and aggressive behaviour in captive Atlantic bottlenosed dolphins *(Tursiops truncatus)*. Zoo Biol 2:93–103

Owren MJ, Bachorowski J (2007) Measuring emotion-related vocal acoustics. In: Coan JA, Allen JJ (eds) Handbook of emotional elicitation and assessment. Oxford University Press, New York, pp 239–266

Palacios DM, Day D (1995) A Risso's dolphin (*Grampus griseus*) carrying a dead calf. Mar Mamm Sci 11:593–594

Panksepp J (1998) Affective neuroscience. Oxford University Press, New York

Panksepp J (2003) At the interface between the affective, behavioral, and cognitive neurosciences: decoding the emotional feelings of the brain. Brain Cogn 52:4–14

Panksepp J (2007) Neurologizing the psychology of affects: how appraisal-based constructivism and basic emotion theory can coexist. Perspect Psychol Sci 2:281–296

Panksepp J, Burgdorf J (2003) "Laughing" rats and the evolutionary antecedents of human joy? Physiol Behav 79:533–547

Parr LA, Gothard KM (2007) Studying emotion in animals. In: Coan JA, Allen JJ (eds) Handbook of emotional elicitation and assessment. Oxford University Press, New York, pp 379–397

Parr LA, Waller BM, Fugate J (2005) Emotional communication in primates: implications for neurobiology. Curr Opin Neurobiol 15:716–720

Paul ES, Harding EJ, Mendl M (2005) Measuring emotional processes in animals: the utility of a cognitive approach. Neurosci Biobehav Rev 29:469–491

Paulos R, Dudzinski K, Kuczaj SA II (2008) The role of touch in select social interactions of Atlantic spotted dolphin (*Stenella frontalis*) and Indo-Pacific bottlenose dolphin (*Tursiops aduncus*). J Ethol 26:153–164

Pinel J (2009) Biopsychology, 7th edn. Pearson Educational, Boston

Plutchik R (2001) The nature of emotions. Am Scientist 89:344

Preuschoff S, van Hooff JARAM (1997) The social function of "smile" and "laughter": variations across primate species and societies. In: Segerstrale U, Molnar P (eds) Nonverbal communication: where nature meets culture. Lawrence Erlbaum Associates, Mahwah, pp 171–190

Puppe B, Schön PC, Tuchscherer A, Manteuffel G (2005) Castration-induced vocalization in domestic piglets, *Sus scrofa*: complex and specific alterations of the vocal quality. Appl Anim Behav Sci 95:67–78

Quaranta A, Siniscalchi M, Vallortigara G (2007) Asymmetric tail-wagging responses by dogs to different emotive stimuli. Curr Biol 20:R199–R201

Reefmann N, Kaszàs FB, Wechsler B, Gygax L (2009) Ear and tail postures as indicators of emotional valence in sheep. Appl Anim Behav Sci 118:199–207

Reeve J (2009) Understanding motivation and emotion. Wiley, Hoboken

Rehn N, Filatova OA, Durban JW, Foote AD (2010) Cross-cultural and cross-ecotype production of a killer whale "excitement" call suggests universality. Naturwissenschafte. doi:10.1007/s00114-010-0732-5

Ridgway S, Carder D, Finneran J, Keogh M, Kamolnick T, Todd M, Goldblatt A (2006) Dolphin continuous auditory vigilance for five days. J Exp Biol 209:3621–3628

Ridley M (1998) The origins of virtue: human instincts and the evolution of cooperation. Penguin, New York

Ritter F (2007) Behavioural responses of rough-toothed dolphins to a dead newborn calf. Mar Mamm Sci 23:429–433

Romanes GJ (1881) Animal intelligence. Kegan Paul, London

Rowell TE, Hinde RA (1962) Vocal communication by the rhesus monkey (*Macaca mulatta*). Proc Zool Soc Lond 138:279–294

Ruys KI, Stapel DA (2008) The secret life of emotions. Psychol Sci 19:385–391

Sakai M, Hishii T, Takeda S, Kohshima S (2006) Laterality of flipper rubbing behaviour in wild bottlenose dolphins (*Tursiops aduncus*): caused by asymmetry of eye use? Behav Brain Res 170:204–210

Samuels A (1996) A systematic approach to measuring the social behavior of bottlenose dolphins. Dissertation, Woods Hole Oceanographic Institution

Sandem AI, Braastad BO, Bøe KE (2002) Eye white may indicate emotional state on a frustration-contentedness axis in dairy cows. Appl Anim Behav Sci 79:1–10

Siebert ER, Parr LA (2003) A structural and contextual analysis of chimpanzee screams. Ann N Y Acad Sci 1000:104–109

Skinner BF (1938) The behavior of organisms: an experimental analysis. D. Appleton-Century Company, New York

Smolker R, Mann J, Smuts B (1993) Use of signature whistles during separations and reunions by wild bottlenose dolphin mothers and infants. Behav Ecol Sociobiol 33:393–402

Suomi SJ (1997) Nonverbal communication in nonhuman primates: implications for the emergence of culture. In: Segerstrale U, Molnar P (eds) Nonverbal communication: where nature meets culture. Lawrence Erlbaum Associates, Mahwah, NJ, pp 131–150.

Svartberg K, Forkman B (2002) Personality traits in the domestic dog (*Canis familiaris*). Appl Anim Behav Sci 79:133–155

Takamatsu H, Noda A, Kurumaji A, Murakami Y, Tatsumi M, Ichise R, Nishimura S (2003) A PET study following treatment with a pharmacological stressor, FG7142, in conscious rhesus monkeys. Brain Res 980:275–280

Talan J (2006) Do animals have feelings? Scientific American Mind 17:26–29

Tamaki N, Tadamichi M, Michihiro T (2006) Does body contact contribute towards repairing relationships?: the association between flipper-rubbing and aggressive behavior in captive bottlenose dolphins. Behav Processes 73:209–215

Thieltges H, Lemasson A, Kuczaj SA II, Böye M, Blois-Heulin C (2011) Visual laterality in dolphins when looking at (un)familiar humans. Anim Cogn 14(2):303–308

Tinbergen N (1951) The study of instinct. Clarendon, Oxford

Tolman EC (1932) Purposive behaviors in animals and men. Century, New York

Tyack P (1986) Whistle repertoires of two bottlenosed dolphins, *Tursiops truncatus*: mimicry of signature whistles? Behav Ecol Sociobiol 18:251–257

Tyack P (2000) Functional aspects of cetacean communication. In: Mann J, Connor RC, Tyack PL, Whitehead H (eds) Cetacean societies: field studies of dolphins and whales. University of Chicago Press, Chicago, pp 270–307

Vazire S, Gosling SD (2004) Personality and temperament: a comparative perspective. In: Bekoff M (ed) Encyclopedia of animal behavior. Greenwood Publishing Group, Westport, pp 818–822

Warren-Smith AB, Dunn WL (2006) Epimeletic behaviour toward a seriously injured juvenile bottlenose dolphin (*Tursiops* sp.) in Port Phillip, Victoria, Australia. Aquat Mamm 32:357–362

Washburn MF (1908) The animal mind: a textbook of comparative psychology. MacMillan, New York
Watson JB (1914) Behavior: an introduction to comparative psychology. Holt, New York
Wolf M, van Doorn GS, Leimar O, Weissing FJ (2007) Life-history trade-offs favour the evolution of animal personalities. Nature 447:581–584
Würsing B (2000) In a party mood. In: Bekoff M (ed) The smile of a dolphin. Discovery Books, New York, pp 188–191
Young PT (1943) Emotion in man and animal. Wiley, New York

Chapter 5
Play and Emotion

Stan A. Kuczaj II and Kristina M. Horback

Abstract The most important feature in interpreting observed emotions in both humans and animals is the context in which the assumed emotion occurs. The same can be said for play. Wrestling between two individuals can be interpreted as either aggressive or playful depending on the amount of self-handicapping, mutual pauses,

S.A. Kuczaj II (✉) • K.M. Horback
Department of Psychology, The University of Southern Mississippi,
118 College Dr. #5025, Hattiesburg, MS 39406-5025, USA
e-mail: s.kuczaj@usm.edu

and damaging fighting tactics that occur during the wrestling bout. Play and emotion are intimately intertwined, and this relationship provides an invaluable context in which to study emotions. In this chapter, we discuss the inherent difficulties in the investigation of play and emotion. We also consider the roles of solitary and social play behavior in individuals' acquisition of both physical and mental skills, including the manner in which play may facilitate the manipulation of one's own emotions in pretense and the ability to correctly interpret and respond to a play partner's emotions.

As you read the chapters in this book, you will undoubtedly notice that "emotion" is very difficult to define. This has long been the case, and a variety of scholars have produced thoughtful and important considerations of emotions without precisely defining the term (e.g., Bain 1859; Cacioppo and Gardner 1999; Darwin 1872). Moreover, attempts to define emotion have typically been unsatisfactory, resulting in a somewhat dizzying array of definitions, many of which contradict one another (e.g., Arnold 1960; Averill 1980; Bekoff 2007; Cornelius 1996; Diener 1999; Ekman and Davidson 1994; James 1884; Kagan 2007; Kuczaj et al. 2012; Young 1943). The elusive nature of emotions sometimes results in overly narrow definitions, and at other times, overly broad definitions. There is no consensus for any single definition that includes all emotions but excludes all non-emotions, most likely because there is little agreement about what should and should not count as instances of an emotion. Despite the difficulties inherent in defining emotions, animals appear to experience both negative emotions such as fear and positive emotions such as happiness (Balcombe 2009; Bekoff 2000; Berridge and Kringelbach 2008; Panksepp 1998, 2007; Sankey et al. 2010). The majority of vertebrate animals share the same dopaminergic pathways and neurological structures that are activated during a pleasurable experience. Additional support for the notion that non-human animals have the capacity to experience positive emotions comes from physiological reactions to social grooming. Such communal physical contact reduces heart rate (Aureli et al. 1999) and releases endorphins (Keverne et al. 1989) for all individuals involved, not just the individual being groomed. Rats approach researchers' hands four times more quickly when reinforced with belly-tickles rather than basic dorsal strokes, the belly-tickles being more similar to the rats' rough-and-tumble play than are the dorsal strokes (Burgdorf and Panksepp 2001). Tickled rats even produce an ultrasonic "laugh-like" repetitive chatter (Panksepp and Burgdorf 2000); a behavioral response previously thought to be present only in humans and great apes (Provine 2001). It seems likely that positive emotions that result from tactile interactions facilitate social bonding and social interactions. In fact, there may be a common set of neurological pathways in many mammalian species that regulate touch-induced mood changes (Pellis and Pellis 2009). Rough and tumble play may allow participants to learn more about how touch affects their own mood as well as that of others. If this is the case, then rough and tumble play may facilitate the ability to use touch to produce particular emotions in others.

The definitional difficulties that plague the study of emotions are also evident in the study of play. The literature contains numerous reports of play in children and young animals, but no universally accepted definition of play (Bekoff and Allen 1998; Bekoff and Byers 1981; Bel'kovich et al. 1991; Bretherton 1984; Bruner et al. 1976; Burghardt 2005; Caro 1995; Coelho and Bramblett 1982; Delfour and Aulagnier 1997; Fagen 1993; Ficken 1977; Gewalt 1989; Groos 1898; Guinet 1991; Harcourt 1991; Heinrich 1999; Kuczaj et al. 2006; Kuyk et al. 1976; Markus and Croft 1995; McCowan et al. 2000; Muller-Schwarze et al. 1982; Ortega and Bekoff 1987; Paulos et al. 2010; Pellis and Pellis 2009; Power 2000; Scarlett et al. 2005; Singer and Singer 1990; Symons 1978; Thompson 1996; West 1974; Würsig et al. 1989). As Wilson (1975) noted, "no behavioral concept has proved more ill-defined, elusive, controversial and even unfashionable than play" (p. 164). The nature of play itself contributes to this definitional confusion. Play can occur in many forms and oftentimes resembles non-play behaviors (e.g., fight behaviors during play-fights and flight behaviors during play chases). Consequently, interpretations of the context in which behaviors are observed are crucial for decisions concerning whether or not the observed behaviors are instances of play (Beach 1945; Bekoff and Byers 1998; Fagen 1981; Kuczaj and Makecha 2008; Paulos et al. 2010). Given that observers can make overly narrow or overly rich interpretations of context, deciding whether or not a behavior should be considered an instance of play is not always straightforward.

For example, consider definitions offered by two of the leading authorities in the field of animal play. More than thirty years ago, Bekoff and Byers (1981) suggested that play consists of "all motor activity performed postnatally that appears to be purposeless, in which motor patterns from other contexts may often be used in modified forms and altered temporal sequencing" (pp. 300–301). The phrase "appears to be purposeless" places the burden of decision on the observer, and is sufficiently vague to cause problems with reliability. Moreover, as Heinrich and Smolker (1998) noted, behavior may appear purposeless for a number of reasons. For example, the observer may not recognize the purpose of the play behavior even though the behavior may in fact be purposeful. Or the purpose may be not be revealed until long after the play bout has ended.

More recently, in his excellent review of animal play, Burghardt (2005, pp. 70–78) proposed five criteria for a behavior to be considered play: (1) the behavior is not fully functional, (2) the behavior is spontaneous, voluntary, pleasurable, or autotelic ("done for its own sake"), (3) the behavior differs from "serious" behavior in terms of its form, duration, and/or frequency, (4) the behavior is repeatedly performed, but not stereotyped, and (5) the behavior occurs when the animal is healthy and free from stress. As was the case with the Bekoff and Byers (1981) definition, Burghardt's criteria require the observer to make some difficult decisions. The observer must decide if a behavior is "not fully functional", "pleasurable", "non-stereotypic", etc. However, the burden on observers may not be as great as it appears. As many authors have noted, one of the curious things about play is that, despite these definitional difficulties, most of us recognize play when we see it (even if we are hard-pressed to state precisely why this is the case).

Fig. 5.1 Dolphins play with a variety of objects

Following Bekoff and Byer's (1981) consideration of play, and consistent with Burghardt's (2005) notion that play is repeatedly performed in some non-stereotypic manner, Kuczaj (1998) noted that play typically involves one or more of the following types of dynamic behaviors: (1) modification of a produced or observed behavior, (2) imitation of another's behavior, and (3) repetition of a behavior produced by the self. Modification occurs during play when a preceding behavior (produced by the self or others) is transformed in some way. Imitation involves the (sometimes partial) reproduction of another's behavior, while repetition occurs when the model behavior is provided by the self. The fact that play often involves both modifications and imitation/repetitions has led numerous scholars to suggest that play is important for the ontogeny and evolution of flexible thought and flexible communication (Fagen 1974; Kuczaj 1982, 1983, 1998; Kuczaj et al. 2006; Kuczaj and Makecha 2008; Piaget 1952; Spinka et al. 2001; Weir 1962) (Fig. 5.1).

5.1 Types of Play

Play behavior can be categorized in many ways, including the extent to which it is social. While solitary play is common in the young of both social and solitary species, social play is particularly important for the behavioral development of social species (Bekoff 2007; Biben 1998; Kuczaj et al. 2006; Špinka et al. 2001). Social play requires more than the simple proximity of playing individuals. For example, a playground might contain ten children playing on swing sets, slides and monkey bars, but no interaction among the playing children. Playing in the same area as others is insufficient to constitute social play. Instead, when children or animals are playing in close proximity but actually playing alone, they are engaging in solitary play in a social context, a form of solitary play called parallel play. True social play involves *interaction* among the players, and it is interaction that is lacking in parallel play. Humans are not the only species to produce solitary, parallel, and social play, but general developmental trends for these forms of play have been specified for only a few non-human animals (see Burghardt 2005 for a general review). Panksepp (1998) noted that mother–offspring play is common throughout infancy in many species (see also Burghardt 2005; Kuczaj et al. 2006). Given that the mother–infant bond is typically the strongest bond in a young animal's life, and that social play is most likely to occur among animals that are familiar with one another (Panksepp 1998), social play among mothers and their infants probably serves multiple functions, including strengthening the mother–infant bond and establishing the foundation for social play with others besides the mother. As important as mother–infant social play appears to be, social play among peers is a critical component of social and cognitive development for many species. For example, it is clear that animal social play involving peers increases the complexity of play behavior (Kuczaj et al. 2006). Such play may also facilitate the imitation of novel behaviors (Kuczaj et al. 2006; Miklosi 1999; Reynolds 1976; Špinka et al. 2001) and provide relatively non-threatening contexts in which children and young animals can learn to express and control emotions.

Social play in both human children and animal youngsters often involves physical interactions. For example, lamb social play contains lots of head butts and other forms of contact (Berger 1980). In fact, rough-and-tumble play is the predominant form of social play seen in both human and non-human animals (Boulton and Smith 1992). Social play of this sort allows the young to practice and perfect skills that might be beneficial in later life—foraging, predator avoidance, hunting, mating, and dominance/territory disputes (Vieira and Sartorio 2002). In this sense, physical play interactions benefit individuals in both their social interactions with members of their own social group (mating, dominance) and their interactions with members of other species (prey or predator) or other social groups (territorial disputes, intergroup mating). Rough-and-tumble play is a good example of the emotional check and balance system necessary for a successful social play bout. Individuals must control their own emotions during play fights and play chases, but they must also be able to read the emotional state of their play partner, including any changes that occur during the play bout. Such appraisals are necessary in order to produce appropriate play behaviors.

In addition to the distinction between solitary and social play, it is possible to group play behaviors according to the extent to which they involve locomotor activities, object manipulation, and pretense (Burghardt 2005; Fagen 1981). Although these categories are not mutually exclusive, they provide a useful framework for any consideration of play. Locomotor play corresponds to the isolated, sporadic movement patterns most often seen in young animals. This type of play is characterized by active movements such as scampering or leaping, and is distinctive from overt "avoidance of a predator or conspecifics…by its non-threatening context…player's loose body tone…and by the repetition and ease of interruption of its activity" (Fagen 1981, p. 8). This type of play seems to result from general exuberance, and is thought to be beneficial in that it facilitates the development of muscle and cardiopulmonary capability as well as providing opportunities for predators to perfect their hunting skills and for prey to practice anti-predator escape behaviors in a variety of contexts (Byers and Walker 1995; Power 2000; Špinka et al. 2001). These practice play bouts are *relatively* safe because they occur in a play context that lacks the normal consequences for failure, but the unwary player may nonetheless be injured or even killed if it does not remain aware of the reality in which its play behavior is embedded.

Object play involves interactions with inanimate things (i.e., toy play). Object play involves physical manipulation of objects that provides no immediate obvious benefit (Burghardt 2005; Hall 1998). Such play aids in the development of manipulative-cognitive skills and exploratory behaviors (Burghardt 2005; Kuczaj et al. 2006; Kuczaj and Makecha 2008). Examples include play with self-made bubble toys by dolphins (Kuczaj et al. 2006; McCowan et al. 2000), object kicking and carrying by young cheetahs (Caro 1995), the capture of sea gulls by killer whales or dolphins to use them as play objects (Kuczaj and Makecha 2008; Kuczaj and Walker 2006; Würsig 2002), object manipulation by gulls and ravens (Heinrich 1999; Heinrich and Smolker 1998; Wheeler 1943), and the exploration of the myriad array of toys with which human children play.

Pretend play "involves a pretender, a reality, a mental representation, an awareness of the divergence between what is represented and reality, and intentionally projecting the representation onto reality" (Lillard 2002, p. 104). Put a bit more simply, pretend play occurs when the player either has an object represent something else (e.g., a child pretending that a block of wood is a dinosaur) or imagines that he is playing with an imaginary object, perhaps even an imaginary friend or foe (e.g., a child sword-fighting with an imaginary sword and an imaginary opponent). Pretend play requires the capacity to mentally represent and manipulate alternative possibilities, abilities that are common in human children (Kuczaj and Daly 1979; Kuczaj and Makecha 2008; Piaget 1951; Singer and Singer 1990). Children with imaginary friends exhibit emotional involvement with these friends but also recognize that the friends are not real (Taylor and Carlson 2002). Outside of humans, make-believe play is thought to only be possible in more cognitively advanced primates, although the evidence for such play is somewhat controversial (see Mitchell 2002). For instance, there are rare anecdotal reports of captive great apes engaging in forms of make-believe object play, such as pretending to eat fake food, pretending to feed a doll, and hiding pretend objects (Parker and McKinney 1999; Pellegrini and Smith 2005). The problem, of course, lies not with the observations themselves but with the range of possible interpretations. It is possible that an ape that appears to be pretending to feed a doll is in fact pretending to do so. But it is also possible that the ape is reproducing its own previous behaviors or mimicking behaviors of others it has witnessed without pretense being involved at all. This is also true for children's pretend play, the difference being that children's spontaneous speech and/or answers to adult inquiries provides valuable clues about the extent to which children are pretending. Such information is lacking in almost all of the anecdotes on ape pretend play, and so it is difficult to determine whether or not the apes are actually pretending. For example, consider Matsuzawa's (reported in Parker and McKinney 1999) observation of a juvenile female carrying a stick in the same manner that her mother was carrying her sick infant sibling. This is a most intriguing observation. But it is not clear if the juvenile was actually pretending that the stick was an infant, mimicking the behavior of the mother with an object that happened to be convenient (but not pretending the object was an infant), or if this was simple a coincidence of some sort. And since there was no way to ask the juvenile what she was doing, it is difficult to eliminate enough possibilities to arrive at a single most reasonable conclusion. However, it has been reported that a bonobo pretended monsters were in another room by indicating the sign "monster" on her keyboard, and so seems to have used a symbol to communicate the content of her pretense to others (Ingmanson 2002).

As noted above, the pretend play of human children may occur during solitary play (e.g., a child playing by herself texting on an imaginary phone) or social play (e.g., children assuming imaginary roles in a pirate attack on a merchant vessel). Social make-believe play is sometimes called sociodramatic play (Singer and Singer 1990). During such play, children often feign an emotion as a part of the play frame that they are creating (Burghardt 2005; Mitchell 2002; Singer and Singer 1990). These virtual emotions generated by children during sociodramatic play sometimes result in an individual experiencing the negative emotion that they were initially pretending to experience (Sutton-Smith 2003). Thus, as noted earlier, play and emotions are intimately intertwined. Even though the play context may provide a means for players to express and understand undesirable emotions, the fragility of the play context can result in an imagined emotion being transformed into an actual emotion. Nonetheless, make-believe play may be an important aspect of children's emotional development in that such play can help children learn to cope with emotional impulses in the relative safety of pretense, and so learn to control emotions both during play and more serious contexts (Bornstein and O'Reilly 1993; Bretherton 1989; Fein 1987; Singer and Singer 1990; Sutton-Smith 1980, 2003). Fein (1987) noted that pretense provided children with an "unusual opportunity" to learn to maintain "comfortable and stimulating" levels of emotional arousal. The success of such unusual opportunities may be reflected in happier children. Singer and Singer (1990) reported that children who engaged in more imaginative play were more likely to experience positive emotions such as joy and liveliness, and were more likely to cooperate with their peers. If Plutchik (2001) is correct in his assertion that emotions help humans and animals to regain a state of equilibrium, play may provide young animals and children with necessary opportunities to learn the usefulness of emotions by manipulating particular emotions in various make-believe scenarios. Animals, like children, may pretend about their emotions during play (Mitchell 2002). For example, Eibl-Eibesfeldt (1978) reported that a pet badger

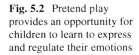

Fig. 5.2 Pretend play provides an opportunity for children to learn to express and regulate their emotions

pretended to be afraid during play that involved chasing the badger. He also observed a dog pretending to be angry (growling while playing with an object), but at the same time wagging its tail in a friendly manner (Fig. 5.2).

In their study of the development of hypothetical reference in young children's speech, Kuczaj and Daly (1979) found that children's discussions of hypothetical situations often contained references to emotions. Oftentimes, the reference was to the emotions the child might experience in a hypothetical situation. For example, when a mother asked her three and one-half year old daughter "What would happen if the girl didn't want her ears pierced?" the daughter replied "then she won't get pierced ears. Or she will cry". Similarly, when a three and one-half year old boy was asked "What would happen if a butterfly would have sat on your nose?" his response was "I would be not angry. I would be happy". Children were also able to speculate about the emotions of others; a three-year-old stated that "It break and daddy get angry" when asked "What if I put this under Daddy's pillow?". In addition, a 4-year-old speculated about the feelings a dinosaur might experience in the following exchange:

Adult: What would happen if we brought a dinosaur home?
Child: He would like us and be happy to live with us. But he might smash our house by accident and be sorry.
Adult: Why would he be sorry?
Child: Because it would be bad to smash our house. And we would cry. The dinosaur too.
Adult: He would cry?

Child: Yeah. But not as much as us.
Adult: Why not?
Child: Because dinosaurs don't like to cry. When they cry too much, they get mad.
Adult: That doesn't sound good.
Child: No. If he gets mad, he might eat us.
Adult: Oooh. Maybe we shouldn't bring him home then.
Child: No. He would be friendly. No mad.

5.2 Emotions and Play

Despite the lack of consensus concerning definitions of play and emotion, many scholars have noted that positive emotions and play are highly correlated. For example, Darwin (1872) observed that "happiness is never better exhibited than by young animals, such as puppies, kittens, lambs, etc., when playing together, like our own children" (p. 448). Following Darwin, Groos (1898) proposed that "there are two quite different popular ideas of play. The first is that the animal (or man) begins to play when he feels particularly cheerful, healthy, and strong: the second—which I found even entertained by a forester—that the play of young animals serves to fit them for the tasks of later life" (p. xix). In his consideration of emotions and play, Young (1961) noted that although play behaviors may occur in many forms, "from the point of view of affective psychology, …play behavior is typically pleasing" (p. 49). The notion that play is pleasurable, perhaps even joyful, is common in the literature on play (Bekoff 2007; Csikszentmihalyi 1990; Kuczaj and Makecha 2008; Špinka et al. 2001; Sutton-Smith 2003), leading Panksepp (1998) to speculate that the experience of joy occurs when play circuits in the brain are aroused. Panksepp (1998) reported that high levels of acetylcholine, glutamate and opioid neurotransmitters were correlated with play behavior in numerous species. In contrast, he found that serotonin, norepinephrine and gamma-aminobutyric acid (GABA) appeared to reduce the play drive. Thus, not only is play associated with positive emotional experiences, but the lack of play seems to reflect the absence of such states.

Bekoff (2007) suggested that one of the reasons that play is contagious is that others can "feel" the joy and glee when observing animals at play. Perhaps the expression of joy during play is in of itself a signal that the individual is playing, which may also serve as an invitation for others to join in the play bout. Emotions convey information (Mayer et al. 2008), and the positive emotions that occur during play may be one reason that specific play signals are relatively rare (see discussion below on play signals). Species that are adept at reading the emotional states of others need not have evolved specific signals for play—the emotions associated with play may be the only signals that are needed. Kano and Tomonaga (2010) found that chimpanzees appear to pay more attention to the contextual information provided in emotional scenarios than to specific cues such as facial expression, and it seems likely that many social animals that engage in social play also use a variety of cues rather than a single signal to determine the extent to which an interaction is playful.

Part of the evidence for the strong correlation between play and positive affect comes from repeated observations that animals that are depressed, stressed in some way, or ill do not play (Burghardt 2005; Carlstead et al. 1993; Fagen 1981; Martin and Caro 1985; Oliveira et al. 2010; Rensch 1973). For example, fear and hunger can eliminate play behavior in human children and young animals (Panksepp 1998). The fact that sick or stressed individuals rarely play may reflect the compromised individual's lack of the requisite physical energy to play and/or an inability to generate the requisite positive emotional energy. For example, a child suffering from a fever will be too lethargic to play. But it is not just physical conditions that limit play. A child or a juvenile animal that is placed in a strange situation is likely to inhibit its play behavior until it grows comfortable with the novel situation. If the youngster has some basis from which to interpret the novel situation, such as a secure base provided by the mother, then play will slowly emerge as the frightening strange situation becomes more familiar. The transformation of familiar or novel events into ones that are moderately discrepant events is one of the reasons that players find play so enjoyable. A situation that is infused with negative affectivity is incompatible with play, but as a familiar situation is made more interesting by altering one's play behaviors or a scary situation is made less threatening by daring to do something such as peering from behind a mother's shoulder, the youngster modifies its emotional state and continues to do so as the play bout evolves (Kuczaj et al. 2006; Kuczaj and Makecha 2008; Piaget 1951).

Although one of the characteristics of play is that it is pleasurable, Burghardt (2005) pointed out that play is not always fun. Play bouts that begin as fun or as friendly interactions may quickly change into contexts that produce anger, fear, surprise, or even grief (Burghardt 2005). This can occur in solitary play or social play. We witnessed an example of solitary play changing from fun to surprise when a young dolphin playing with a large crab had the tables turned on him by the crab. The dolphin was grabbing the crab with its mouth, carrying the crab for a distance, releasing the crab, and then catching the crab before the crab settled to the bottom. This behavior continued until the overly confident dolphin nonchalantly mouthed the crab and received a pinch on its tongue for its efforts. This was immediately followed by a dolphin yelp and the dolphin hurrying back to its mother. What had started out as a pleasurable experience for the dolphin (but perhaps not for the crab) became a painful one, at which time the play bout ended.

Social play can also start off as fun but turn more serious. Some of the best examples of this involve social play fights. What begins as playful jousting among peers sometimes escalates into true aggressive behaviors, such escalation being more likely to occur with increasing age (Burghardt 2005; Burghardt and Burghardt 1972; Pellis and Pellis 1996; Power 2000; Sommer and Mendoza-Granados 1995). During a play bout, individuals experience both the autonomic physical reactions to and cognitive appraisals of specific play cues and behaviors. According to Sutton-Smith (2003), individuals trigger "virtual emotions" during play, such as aggressive body postures and elevated heart rates. The cognitive appraisal of the play context reassures the individual that it is safe to display these virtual emotions. The stimulation of these virtual emotions, however, allows the actual negative emotion to be more readily triggered (Sutton-Smith 2003). Consequently, social rough-and-tumble play can quickly transition from play to aggression. This is why Burghardt (2005) suggested that "play may be proximally controlled by a broad array of emotions, not just fun" (p. 140). Of course, if the behavior ceases to be play, as happens when play fights turn brutal or a child becomes frustrated with some play activity, then it is not really the case that anger or aggression accompanies play. Instead, the once playful experience has become something quite different, and this difference is reflected in the nature of the behavior as well as the emotions that are involved. Nonetheless, it does seem to be the case that the play context may allow players to experience a variety of emotions in a non-threatening setting (Burghardt 2005; Singer and Singer 1990; Sutton-Smith 2003). Despite those occasions in which play is transformed into a more serious situation, social play appears to reduce the frequency of negative emotions among animals in a variety of contexts, not just those involving play. Social play helps animals and children learn appropriate responses to other's behaviors, including how to adjust one's emotional response so that it fits the context (Bekoff 2002; Fagen and Fagen 2004; Pellis and Pellis 2009; Singer and Singer 1990). Depriving animals of the opportunity for social play can have dire consequences. For example, rats that were not allowed to play fight overreacted when touched by other rats, these extreme reactions resulting in more frequent attacks on the deprived rats (Pellis and Pellis 2009). Play, then, may help animals and children learn to regulate their emotions in socially appropriate ways.

Human beings can verbally communicate their subjective experiences of visceral reactions such as changes in heart rate as well as their cognitive appraisals of such changes. In this sense, humans can directly communicate their emotional states to others (Cosmides and Tooby 2000; Darwin 1872; Oatley and Jenkins 1998), although the recipient must decide if the message is truthful or deceptive. It appears that the human ability to identify emotions in others relies on our autonomic responses, others' physical expressions, and others' verbal reports (Oatley and Jenkins 1998). The extent to which we rely on each type of information depends on the context and on our previous experience with the individual. Nonetheless, correctly identifying emotions in others of our own species can be challenging. For example, many couples argue about the validity of one partner's assessment of the other's emotional state. In addition to hazarding guesses about other humans' emotional states, we also infer the affective states of non-human animals by assessing body posture, facial expression and behavioral context (Darwin 1872; Ekman 1999). Of the six primary emotions proposed by Ekman et al. (1969), fear, surprise and anger seem to be the easiest for humans to recognize in a variety of species, most likely because the behaviors that accompany such emotions are more apparent than those associated with emotions such as happiness, sadness and disgust (the other three primary emotions proposed by Ekman et al. 1969).

We, like many authors before us (see also Kuczaj et al. (2012) for a more thorough consideration of this possibility), assume that animals are able to recognize the emotional states of other members of their own species, particularly members with which they have had previous social experience. This ability is of particular importance when a young animal wants to play with another. If the youngster misinterprets the willingness of another to play, she runs the risk of being rebuffed and perhaps even injured. Even if the initial interpretation is correct, the playing animal must continually monitor and successfully interpret the emotions of her play partner. Misinterpreting another's facial and postural cues can result in a playful wrestling escalating into a costly physical fight.

5.3 Play Signals

Social play may facilitate the development and maintenance of social bonds and flexible social interaction skills (Baldwin and Baldwin 1974; Bekoff 1984; Byers 1984; Colvin and Tissier 1985; Fairbanks 1993; Kuczaj et al. 2006; Kuczaj and Makecha 2008; McCowan and Reiss 1997; Snowdon et al. 1997). Creels and Sands (2003) suggested that animals that find it difficult to interpret another's social signals have higher chronic levels of stress than animals that have such skills. Play may help animals learn to assess social situations and communicative cues. If so, play may contribute to developing organisms learning to be more comfortable in social situations and so reduce the incidence of social stress in later life (Fig. 5.3).

The notion that play is important in the ontogeny and evolution of communication was advanced by Bateson (1955, 1972), and has since been strongly advocated by others (see Kuczaj and Makecha 2008). Lancy (1980) argued that "to the extent that communication systems are vital to the creation and maintenance of social systems, play appears to be the medium through which the young learn to use and

Fig. 5.3 Play provides opportunities for social learning. Here a dolphin watches another dolphin play with a bubble ring

understand these systems" (p. 489). Around the same time, Sutton-Smith (1980) suggested that play is "first and foremost a kind of communication" (p. 11). Bateson's work was pivotal in a number of ways. First, he noted that all play requires an ability to distinguish pretense and reality, and so the type of play we have termed "pretend play" is in fact a form of play that involves additional components of pretense beyond those involved in normal play. Second, Bateson also emphasized that social play requires players to communicate that they are playing in order for common "play frames" to be established. Bateson suggested that play frames are established and maintained via "play signals", signals that communicate that behaviors produced during play are not as serious as they normally would be in other contexts. The use of play signals likely evolved to decrease the likelihood that playing animals misinterpret each other's intent. After all, such misinterpretations could have serious consequences. Given the ubiquitous nature of play and its roles in social development, it makes sense that evolution would have selected for individuals that were able to successfully communicate playful intent.

As Bateson suspected, play signals have been found in a number of species that engage in social play, albeit far from all such species. The communicative significance of play signals is supported by the fact that play signals are more likely to occur during social play than solitary play (Biben and Symmes 1986). Acoustic play signals have been found in the social play of chimpanzees (*Pan troglodytes*) (Matsusaka 2004), cotton-top tamarins (*Saguinus oedipus*) (Goedeking and Immelman 1986), bottlenose dolphins (*Tursiops truncatus*) (Blomqvist et al. 2005), dwarf mongooses (*Helogale parvula*) (Rasa 1984), harp seals (*Pagophilus groenlandicus*) (Kovacs 1987), rhesus monkeys (*Macaca mulatta*) (Gard and Meier 1977), and squirrel monkeys (*Saimiri sciureus*) (Biben and Symmes 1986). However, play signals are not always vocalizations. Australian magpies display a "bouncy" gait prior to playing (Pellis 1983). Adult bonobos use play faces to solicit and facilitate play from/with others, and so in a sense choose their play partners by signaling them (Palagi 2008). Some play signals mimic authentic aggressive postures and behaviors, and therefore require subtle additions in order to be received as playful. Baboons (*Papio* sp.) display a relaxed open mouth during play that is very similar to an aggressive open mouth threat (Leresche 1976). Dolphins often display bodily *S-postures* during aggressive contexts (Horback et al. 2010; Pryor 1990), but Dudzinski (1998) reported that S-postures performed with an oblique angle of approach by juvenile spotted dolphins often occurred during playful contexts. Chimpanzees sometimes use head tilts (Sade 1973) and/or play faces (van Hoof 1967) to indicate playful intent. Palagi and Mancini (2011) found that the type of play face used changed as a function of age in gelada monkeys. Young monkeys used a partial play face in which the mouth was opened with only the lower teeth exposed. Adults rarely used the partial play face, but instead used a full play face in which both lower and upper teeth were exposed. This suggests that the full play face is less ambiguous than the partial play face, and so replaces it as a more efficient signal. The use of play postures and/or play faces as play signals has also been observed in dogs (*Canis* sp.) (Bekoff 1995), spotted hyenas (*Crocuta crocuta*) (Drea et al. 1996), and squirrel monkeys (Baldwin and Baldwin 1974). Kuczaj and

Makecha (2008) reported a possible play signal for bottlenose dolphin calves that involved positioning themselves on the surface of the water perpendicular to a playmate and remaining stationary. This behavior usually preceded another juvenile dolphin approaching the stationary animal and pushing it sideways through the water. Kuczaj and Makecha speculated that the stationary perpendicular posturing was a play signal indicating a willingness to be pushed, noting that this behavior was only produced during play contexts. Moreover, it always resulted in one of two outcomes. Other dolphins either ignored the posing dolphin or the posing dolphin was pushed sideways along the surface of the water. Bouts in which the posing dolphin was pushed sometimes resulted in the dolphin posing again in order to solicit another push. However, in some play bouts two dolphins alternated between being the "pusher" and the "pushee". This turn-taking behavior suggested to Kuczaj and Makecha that the young dolphins were engaging in a form of cooperative play and that the play signal used to indicate the willingness to be the "pushee" was a critical aspect of this game.

Although specific play signals are relatively rare (or are still remaining to be discovered), species that use play signals appear to use them throughout play bouts. Play signals are used to indicate a willingness to play, to assess another's willingness to play, and as a means to check that an interaction is still playful (Bekoff 1975; Bekoff and Byers 1981; Fagen 1981; Feddersen-Petersen 1991; Hailman 1977; Hailman and Dzelzkalns 1974; Kuczaj and Makecha 2008; Loeven 1993). Individuals engaging in play may help to maintain the playful temperament by producing species-specific play signals, although such signals do not appear to be necessary for either play solicitation or play maintenance in all social species. Play signals may also help prevent simulated negative emotions from escalating into genuine aggression or fear. They also affect the manner in which other signals are interpreted (Bekoff and Allen 1998; Hailman 1977). For example, coyotes' responses to threat gestures preceded by a play signal differ from the responses to threat gestures that are not related to a play signal (Bekoff 1975) .

During social play bouts that involve play signals, all participants may produce play signals, albeit at different times and in different contexts. For example, chimpanzees taking the active role in a play bout are more likely to display a "wide-open" mouth (van Hoof and Preuschoft 2003). In contrast, chimpanzees in the passive role are more likely to produce rapid "play chuckles". In a similar vein, Matsusaka (2004) reported that social play was the context most likely to result in chimpanzee "play pants", and that the recipient of the interaction (e.g., the one being chased or tickled) was more likely to be the individual to produce "play pants". Matsusaka also found that play pants were more likely to be followed by bouts of "play aggression" than by any other forms of behavior. It is not clear if these differences reflect different emotions that are being experienced by the active and passive players, but the relationship between type of play signal and emotion warrants additional investigation. Moreover, even though certain forms of play signals are correlated with certain play contexts, play signals are not automatically produced in response to other's behaviors. Play signals are not reflexive responses to external stimuli, but are instead "flexibly related to the occurrence of events in a play sequence" (Bekoff and Allen 1998, p. 109).

The flexibility of play signals may depend at least in part on when they occur during play. Bekoff's (1977) investigation of the play bows produced by dogs revealed that the form of play bows was relatively constant throughout play bouts. However, the length of time a dog stayed in the play bow position was not constant, but instead changed depending upon where the play bows occurred during the play bout. The duration of play bows was more constant at the beginning of a play bout and more variable during ongoing play interactions. These results suggest that unambiguous signals are very important in order to signal the intent to play prior to the onset of a play bout, but that more flexibility in play signals is possible once play is underway. It may be that the already established play context enables less well formed play signals to nonetheless convey the continuation of playful intent. Alternatively, it may be the case that the playing animals are too busy playing to produce well formed signals. Or perhaps it is a combination of these two factors. Furthermore, it is far from clear exactly how much flexibility exists in species-specific play signals. Bekoff's (1977) analysis of dog play bows revealed variability in the duration but not the form of play bows. However, duration was more likely to vary during the middle of a play bout than during its initiation.

Despite the fact that play signals are not universal across species (or even within a play context, as in the case of dog play bows), the behavior of playing individuals must somehow convey information that allows both participants and observers to recognize that the social interaction is playful. We have watched young dolphins play chase and play fight while their mothers keep a watchful eye. As long as the interactions remain playful, the dolphin mothers keep their distance and do not interfere in the play interaction. But as soon as the play context is transformed, mothers charge in and intervene. It appears the mothers somehow recognize when play behaviors cross the line into something potentially more serious, and conscientious mothers act to prevent something bad from happening.

Humans are good at recognizing play behaviors produced by members of their own species, and are also capable of recognizing play behaviors in other species (otherwise, there would be no comparative psychology of play). Some non-human species play with members of species other than their own, suggesting that the ability to recognize playful intent and play signals in other species may not be unique to humans (Brown 1994; Mitchell and Thompson 1990). But these abilities need not rely on the interpretation of discrete play signals. Griebel (personal communication, cited in Kuczaj and Makecha 2008) questioned the general significance of play signals given that relatively few discrete play signals have been discovered in other species (including humans). She suggested that the "intent to play" may be coded in rhythmic and temporal behavioral patterns that distinguish play behaviors from the "real thing", the existence of such patterns making discrete play signals unnecessary. And as we noted above, the emotions associated with play may be sufficient to signal play if participants are adept at "reading" another's emotions. The notion that the emotional context is an important cue for playing animals gains support from a recent article by Pellis and Pellis (2011). They studied the use of a "headshaking" play signal in spider monkeys. Although the monkeys used the signal to facilitate play, they also produced headshakes in many non-social contexts, demonstrating that headshakes are not exclusively play signals. Given this, the use of headshakes as play signals must depend on information outside of the signal itself. We suspect that the signaling animal's emotional state is an important piece of that additional information.

5.3.1 Play Signals Vs. Playing with Signals

Although human language appears to lack specific play signals, human children do use language to communicate the intent to play and the nature of play in which they wish to engage. However, human children also play *with* language, a phenomenon that occurs in both solitary and social contexts (Kuczaj 1983). Language play typically involves the manipulation of sounds, words and linguistic structures in young children, and evolves into puns and verbal riddles in older children.

The emotional content of language play is overwhelmingly positive, as evidenced by the laughter that often accompanies such play. For example, "expressive sound play" (playing with the sounds of words) is most common between 2 and 5 years of age, and is readily apparent to even casual observers because the playing children often laugh as they play with language sounds (Ferguson and Macken 1983; Keenan 1974; Kuczaj 1983). The following 2-year-old child's sound play illustrates the joyous nature of this type of play (Kuczaj 1983):

Nice doggy
Noce doggy (laughs)
Nose piggy (laughs)
Dose diggy (laughs)
Dose diggy (laughs)
Dose diggy, dose piggy, dose piggy (laughs)

In addition to playing with the sounds of words, children may play with the order of sounds in songs, chants, and rhymes. Nonsensical rhyming patterns seem to provide young children with ample amounts of enjoyment, as does altering their voice or accent, resulting in "funny" talk that children find amusing. Although one might expect animals with sophisticated vocal repertoires to engage in some sort of sound play, such play is rare in non-human animals, and so it is difficult to determine if the joy that accompanies such play in young children also exists in other animals that play with sounds (Kuczaj 1998).

5.4 Benefits of Play

Humans are the most advanced of animals—although a case could be made for dolphins—because they seldom grow up. Behavioral traits such as curiosity about the world, flexibility of response, and playfulness are common to practically all young mammals but are usually rapidly lost with the onset of maturity in all but humans.

Humanity has advanced, when it has advanced, not because it has been sober, responsible, and cautious, but because it has been playful, rebellious, and immature (Robbins 1980, p. 19).

Regardless of whether one agrees with Robbins' assessment of the significance of play in the evolution of the human species, even a casual perusal of the literature on play reveals a myriad array of possible advantages that play provides those species that have evolved the capacity for play. The hypothesized benefits of play are as diverse as the many definitions of play, and range from a mechanism in which organisms can release surplus energy in relative safety to an essential developmental phenomenon that enables animals (human and non-human) to develop and maintain flexible communication and problem solving skills (Beach 1945; Caro 1988; Fagen 1981; Kuczaj and Makecha 2008; Kuczaj et al. 2006; Kuczaj and Walker 2006; Lancy 1980; Špinka et al. 2001; Vieira and Sartorio 2002). Theorists may advocate play as youthful practice of physical skills and subsequent adult skills, a means of exploration, a process to facilitate socialization and self expression, or an essential aspect of cognitive growth, but there is a general consensus that play is an important aspect of development. For example, Einon et al. (1978) found that experimental rats deprived of juvenile social play were slower to learn complex motor tasks and exhibited reduced behavioral flexibility in later life. These results are consistent with the notion that the positive emotions engendered in play provide important foundations for later cognitive and social skills, and perhaps even improve the capacity to cope with the difficult situations that pepper most individuals' lives (Frederickson 2001). Špinka et al. (2001) hypothesized that one of the reasons that play is found in so many species concerns the role play has in conditioning animals to be physically and emotionally ready for unexpected events. The basic idea is that play actually increases an individual's behavioral plasticity, which in turn increases survival and inclusive fitness (Kuczaj and Makecha 2008; Kuczaj et al. 2006; Špinka et al. 2001; Vieira and Sartorio 2002). Given that play behavior is typically more variable than non-play behavior, play provides opportunities for exploration in the absence of normal consequences, which in turn allows players to learn how changes in their behavior affect various sorts of outcomes in various sorts of environmental contexts (Fagen 1974). The result is an increased ability to cope with unpredictable environments, that is, to survive change. And survival, of course, is essential for evolutionary success. Thus, play may facilitate an individual's ability to adapt to changing environments and so contribute to the continued survival of species (Špinka et al. 2001; Vieira and Sartorio 2002).

One of the challenges for any player involves keeping the play interesting and thereby positive emotionally. Children and animals sometimes do this by handicapping themselves to make the play situation more novel, complex and unpredictable (Pellegrini and Bjorklund 2004; Piaget 1952). For example, Kuczaj et al. (1998) observed killer whales using bits of fish to bait sea gulls, the goal being to capture a live sea gull (which the whales seemed to view as interesting toys). As they learned to bait and catch sea gulls, each young whale developed its own technique via trial and error learning. The initial observations of killer whale gull baiting led the authors to believe that the ultimate goal of gull baiting was the capture of a live gull since

Fig. 5.4 Dolphins sometimes use objects to play with one another. Here a dolphin tows another dolphin using a hoop

successful captures resulted in interesting play objects for the whales. However, they later learned that catching a gull was not the primary goal of gull baiting. Once whales became proficient at gull baiting, they modified their behaviors to make catching gulls more difficult! Kuczaj and Makecha (2008) provide the following example:

One whale had perfected its technique of lurking below the surface as a piece of fish floated on the surface and catching the gull the instant that it touched the fish. This whale subsequently modified its behavior to try to catch the gulls approximately 4–5 feet *above* the surface of the water as the gull swooped down towards the fish. This change resulted in the whale catching far fewer gulls, but the whale nonetheless persisted at this new behavior. Clearly, the whale was motivated by the challenge rather than by the outcome itself. If catching gulls was all that was important, the whale should have maintained the behavior that resulted in more gull captures rather than modifying its behavior to make it more difficult to catch gulls (p. 256).

The whales' behaviors as they were learning to catch gulls and their subsequent modifications of successful behavior demonstrate that the play *process* is just as, if not more, important than the play product. The process is what makes play interesting, and whales, like children and other animals, do what they can do keep the process entertaining. In a very real sense, players insure that play bouts are optimal learning environments—ones that are challenging but at the same time emotionally positive. This is possible when players have sufficient control over the situation to avoid becoming overly frustrated or anxious. Otherwise, the play bout can turn ugly. Play, then, provides an ideal context for moderately discrepant events, such events being ideal breeding grounds for cognitive and emotional growth (Fig. 5.4).

5.5 Conclusions

Despite the fact that both play and emotion are difficult to define, we concur with Burghardt (2005) that a more complete understanding of play must consider the emotional aspects of play. For example, one of the common notions in the play literature concerns the use of play as practice, a means by which young children and animals perfect both physical and mental skills. However, the practice that occurs during play is neither mundane nor boring. Instead, the play context provides an intrinsically reinforcing scenario in which children and young animals enjoy the activities in which they are engaging, the result being a much more efficient learning session than one in which youngsters attention is being directed by others. If emotional experience is closely related to "instinctual" neurological processes (Panksepp 2007; James 1884), then it may be the case that evolution has selected for individuals who enjoy the challenges and learning opportunities afforded by play.

Social play is impossible unless the players communicate and cooperate with one another. Such play seems ideally suited to help children and young animals learn appropriate interaction strategies, including the importance of taking turns, one of the most fundamental aspects of cooperation. Some forms of social play might also require players to assume the role of another player and so enhance perspective-taking, an essential aspect of theory of mind. Play, then, may facilitate the development and maintenance of social bonds, sometimes directly through play interactions but more often indirectly though the more general acquisition of social interaction skills, including communicative skills. Play may result in flexible social abilities, and so allow individuals to thrive (or at least survive) in a variety of social contexts. Learning to recognize and control one's emotions, and to recognize and respond appropriately to the emotions of others, is critical for successful social interactions, and social play provides ample opportunities for such understandings to flourish.

The notion that the positive affective nature of play may enhance well-being, learning, and cognition gains is consistent with the view that emotions may be essential aspects of cognition and behavior (Damasio 1994; Forgas 2008; Griffin 2001). For example, horses that had positive interactions with humans remembered these interactions and even generalized them to novel humans, demonstrating the long-lasting effects of such interactions (Sankey et al. 2010). These findings are consistent with Piaget's (1952) hypothesis that assimilation and accommodation are critical components of all organisms' interpretation of their experiences. Assimilation occurs when new information is incorporated into existing concepts, beliefs and knowledge, whereas accommodation involves changes in existing knowledge and cognitive structures as a result of new experiences. These two processes are complementary, but seem to be influenced at least in part by emotional state. Positive affect is related to a more assimilative processing style, while negative affect seems to result in a more accommodative cognitive style (Bless and Fielder 1995; Fielder 2001; Forgas 2008). Given that affect may influence both the content and process of how people and animals think (Harding et al. 2004; Paul et al. 2005), and that positive affect may enhance cooperation, creativity, and flexibility (Frederickson 2001; Isen et al. 1987), the positive emotions associated with play seem likely to promote the ontogeny of cognitive abilities that will facilitate the playing individual's ability to successfully navigate its way through its physical and social worlds.

References

Arnold MB (1960) Emotion and personality. Columbia University Press, New York
Aureli F, Preston SD, de Waal FBM (1999) Heart rate responses to social interactions in free-moving rhesus macaques (*Macaca mulatta*): a pilot study. J Comp Psychol 113:59–65
Averill JR (1980) A constructivist view of emotion. In: Plutchik R, Kellerman H (eds) Emotion: theory, research and experience, vol I, Theories of emotion. Academic, New York, pp 305–339
Bain A (1859) The emotions and the will. Longmans, London
Balcombe J (2009) Animal pleasure and its moral significance. Appl Anim Behav Sci 118:208–216
Baldwin JD, Baldwin JI (1974) Exploration and social play in squirrel monkeys (*Saimiri*). Am Zool 14:303–315
Bateson G (1955) A theory of play and fantasy. Psychiatr Res Rep 2:39–51
Bateson G (1972) Steps to an ecology of mind. Ballantine Books, New York
Beach FA (1945) Current concepts of play in animals. Amer Nat 79:523–541
Bekoff M (1975) The communication of play intention: are play signals functional? Semiotica 15:231–239
Bekoff M (1977) Social communication in canids: evidence for the evolution of a stereotyped mammalian display. Science 197:1097–1099
Bekoff M (1984) Social play behavior. Bioscience 34:228–233
Bekoff M (1995) Play signals as punctuation: the structure of social play in canids. Behavior 132:419–429
Bekoff M (2000) Animal emotions: exploring passionate natures. Bioscience 50:861–870
Bekoff M (2002) Minding animals: awareness, emotions and heart. Oxford University Press, New York
Bekoff M (2007) The emotional lives of animals. New World Library, Novato
Bekoff M, Allen C (1998) Intentional communication and social play: how and why animals negotiate and agree to play. In: Bekoff M, Byers JA (eds) Animal play: evolutionary, comparative, and ecological perspectives. Cambridge University Press, pp 97–114
Bekoff M, Byers JA (1981) A critical reanalysis of the ontogeny of mammalian social and locomotor play: an ethological hornet's nest. In: Immelman K, Barlow GW, Petrinovich L (eds) Behavioral development: the Bielefeld interdisciplinary project. Cambridge University Press, New York, pp 296–337
Bekoff M, Byers JA (1998) Animal play: evolutionary, comparative, and ecological perspectives. Cambridge University Press, Cambridge
Bel'kovich VM, Ivanova EE, Kozarovitsky LB, Novikova EV et al (1991) Dolphin play behavior in the open sea. In: Pryor K, Norris KS (eds) Dolphin societies. University of California Press, Los Angeles, pp 67–77
Berger J (1980) The ecology, structure and functions of social play in bighorn sheep (*Ovis canadensis*). J Zool 192:531–542
Berridge KC, Kringelbach ML (2008) Affective neuroscience of pleasure: reward in humans and animals. Psychopharmacology 199:457–480

Biben M (1998) Squirrel monkey playfighting: making the case for a cognitive training hypothesis. In: Bekoff M, Byers JA (eds) Animal play: evolutionary, comparative and ecological approaches. Cambridge University Press, New York

Biben M, Symmes D (1986) Play vocalizations of squirrel monkeys (*Saimire sciureus*). Folia Primatol 46:173–182

Bless H, Fielder K (1995) Affective states and the influence of activated general knowledge. Pers Soc Psychol Bull 21:766–778

Blomqvist C, Mello I, Amundin M (2005) An acoustic play-fight signal in bottlenose dolphins (*Tursiops truncatus*) in human care. Aquat Mamm 31:187–194

Bornstein MH, O'Reilly AW (1993) The role of play in the development of thought. Jossey-Bass Publishers, San Francisco

Boulton MJ, Smith PK (1992) The social nature of play fighting and play chasing: mechanisms and strategies underlying cooperation and compromise. In: Barkow JH, Cosmides L, Tooby J (eds) The adapted mind: evolutionary psychology and the generation of culture. Oxford University Press, New York, pp 429–444

Bretherton I (1984) Representing the social world in symbolic play: reality and fantasy. In: Bretherton I (ed) Symbolic play: the development of social understanding. Academic, New York, pp 3–41

Bretherton I (1989) Pretense: the form and function of make-believe play. Dev Rev 9:383–401

Brown SL (1994) Animals at play. Natl Geogr 186:2–35

Bruner J, Jolly A, Sylva K (1976) Play—its role in development and evolution. Basic Books, New York

Burgdorf J, Panksepp J (2001) Tickling induces reward in adolescent rates. Physiol Behav 72:167–173

Burghardt GM (2005) The genesis of animal play. MIT Press, Cambridge

Burghardt GM, Burghardt LS (1972) Notes on behavioral development of two female black bear cubs: the first eight months. In: Herrero SM (ed) Bears—their biology and management. IUCN N. S., Morges, pp 207–220

Byers JA (1984) Play in ungulates. In: Smith PK (ed) Play in animals and humans. Basil Blackwell, Oxford, pp 43–65

Byers JA, Walker C (1995) Refining the motor training hypothesis of play. Amer Nat 146:25–40

Cacioppo JT, Gardner WL (1999) Emotion. Annu Rev Psychol 50:191–214

Carlstead K, Brown JL, Strawn W (1993) Behavioral and physiological correlates of stress in laboratory cats. Appl Anim Behav Sci 38:143–158

Caro TM (1988) Adaptive significance of play: are we getting closer? Trends Ecol Evol 3:50–54

Caro TM (1995) Short-term costs and correlates of play in cheetahs. Anim Behav 49:333–345

Coelho AM, Bramblett CA (1982) Social play in differentially reared infant and juvenile baboons (Papio sp.). Am J Primatol 3:153–160

Colvin JD, Tissier G (1985) Affiliation and reciprocity in sibling and peer relationships among free-ranging immature male rhesus monkeys. Anim Behav 33:959–977

Cornelius RR (1996) The science of emotion. Research and tradition in the psychology of emotion. Prentice-Hall, Upper Saddle River

Cosmides L, Tooby J (2000) Evolutionary psychology and the emotions. In: Lewis M, Haviland-Jones JM (eds) Handbook of emotions, 2nd edn. Guilford, New York, pp 91–115

Creels S, Sands JL (2003) Is social stress a consequence of subordination or a cost of dominance? In: de Waal FBM, Tyack P (eds) Animal social complexity. Harvard University Press, Cambridge, pp 153–179

Csikszentmihalyi M (1990) Flow: the psychology of optimal experience. HarperCollins, New York

Damasio AR (1994) Descartes' error: emotion, reason, and the human brain. Penguin, New York

Darwin C (1872) The expression of emotions in man and animals. Murray, London

Delfour F, Aulagnier S (1997) Bubbleblow in beluga whales (*Delphinapterus leucas*): a play activity? Behav Processes 40:183–186

Diener E (1999) Introduction to the special section on the structure of emotion. J Pers Soc Psychol 76:803–804

Drea CM, Hawk JE, Glickman SE (1996) Aggression decreases as play emerges in infant spotted hyaenas: preparation for joining the clan. Anim Behav 51:1323–1336

Dudzinski KM (1998) Contact behavior and signal exchange in Atlantic spotted dolphins (*Stenella frontalis*). Aquat Mamm 24:129–142

Eibl-Eibesfeldt I (1978) On the ontogeny of behavior of a male badger (*Meles meles* L.) with particular reference to play behavior. In: Mueller-Schwarze D (ed) Evolution of play behavior. benchmark papers in animal behavior, vol 10. Dowden Hutchins & Ross, Stroudsburg

Einon DE, Morgan MJ, Kibbler CC (1978) Brief periods of socialization and later behavior in the rat. Dev Psychobiol 11:213–225

Ekman P (1999) Basic emotions. In: Dalgleish T, Power M (eds) Handbook of cognition and emotion, 2nd edn. Wiley, Sussex, pp 45–60

Ekman P, Davidson RJ (eds) (1994) The nature of emotion: fundamental questions. Oxford University Press, New York

Ekman P, Sorenson ER, Friesen WV (1969) Pan-cultural elements in facial displays of emotions. Science 164:86–88

Fagen RM (1974) Selective and evolutionary aspects of animal play. Amer Nat 108:850–858

Fagen RM (1981) Animal play behavior. Oxford University Press, New York

Fagen RM (1993) Primate juveniles and primate play. In: Pereira ME, Fairbanks LA (eds) Juvenile primates: life history, development, and behavior. Oxford University Press, Oxford, pp 182–196

Fagen RM, Fagen J (2004) Juvenile survival and benefits of play behavior in brown bears, *Ursus arctos*. Evol Ecol Res 6:89–102

Fairbanks LA (1993) Juvenile vervet monkeys: establishing relationships and practicing skills for the future. In: Pereira ME, Fairbanks LA (eds) Juvenile primates: life history, development and behavior. Oxford University Press, New York, pp 211–227

Feddersen-Petersen D (1991) The ontogeny of social play and angonistic behavior in selected canid species. Bonner Zoologische Beitraege 42:97–114

Fein G (1987) Pretend play: creativity and consciousness. In: Gorlitz P, Wohlwill J (eds) Curiosity, imagination and play. Lawrence Erlbaum Associates, Hillsdale, pp 281–304

Ferguson C, Macken M (1983) The role of play in phonological development, vol 4. Erlbaum, Hillsdale

Ficken MS (1977) Avian play. Auk 94:573–582

Fielder K (2001) Affective states trigger processes of assimilation and accommodation. In: Martin LL, Clore GL (eds) Theories of mood and cognition: a user's guidebook. Lawrence Erlbaum Associates, Mahwah, pp 85–98

Forgas JP (2008) Affect and cognition. Perspect Psychol Sci 3:94–101

Frederickson BL (2001) The role of positive emotions in positive psychology: the broaden-and-build theory of positive emotions. Am Psychol 56:218–226

Gard GC, Meier GW (1977) Social and contextual factors of play behavior in sub-adult rhesus monkeys. Primates 18:367–377

Gewalt W (1989) Orinoco freshwater dolphins (*Inia geoffrensis*) using self-produced air bubble rings as toys. Aquat Mamm 15:73–79

Goedeking P, Immelman K (1986) Vocal cues in cotton-top tamarin play vocalizations. Ethology 73:219–224

Griffin DR (2001) Animal minds, beyond cognition to consciousness. University of Chicago Press, Chicago

Groos K (1898) The play of animals. Appleton & Company, New York

Guinet C (1991) Intentional stranding apprenticeship and social play in killer whales (*Orcinus orca*). Can J Zool 69:2712–2716

Hailman JP (1977) Optical signals: animal communication and light. Indian University Press, Bloomington

Hailman JP, Dzelzkalns JJI (1974) Mallard tail-wagging: punctuation for animal communication? Amer Nat 108:236–238

Hall SL (1998) Object play by adult animals. In: Bekoff M, Byers JA (eds) Animal play. Cambridge University Press, Cambridge, pp 27–44
Harcourt R (1991) The development of play in the South American fur seal. Ethology 88:191–202
Harding EJ, Paul ES, Mendl M (2004) Cognitive bias and affective state. Nature 427:312
Heinrich B (1999) Mind of the raven. Cliff Street Books, New York
Heinrich B, Smolker R (1998) Play in common ravens (*Corvus corax*). In: Bekoff M, Byers JA (eds) Animal play: evolutionary, comparative, and ecological perspectives. Cambridge University Press, Cambridge, pp 27–44
Horback KM, Friedman WR, Johnson CM (2010) The occurrence and context of S-posture display in captive belugas (*Delphinapterus leucas*). Int J Comp Psychol 23:689–700
Ingmanson E (2002) Empathy in a bonobo. In: Mitchell RW (ed) Pretending and imagination in animals and children. Cambridge University Press, Cambridge, pp 280–284
Isen AM, Daubman KA, Nowicki GP (1987) Positive affect facilitates creative problem solving. J Pers Soc Psychol 52:1122–1131
James W (1884) What is an emotion? Mind 19:188–205
Kagan J (2007) What is emotion? history, measures, and meanings. Yale University Press, New Haven
Kano F, Tomonaga M (2010) Face scanning in chimpanzees and humans: continuity and discontinuity. Anim Behav 79:227–235
Keenan E (1974) Conversational competence in children. J Child Lang 1:163–183
Keverne EB, Martensz ND, Tuite B (1989) Beta-endorphin concentrations in cerebrospinal fluid of monkeys are influenced by grooming relationships. Psychoneuroendocrinology 14:155–161
Kovacs KM (1987) Maternal behavior and early behavioural ontogeny of harp seals, *Phoca groenlandica*. Anim Behav 35:844–855
Kuczaj SA II (1982) Language play and language acquisition. Adv Child Dev Behav 17:197–232
Kuczaj SA II (1983) "I mell a kunk!" Evidence that children have more complex representations of word pronunciations which they simplify. J Psycholinguist Res 12:69–73
Kuczaj SA II (1998) Is an evolutionary theory of language play possible? Curr Psychol Cogn 17:135–154
Kuczaj II SA, Lacinak CT, Garver A, Scarpuzzi M (1998) Can animals enrich their own environment? In: Hare VJ, Worley KE (eds) Proceedings of the third international conference on environmental enrichment, The Shape of Enrichment, Orlando, pp 168–170
Kuczaj SA II, Daly MJ (1979) The development of hypothetical reference in the speech of young children. J Child Lang 6:563–579
Kuczaj SA II, Makecha R (2008) The role of play in the evolution and ontogeny of contextually flexible communication. In: Oller DK, Griebel U (eds) Evolution of communicative flexibility: complexity, creativity, and adaptability in human and animal communication. Cambridge, MIT Press, pp 253–277
Kuczaj SA II, Highfill LE, Makecha R, Byerly HC (2012). Why do dolphins smile? A comparative perspective on dolphin emotions and emotional expressions. In: Watanabe S, Kuczaj S (eds), Emotions of animals and humans. Springer
Kuczaj SA II, Walker RT (2006) How do dolphins solve problems? In: Wasserman EA, Zentall TR (eds) Comparative cognition. University Press, Oxford, pp 580–601
Kuczaj SA, Makecha R, Trone M, Paulos RD, Ramos JA (2006) Role of peers in cultural innovation and cultural transmission: evidence from the play of dolphin calves. Int J Comp Psychol 19:223–240
Kuyk K, Dazey J, Erwin J (1976) Play patterns of pigtail monkey infants: effects of age and peer presence. J Biol Psychol 18:20–23
Lancy DF (1980) Play in species adaptation. Annu Rev Anthropol 9:471–495
Leresche LA (1976) Dyadic play in Hamadryas baboons. Behaviour 57:190–205
Lillard AS (2002) Pretend play and cognitive development. In: Goswami U (ed) Handbook of cognitive development. Blackwell, London, pp 188–205

Loeven J (1993) The ontogeny of social play in timber wolves, Canis lupus. Masters Thesis, Dalhousie University

Markus N, Croft DB (1995) Play behavior and its effects on social development of common chimpanzees (*Pan troglodytes*). Primates 36:213–225

Martin P, Caro TM (1985) On the functions of play and its role in behavioral development. In: Rosenblatt JS, Beer C, Busnel MC, Slater PJB (eds) Advances in the study of behavior, vol 15. Academic, Orlando, pp 59–103

Matsusaka T (2004) When does play panting occur during social play in wild chimpanzees? Primates 45:221–229

Mayer JD, Salovey P, Caruso DR (2008) Emotional intelligence: new ability or eclectic traits? Am Psychol 63:503–517

McCowan B, Reiss D (1997) Vocal learning in captive bottlenose dolphins: a comparison with humans and non-human animals. In: Snowdon CT, Hausberger M (eds) Social influences on vocal development. Cambridge University Press, Cambridge, pp 178–207

McCowan B, Marino L, Vance E, Walke L, Reiss D (2000) Bubble ring play of bottlenose dolphins (Tursiops truncatus): implications for cognition. J Comp Psychol 114:98–106

Miklosi A (1999) The ethological analysis of imitation. Biol Rev 74:347–374

Mitchell RW (2002) Imitation as a perceptual process. In: Dautenhahn K, Nehanivc L (eds) Imitation in animals and artifacts. MIT Press, Cambridge, pp 441–469

Mitchell RW, Thompson NS (1990) The effects of familiarity on dog-human play. Anthrozoös 4:24–43

Muller-Schwarze D, Stagge B, Muller-Schwarze C (1982) Play behaviour: persistence, decrease and energetic compensation during food shortage in deer fawns. Science 215:85–87

Oatley K, Jenkins JM (1998) Understanding emotions. Blackwell, Oxford

Oliveira A, Rossi AO, Silva L, Lau MC, Barreto RE (2010) Play behavior in nonhuman animals and the animal welfare issue. J Ethol 28:1–5

Ortega JC, Bekoff M (1987) Avian play: comparative evolutionary and developmental trends. Auk 104:338–341

Palagi E (2008) Sharing the motivation to play: the use of signals in adult bonobos. Anim Behav 75:887–896

Palagi E, Mancini G (2011) Playing with the face: playful facial "chattering" and signal modulation in a monkey species (*Theropithecus gelada*). J Comp Psychol 125:11–21

Panksepp J (1998) Attention deficit hyperactivity disorders, psychostimulants, and intolerance of childhood playfulness: a tragedy in the making. Curr Dir Psychol Sci 7:91–98

Panksepp J (2007) Can PLAY diminish ADHD and facilitate the construction of the social brain? J Can Acad Child Adolesc Psychiatry 16:57–66

Panksepp J, Burgdorf J (2000) 50 k-Hz chirping (laughter?) in response to conditioned and unconditioned tickle-induced reward in rats: effects of social housing and genetic variables. Behav Brain Res 115:25–38

Parker ST, McKinney ML (1999) Origins of intelligence: the evolution of cognitive development in monkeys, apes and humans. Johns Hopkins Press, Baltimore

Paul ES, Harding EJ, Mendl M (2005) Measuring emotional processes in animals: the utility of a cognitive approach. Neurosci Biobehav Rev 29:469–491

Paulos RD, Trone M, Kuczaj SA II (2010) Play in wild and captive cetaceans. Int J Comp Psychol 23:701–722

Pellegrini AD, Bjorklund DF (2004) The ontogeny and phylogeny of children's object and fantasy play. Human Nature 15:23–43

Pellegrini AD, Smith PK (eds) (2005) The nature of play: great apes and humans. Guilford, New York

Pellis SM (1983) Development of head and foot coordination in the Australian magpies *Gymnorhina tibicen*, and the function of play. Bird Behav 4:57–62

Pellis SM, Pellis VC (1996) On knowing it's only play: the role of play signals in play fighting. Aggress Violent Behav 1:249–268

Pellis S, Pellis V (2009) The playful brain: venturing to the limits of neuroscience. Oneworld Publications, Oxford

Pellis SM, Pellis VC (2011) To whom the play signal is directed: a study of headshaking in black-handed spider monkeys (*Ateles geoffroyi*). J Comp Psychol 125:1–10
Piaget J (1951) Play, dreams and imitation in childhood. Norton, New York
Piaget J (1952) The origins of intelligence in children. W.W. Norton & Company, New York
Plutchik R (2001) The nature of emotions. Am Sci 89:344–350
Power TG (2000) Play and exploration in children and animals. Erlbaum, Mahwah
Provine RR (2001) Laughter. Penguin, New York
Pryor KW (1990) Non-acoustic communication in small cetaceans: glance, touch, position, gesture, and bubbles. In: Thomas JA, Kastelein R (eds) Sensory abilities of cetaceans. Plenum, New York, pp 537–544
Rasa OAE (1984) A motivational analysis of object play in juvenile dwarf mongooses (*Helogale undulate rufula*). Anim Behav 32:579–589
Rensch B (1973) Play and art in apes and monkeys. In: Symposia of the 4th international congress of primatology, vol 1, pp 102–123
Reynolds PC (1976) Play, language, and human evolution. In: Bruner J, Jolly A, Sylva K (eds) Play. Basic Books, New York, pp 621–637
Robbins T (1980) Still life with woodpecker. Bantam, New York
Sade DS (1973) An ethogram for rhesus monkeys: I. Antithetical contrasts in posture and movement. Am J Phys Anthropol 38:537–542
Sankey C, Richard-Yris MA, Leroy H, Henry S, Hausberger M (2010) Positive interactions lead to lasting positive memories in horses, *Equus caballus*. Anim Behav 79:869–875
Scarlett WG, Naudeau S, Salonius-Pasternak D, Ponte I (2005) Children's play. Sage Publications, Thousand Oaks
Singer DG, Singer JL (1990) The house of make-believe: children's play and the developing imagination. Harvard University Press, Cambridge
Snowdon CT, Elowson AM, Roush RS (1997) Social influences on vocal development in New World Primates. In: Snowdon CT, Hausberger M (eds) Social influences on vocal development. Cambridge University Press, Cambridge, pp 234–248
Sommer V, Mendoza-Granados D (1995) Play as indicator of habitat quality: a field study of langur monkeys (*Presbytis entellus*). Ethology 99:177–192
Špinka M, Newberry RC, Bekoff M (2001) Mammalian play: training for the unexpected. Q Rev Biol 76:141–168
Sutton-Smith B (1980) Conclusion: the persuasive rhetorics of play. In: Pellegrini AD (ed) The future of play theory: a multidisciplinary inquiry into the contributions of Brian Sutton-Smith. State University of New York Press, Albany, pp 275–295
Sutton-Smith B (2003) Play as a parody of emotional vulnerability. In: Lytle D (ed) Play and culture studies, vol 5. Praeger, Westport, pp 3–17
Symons D (1978) Play and aggression. Columbia UP, NY
Taylor M, Carlson SM (2002) Imaginary companions and elaborate fantasy in childhood: discontinuity with nonhuman animals. In: Mitchell RW (ed) Pretending and imagining in animals and children. Cambridge University Press, New York, pp 167–180
Thompson KV (1996) Behavioral development and play. In: Kleiman D, Allen ME, Thompson KV, Lumpkin S (eds) Wild mammals in captivity: principles and techniques. University of Chicago Press, Chicago, pp 352–371
van Hoof JARAM (1967) The facial displays of the catarrhine monkeys and apes. In: Morris D (ed) Primate ethology. Weidenfeld & Nicholson, London, pp 7–68
van Hoof JARAM, Preuschoft S (2003) Laughter and smiling: the intertwining of nature and culture. In: de Waal FBM, Tyack P (eds) Animal social complexity. Harvard University Press, Cambridge, pp 153–179
Vieira ML, Sartorio R (2002) Motivational, causal and functional analysis of play behavior in two rodent species. Estudos de Psicologia 7:189–196
Weir R (1962) Language in the crib. Mouton, The Hague
West M (1974) Social play in the domestic cat. Am Zool 14:427–436
Wheeler R (1943) Pacific gull at play? Emu 42:181

Wilson EO (1975) Sociobiology: the new synthesis. Harvard University Press, Cambridge
Würsig B (2002) Playful behavior. In: Perrin WF, Wursig B, Thewissen JGM (eds) Encyclopedia of marine mammals. Academic, San Diego, pp 942–945
Würsig B, Dorsey EM, Richardson WJ, Wells RS (1989) Feeding, aerial and play behaviour of the bowhead whale, *Balaena mysticetus*, summering in the Beaufort Sea. Aquat Mamm 15:27–37
Young PT (1943) Emotion in man and animal: its nature and relation to attitude and motive. Wiley, New York
Young PT (1961) Motivation and emotion: a survey of the determinants of human and animal activity. Wiley, New York

Chapter 6
The Use of Emotion Symbols in Language-Using Apes

Heidi Lyn and Sue Savage-Rumbaugh

Courtesy of the Great Ape Trust of Iowa

H. Lyn (✉)
University of Southern Mississippi, Gulf Coast; College of Education & Psychology,
730 East Beach Blvd., Long Beach, MS 39560, USA
e-mail: heidi.lyn@usm.edu

S. Savage-Rumbaugh
Great Ape Trust of Iowa, Des Moines, IA, USA

Abstract There has been a long history of scientific disagreement about the ability of nonhumans to feel complex or even simple emotions. One of the stumbling blocks for recognition of emotions has been the inability of animals to communicate about emotions or any internal state. Here we look at the internal state utterances of language competent apes. In particular, four internal state symbols are explored: mad, happy, scared, and hurt. We find that these apes use these words appropriately and that developmentally, their use of internal state words occurs after they begin to use value based words like good and bad. We also find that there is less co-construction of internal state words as opposed to the more culturally changeable value based words.

6.1 Emotion and Language-Using Apes

> Now by these ... means one can also know the difference between men and beasts. For it is rather remarkable that there are no men so dull and so stupid (excluding not even the insane), that they are incapable of arranging various words together and of composing from them a discourse by means of which they might make their thoughts understood, and that, on the other hand, there is no other animal at all, however perfect and pedigreed it may be, that does the like.—Descartes, *Discourse on the Method for Conducting One's Reason*

Descartes famously expressed the opinion that nonhuman animals were without reason, driven entirely by biological urges with no thought (and consequently no emotion) behind them (e.g. Descartes 1637). As can be gleaned from the quotation above, his main argument for this belief was the lack of language capacity outside of the human species. In contrast, Charles Darwin in The Expression of Emotions in Man and Animals (Darwin 1872) wrote of the fundamental *continuities* between basic emotions in humans and nonhumans, pointing out that, "...the young and the old of widely different races, both with man and animals, express the same state of mind by the same movements" (p. 352). This push and pull in scientific opinion continues to today, with most scientists recognizing continuities between human and nonhuman, but also delineating discontinuities (e.g. Penn et al. 2008).

When comparing humans to animals in most physical or psychological realms, including emotion, nonhuman primates are frequently the comparison of choice due to their evolutionary similarity to humans. Among nonhuman primates, there are further preferences for great apes as they are even more similar to humans. For example, in an early study of chimpanzee emotions, Ladygina-Kohts et al. (2002) published a series of photographs depicting chimpanzee emotional facial expressions and comparing them directly to those of human children.

Since Darwin, researchers have continued to explore the similarities between nonhuman primate and human emotional expression through neurological similarities (e.g. Panksepp and Watt 2011), muscle movements (e.g. Parr et al. 2007) and behavior (e.g. Yamanashi and Matsuzawa 2010). While researchers generally agree on the continuity of basic emotions such as rage, joy, and fear, secondary emotions have yet to be scientifically identified in nonhumans (Panksepp and Watt 2011). Secondary emotions are those that require some level of self-consciousness, or even Theory of Mind, such as guilt or shame. While there are no experimental publications, one study found suggestive subjective evidence of one secondary emotion—jealousy—in nonprimates (Panksepp and Watt 2011). Similarly, complex emotional behaviors such as reconciliation and empathy (e.g. Preston and de Waal 2002) have been extensively studied in nonhuman primates. The lack of published reports on other secondary emotions such as shame and guilt in nonhumans is not surprising considering that, because of the highly cultural nature of these emotions, they are developmentally more advanced in humans than the basic emotions. Shame and guilt don't appear until 4–5 years of age and basic emotions are present at or soon after birth (Izard 1971).

This late development of secondary emotions, or even primary emotions, may be driven not just by cultural acquisition, but by other factors such as the simplicity of the concepts themselves, or more direct internal development. Supporting this possibility, Lamb (1991) found that in children, the use of internal state words, or words that represent emotional and psychological states, such as: happy, tired, and hungry, *preceded* the first stages of moral development, while following the development of "awareness of standards". Lamb also found that the caregivers' use of these internal state words was rare and did not seem to drive the children's use of these words, suggesting an internal development of the use of emotion words.

Contrary to Descartes assertion, several great apes have acquired the symbolic capacity to communicate about their internal states (Brakke and Savage-Rumbaugh 1995, 1996; Fouts and Mills 1998; Gardner et al. 1989; Lyn 2012; Lyn et al. 2008; Savage-Rumbaugh 1986; Savage-Rumbaugh et al. 1986). The complex symbolic abilities shown in two bonobos and a chimpanzee allow for an exploration of emotional communication in nonhuman apes; Sue Savage-Rumbaugh et al. (e.g. 1986, 1993) have shown that these bonobos (*Pan paniscus*) and chimpanzee (*Pan troglodytes*), when reared in an environment with English, a keyboard of visual symbols (lexigrams), and cultural emersion, can acquire symbolic capacities without explicit training (Greenfield and Savage-Rumbaugh 1991; Savage-Rumbaugh et al. 1980, 1986, 1993). Further research has shown that these apes are capable of many linguistic feats: using symbols to name objects in double blind studies (Savage-Rumbaugh et al. 1986); associating novel English names with novel objects with very few exposures to both object and word (Lyn and Savage-Rumbaugh 2000); utilizing imitation in an intentionally communicative context (Greenfield and Savage-Rumbaugh 1993); making semantically-based combinations across both lexigram and gestural combinations (Greenfield and Savage-Rumbaugh 1990, 1991; Greenfield and Lyn 2007; Greenfield et al. 2008; Lyn et al. 2011a); mentally representing symbols on several levels at the same time (Lyn 2007); and comprehending English sentences at least at a similar level to a two-and-a-half year old child tested in the same manner as the ape (Savage-Rumbaugh et al. 1993).

Additionally, recent publications have explored the development of cultural scaffolding of cognitive abilities in these apes at Language Research Center and now the Great Ape Trust of Iowa (Savage-Rumbaugh et al. 2004). The intensive 24-h interaction among the apes and humans, sharing daily activities and communicative modes, have produced extended spontaneous cognitive and social abilities not otherwise documented in these species, for example, pretend play (Lyn et al. 2006) and joint attention and imitation (Savage-Rumbaugh et al. 2004). These apes have been directly compared to apes from standard environments and have shown significantly stronger cognitive abilities, particularly abilities that require social interaction and communication (Lyn et al. 2010; Russell et al. 2011).

In a study that focused on the co-reared bonobo, Panbanisha, and chimpanzee, Panpanzee, researchers explored the understanding of morality precursors in the use of the lexigrams "bad" and "good" (Lyn et al. 2008). Within the ape/human social structure, human caregivers utilized the terms "good" and "bad" (in English and/or Lexigram form) as value judgments as they would in ordinary interactions. For example, "Panzee's being bad" or, "This is a good apple." They would also produce interrogatives that requested value judgments from the apes. For example, "Do you think that's a good idea?" or "Do you know how you've been acting?" Accordingly, within the human and ape culture, the apes learned to use the "bad" and "good" lexigrams in many diverse contexts (Lyn et al. 2008). However, the apes also appeared to co-construct their understanding of these terms, using conversational negotiation to clarify meaning and action (Lyn et al. 2008).

Also included in the exploration of Panbanisha and Panpanzee's use of good and bad was a description of the development of the four internal state words that were on their keyboard (happy, mad, scared, and hurt; Lyn et al. 2008). The authors found that, unlike the children in Lamb (1991) the apes used "good" and "bad" before they used internal state words.

Recently, a database of symbol use from a study of language development in human children by Greenfield and Smith (1976) was combined with the database from the Savage-Rumbaugh study of symbol use in apes into a combined data set—comprising both ape and child symbol use. This database is unique—composed of over 100,000 "utterances" of apes using a symbolic keyboard as well as over 4,000 verbal utterances of children at a similar linguistic stage of development (Gillespie-Lynch et al. 2011). Because sets of utterances were recorded according to similar protocols and coded using the same coding scheme regardless of species, the database provides the first truly comparative record of human and ape symbol use.

For this chapter, we will continue our exploration of the apes' use of internal state words. We will examine the development of both comprehension and production of internal state words as well as any social interactions that may have influenced the apes' use of these words. While the meanings of "good" and "bad" were shown to be co-constructed between the apes and the caregivers, one might expect less evidence of co-construction of purely internal concepts like "happy" and "hurt".

6.2 Methods

6.2.1 Participants

6.2.1.1 Apes

Our ape participants were Kanzi, a bonobo (*Pan paniscus*), Panbanisha, his half sister, and Panpanzee, a chimpanzee (*Pan troglodytes*) born within 6 weeks of Panbanisha, and reared alongside her (much of the methodological details included here are copied directly from Lyn et al. 2011b). All three were reared at the Language

Research Center, in Atlanta, Georgia in a within- and cross-species communicative environment. This communicative environment consisted of gesture, speech, and written visual symbols (lexigrams) placed on a keyboard. The lexigram keyboard was made available to the apes at all times and some of the available keyboards emitted the sounds of a computer-synthesized English word when the corresponding key was touched. While utilizing any of these keyboards, the caregivers were also instructed to communicate in English. Therefore, English and lexigram use were paired in the communicative environment of the apes.

Emphasis on keyboard use and spoken English began earlier with Panbanisha and Panpanzee than with Kanzi (before they were 6 weeks old, as opposed to Kanzi's several months old). Unlike Kanzi, who was initially exposed to a keyboard of 6 symbols during reward-based training sessions with his mother (the first keyboard that he used himself contained 12 symbols), Panbanisha and Panpanzee were initially exposed to a keyboard of more than 256 symbols and were expected to learn the use of these symbols without specific training. Like Kanzi, Panbanisha and Panpanzee were in the company of human caregivers and each other 24 h a day, 7 days a week, although this experience began for them at birth and did not begin for Kanzi until he reached 2.5 years of age (Savage-Rumbaugh et al. 1986). Whereas Kanzi produced his first meaningful lexigram at 30 months of age, Panbanisha and Panpanzee produced theirs around age one, a starting point very similar to that of a human child. (For further details of rearing and linguistic capabilities, see Brakke and Savage-Rumbaugh 1995, 1996; Savage-Rumbaugh 1986; Savage-Rumbaugh et al. 1993.)

6.2.1.2 Children

Archived data from two children, NT and MG, were used for comparative purposes. These children were raised in a normal family and home environment, and their language development was tracked by a research assistant through monthly formal observation sessions. Regular observations continued on a monthly basis for MG from 12 to 22 months of age. Regular observations continued on a monthly basis for NT from 18 to 24 months of age.

6.2.2 Data Collection

6.2.2.1 Apes

Panbanisha and Panpanzee were co-reared for almost 5 years. Beginning 1 year into the study, data collection included all uses of the keyboard by Panbanisha and Panpanzee, as well as communicative gestures used in combination with keyboard utterances, all of which were recorded by hand by the caregivers and input into the computer at the end of the day. This procedure provided the researchers with an exhaustive written record of the apes' symbol use. The database contains each observed utterance (which could include lexigram(s), gesture(s), or combination(s) of lexigram with gesture), date, record number, ape, researcher, codes as to pragmatic force of the utterance, behavioral concordance notes, and a short contextual note. Behavioral concordance—the relationship between the utterance and the ape's behavior, particularly his or her behavior subsequent to the utterance, was the main clue as to semantic relations and pragmatic force (request, statement, comment, etc.). These were similar to observational protocols used in classic studies of child language which provided the comparative foundation for this study (Bowerman 1973b; Brown 1973; Greenfield and Smith 1976).

This same procedure was operative for Kanzi over the same 5-year period, but the developmental period differed, as Kanzi was 5 years old at the beginning of the co-rearing study. At the end of this study period, the chimpanzee, Panpanzee, was moved out of the language study and no further utterance data was recorded for her. However, opportunistic data were recorded when staffing and scientific priorities allowed for Panbanisha and Kanzi utilizing these same procedures over the next 6 years, yielding a total of 11 years of data for each of the bonobos, although these data are not exhaustive.

Reliability for Data Collection

As a reliability check, real-time recording was checked against 4.5 h of videotape. Thirty-seven out of 46 utterances, or 80%, were noted by both the real-time and the video observer (the other 11 were noted only by the video observer) and there was 100% agreement on the lexigram that had been used when both observers noted the utterance. Hence, we conclude that our corpus is highly reliable, but an underestimation of quantity.

6.2.2.2 Children

Data from the formal observation sessions with the children were included in the present study. During these observations, all verbal utterances were recorded. Contextual notes were included in the child data as were prior utterances and direct responses to the child's utterance. This data collection procedure occurred prior to and served as the blueprint for the data collection with the apes.

6.2.3 Coding

The data from the children were added to the electronic, searchable database of ape utterances. As each record was added, it was given an unchangeable Record ID number; these record IDs were used to identify particular utterances in the database.

The corpus for the apes consisted of 105,629 utterances and the children's corpus consisted of 4,445 verbal utterances. These quantities differed mainly due to data collection procedures—daily for the apes, monthly for the children; over 12 months for the children as opposed to 11 years for some of the apes. Also, the children's sessions lasted only a few hours, whereas the apes were observed all day. To control for these differences, our analyses use percentages to report on utterance types. Gestures were not noted in the children's utterances, but any gesture that was combined with a verbal utterance was noted in the contextual notes. This allowed for an analysis of gesture/symbol combinations for both children and apes.

6.2.4 Database

The utterance database includes information about the apes' use of the keyboard from November 11, 1985 to January 19, 1997. During intensive study periods (1987–1990), the utterance database included all uses of the keyboard by Kanzi, Panbanisha and Panpanzee, as well as communicative gestures used in combination with keyboard utterances. These were recorded by hand by the caregivers and input into the computer at the end of the day. This procedure provided the researchers with an exhaustive written record of their symbol use. At other times (1985–1987 and 1990–1997), due to caretaker shortages, data recording was more sporadic, including all uses of the keyboard for only certain hours of the day and certain days of the week. At no time were utterances selectively entered into the database.

Each record contained the utterance, date, record number, ape, researcher, codes as to pragmatic force of the utterance, behavioral concordance notes, and a short contextual note. The procedure was similar to observational protocols used in studies of child language which provided the comparative foundation for this study (Bowerman 1973a; Brown 1973). Several studies have been published utilizing this database (Brakke and Savage-Rumbaugh 1995, 1996; Greenfield and Savage-Rumbaugh 1990, 1991; Lyn et al. 2011a; Lyn 2007).

Contextual notes were provided by a second caregiver who functioned as an observer. This contextual description in the utterance database was our basis for coding these combinations as to their semantic meaning. When the ape was with only one caregiver, the data would be excluded from our corpus for lack of contextual information. We estimate that two caregiver/researchers were present with the apes about four and one-half hours per day.

The database contains exhaustive utterance information for Panbanisha and Panpanzee from 1 year of age (1988) though 3 years of age (1990). Kanzi was already over 5 years of age at the initiation of the utterance database. Therefore, all developmental analyses will focus on Panbanisha and Panpanzee.

6.3 Results

6.3.1 Use and Comprehension of Internal State Words

Internal state words made up a small proportion of the apes' utterances, particularly for Kanzi and Panpanzee. Kanzi used only 10 internal state lexigrams in 9,204 utterances (0.11% of utterances) and Panpanzee used 17 internal state lexigrams in 21,351 utterances (0.08% of utterances). Panbanisha used more internal state words (554 in 46,681utterances, or 1.19%), however this proportion was still very low. However, this low proportion of internal state words was also found for the two human children. For both MG, only one utterance of these four internal state words was recorded out of 1,411 utterances (0.07%) and two utterances of "hurt" were recorded for NT out of 3,034 utterances (0.06%), both after he had begun combining multiple words:

> MG: Record 109221: "hurt hurt": His sister had hit him. Age: 1:10[1]
> NT: Record 111543: "ma BandAid hurt Mommy". Age: 2:1
> NT: Record 111680: "my tummy hurts (whine)". Age: 2:2

Like the human children, of the four internal state words (happy, mad, scared, and hurt) the bonobos were most likely to use "hurt". Kanzi's use of "hurt" made up 90.0% of his internal state utterances and Panbanisha's use of "hurt" made up 68.9% of hers. Panpanzee, the chimpanzee, also used "hurt" frequently (29.4% of her internal state utterances), but she was most likely to use "scared" (64.7%).

Developmentally, we can only compare Panbanisha and Panpanzee, but we can see a general trend for more internal state words as the apes aged (see Fig. 6.1).

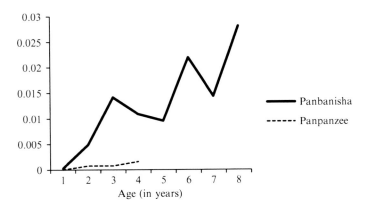

Fig. 6.1 Proportion of utterances that included internal state lexigrams over the first 8 years for Panbanisha and Panpanzee. Systematic observations of Panpanzee's utterances ceased in her 5th year of life, but a general upward trend can still be seen

---

[1] Examples of utterances will be formatted as: Participant: Record number: "utterance": context. Age. Utterances in all CAPS indicate lexigrams glossed as the English word.

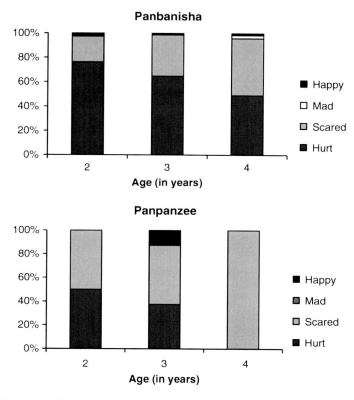

Fig. 6.2 Proportion of internal state utterances of each lexigram type over the first 4 years of life

As reported in Lyn et al. (2008), Panbanisha began using internal state words at one year of age but only used them at extremely low levels for almost 9 months, at which point her internal state word usage increased. The data for Panpanzee show a similar increase for internal state words at three years of age, but the pattern is not strong.

Specific internal state words showed different developmental tracks, as well. The data in Fig. 6.2 show that both Panbanisha and Panpanzee had a general trend of increasing use of "scared" and decreasing use of "hurt", with Panpanzee using the lexigram for "scared" in five out of five utterances that included internal state words when she was 4 years old. Both apes also had a much lower rate of utterances that included "mad" and "happy". Panpanzee never used the lexigram for "mad" and only used the lexigram for "happy" once (while age 3). Similarly, Panbanisha used "happy" a total of five times and "mad" only three times in the 4-year period. Panbanisha only used the lexigram for "mad" after she turned four.

Similarly, the apes showed greater comprehension of the terms "hurt" and "scared" than they did "mad" and "happy" (see Fig. 6.3). Unfortunately, it is impossible to know whether the caregivers simply used "hurt" and "scared" more frequently, artificially inflating the number of items available to be counted as comprehension. However, it is clear that the ape responded with clear understanding more often to these lexigrams.

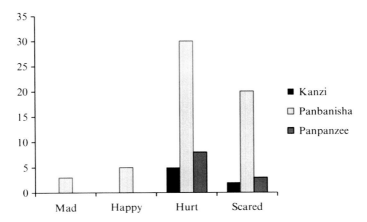

Fig. 6.3 Number of examples in the database of comprehension of internal state words

One important point to make is that both "Hurt" and "Scare(d)" while frequently used to label internal states, also served as a noun (in the case of hurt) and a verb (in the case of scare), both much more concrete meanings. For example, caregivers would frequently suggest that an ape "scare" a person or another ape as part of a social game. For example:

> Panbanisha: Record 29793: "let's GO to the CHILDSIDE and SCARE someone": Panbanisha found a monster mask in the cabinet that we opened. I made this suggestion, referring to the mask as a way to scare people. I put the mask on my head, but Panpanzee got scared herself and didn't really want to go. So I gave it to Panbanisha - she put it over her head, and led us to the childside, wearing the mask the entire way. She stopped by the front door and peered in, apparently looking for someone to scare, but no one was there. Age: 2:8

"Hurt", in contrast, was often used to refer to actual injuries–as the only appropriate word on the keyboard at that time. For example,

> Kanzi: Record 13989: "Show her your hurt in your mouth": Rose tells Kanzi this (english only). Kanzi then comes to Krista and sticks out his lower lip so that she can see his scrape Age: 6:6

These multiple uses for the same symbol are reflected in the apes' keyboard use, although due to the lack of syntactical markers, the exact meaning must be inferred from the context. Examples:

> Kanzi: Record 107637: "LEMON HURT": Kanzi had just had a lemon and again, as I said no to another one, he said hurt. Age: 15:9
>
> Panbanisha: Record 28880: "HURT": Panbanisha had been stirring her hot coffee with a spoon when she accidentally spilled a few drops on her foot. About 15–20 s later she pointed to the HURT lexigram and then asked me to look over her foot. Her foot looked fine. Age: 3:2
>
> Panbanisha: Record 98630: "HURT MILK": It seemed that Panbanisha was using the word 'hurt' to comment on that she felt bad and could not drink the milk even though she wanted it. Age: 7:3
>
> Panbanisha: Record 65916: "HIDE SCARE MONSTER": Panbanisha led Linda G. down the hallway from the observation room. When they came back Panbanisha had a mask on her head. I pretended to be scared of her MONSTER HAT. Age: 4:0

In contrast, "Mad" and "Happy" are more abstract and specifically internally referential, even in comprehension examples:

> Panbanisha: Record 20080: "Are you happy?": When I asked Panbanisha if she was happy about going to the Observation Room she assumed a position for me to slap her (a playful behavior). Age: 2:7
>
> Panbanisha: Record 69906: "If you're bad I'm going to get mad": I let Panbanisha know that if she was bad because I wouldn't give her any milk I was going to get mad at her. Age: 3:9
>
> Panbanisha: Record 97656: "HAPPY": As Panbanisha drank her milk, she commented that she was happy. Age: 7:8
>
> Panbanisha: Record 97656: "MAD": When Liz does not allow Panbanisha to see the vacuum again, Panbanisha states that she is mad. Age: 4:11

6.3.2 *Pragmatic Force and Co-Construction*

The coding system for the apes' utterances included information about the basic pragmatic force of the communication. Interestingly, all three apes showed a different pattern of pragmatic force. As may be expected, the majority of Panbanisha's internal state lexigrams were found within declarative utterances (either comments or statements of intention). Perhaps due to the exceeding rareness of their use of internal state words, Kanzi and Panpanzee did not follow the expected path. Kanzi most frequently used internal state lexigrams in requests (70% of internal state lexigrams). For example:

> Kanzi: Record 28878: "CELERY HURT HURT LEMON": We had been talking about what Kanzi wanted before we worked and he wanted many things we didn't have. Possibly this combination was to tell me he was upset he couldn't have those things and to suggest other possibilities. Age: 15:9

In contrast, Panpanzee used internal state lexigrams often in declaratives (24%), but even more often in imitative utterances (47%). As has been reported elsewhere (Greenfield and Savage-Rumbaugh 1991), the apes frequently use imitations as specific communicative devices—most frequently, agreement:

> Panpanzee: Record 36099: "SCARE": Our play was interrupted by a scary noise (drilling) coming from the other side of the bathroom. I commented on this and Panpanzee deliberately and quietly touched SCARE and kept her finger on the symbol for a few seconds. Age: 3:1
>
> Panbanisha: Record 104370: "GOOD MAD": After I ask Panbanisha if she won't get mad when I take the vacuum away, she points to 'good mad', agreeing to be good and not to get mad. Age: 8:4

6.3.2.1 Self- and Other-Reference

Panbanisha frequently commented on her own hurts as she was self-grooming, e.g.:

> Panbanisha: Record 28878: "HURT": Panbanisha sits before the keyboard grooming a sore on her knuckle. She stops grooming, activates the HURT lexigram 3 or 4 times then continues to groom her sore. Age: 2:12

However, she also commented on other's hurts:

> Panbanisha: Record 107471: "HURT ICE SHOT gesture to arm": Panbanisha pointed to a spot on my arm where I had received a shot a few days ago. She then proceeded to put some ice on it for me. Note: She put ice on one of my sores a couple of weeks ago after making a similar type of combination at the keyboard. Age: 10:8

She also used Internal State lexigrams when communicating with other apes:

> Panbanisha: Record 106979: "HURT GROOM YES": Statement made to Kanzi after he objected to her grooming me (Liz). After this he left us alone. Age: 10:2

6.3.2.2 Past and Future Reference

As with other symbols, the apes used the internal state words to refer to both future and past experiences, indicating some ability to mentally time travel:

> Panbanisha: Record 97654: "HAPPY": After she had her shot, Panbanisha started grinning and commented how happy she was because she was about to get her milk. Age: 7:8
> Panpanzee: Record 85044: "SCARE": We had gotten out to check a cooler by the trail and then got back in car. Panpanzee stared outdoors then touched SCARE SNAKE. Sue said they had seen a snake in that direction during the previous week. Age: 4:3

6.3.2.3 Co-Construction

As hypothesized, there was less evidence of co-construction of the internal state words. Apes might label an internal state to caregivers and the caregiver may then agree, but there were no examples of disagreements or misunderstandings of meaning as there were with the lexigrams "good" and "bad".

> Panbanisha: Record 36095: "SCARE": After Panbanisha said SNAKE, I became alarmed and asked her if she saw a snake somewhere. She said SCARE and I agreed that snakes are scary. Age: 2:12
> Panbanisha: Record 80070: "MAD": When Panbanisha ran over to Linda against my wishes I told Panbanisha that she must come back. My tone of voice was somewhat irritated. Panbanisha came back over and commented that I was mad at her. I definitely agreed with her. This was the first time I've seen her use the "MAD" lexigram so clearly. Age: 4:4

6.4 Discussion

Our findings indicate that apes may not only feel basic emotions like happiness and anger, but that they can communicate symbolically about these feelings as well, negating Descartes' claim for non-rationality in animals (1637). Internal development may drive the acquisition of internal state communications in apes as is theorized in humans (Lamb 1991). In humans, Lamb (1991) found that the use of internal state words in children preceded the first stages of moral development, but was associated with the internal development of empathy, suggesting that the internal development of empathy could first drive the acquisition of internal state words and then, value judgments. However, Lamb also reported that the increase in the children's use of internal state words occurred after an increase in the *awareness* of standards (as measured non-linguistically). Both Panbanisha and Panpanzee spontaneously used "good" and "bad" as value judgments *before* they used internal state words (in contrast to the findings in children), however, this use of "good" and "bad" suggests an awareness of standards, which also preceded the use of internal state words in children, suggesting that the developmental track of "awareness of standards" before internal state words is the same in apes and humans.

However there is less negotiation or co-construction of the meaning of these symbols than in more culturally-mediated value judgments like good and bad, even though the apes parallel their caregivers in using Internal State symbols in multiple forms (like hurt indicating both the internal state of hurting and the physical injury). These findings may suggest that while symbolic use of internal state words and value judgments are environmentally mediated, internal development does indeed drive the understanding of internal states, and value judgments are negotiated on top of that understanding.

It is important to note that internal state symbols made up only a small proportion of the apes' utterances. However, infrequent phenomena, such as declarative communication and use of internal state words in apes, can be significant in evolution because natural selection sometimes utilizes genes for infrequent, but adaptive behavior. As the survival rate of individuals displaying said behavior increases, the behavior becomes more frequent. Similarly, sometimes infrequent behaviors piggyback on genes for other behaviors. Those infrequent behaviors may not be adaptive at all, until an environmental change occurs and those previously infrequent behaviors become more adaptive, and then more frequent.

Given this evolutionary tendency, a behavior, however infrequent (in this case, symbolic communication of internal states), present in all members of an evolutionary clade [all member species diverging from an evolutionary node (like that which gave rise to humans, bonobos, and chimpanzees) (Byrne 1995)] would be assumed to have been present in the evolutionary ancestor. The genes that gave rise to that behavior would therefore be excluded as a candidate for a possible biological change in one of the divergent species. We have put forth similar arguments in earlier papers for hierarchical categorization (Lyn 2007), complex mental representation of symbols (Lyn 2007), word ordering preferences (Lyn et al. 2011a) and pretense (Lyn et al. 2006). However, it is possible that, as has been suggested by others (Moll and Tomasello 2007; Tomasello 2007), use of internal state symbols in apes is so infrequent that one could dismiss it as an artifact of the data collection.

Evidence contradicts this viewpoint, however, with different apes showing different communicative preferences but similar developmental tracks in their use of internal state symbols. It seems more likely that Darwin was correct—emotions are more-or-less evolutionarily continuous in our closest relatives.

Acknowledgements The authors wish to acknowledge the contributions of Patricia Greenfield and Kristen Gillespie-Lynch in the collection and consolidation of the child-ape utterance database. Additional thanks go to the staff of the Language Research Center, Atlanta, GA in collecting the ape data. All applicable guidelines for animal care and use were followed throughout this study. The authors declare that they have no professional or financial affiliations that may be perceived to have biased this study.

References

Bowerman M (1973a) Early syntactic development: a cross-linguistic study with special reference to Finnish. Cambridge University Press, Cambridge

Bowerman M (1973b) Structural relationships in children's utterances: syntactic or semantic? In: Moore TE (ed) Cognitive development and the acquisition of language. Academic, New York, pp 197–213

Brakke KE, Savage-Rumbaugh ES (1995) The development of language skills in bonobo and chimpanzee—1. Comprehension. Lang Commun 15(2):121–148

Brakke KE, Savage-Rumbaugh ES (1996) The development of language skills in *Pan*—II. Production. Lang Commun 16(4):361–380

Brown R (1973) A first language. Harvard University Press, Cambridge

Byrne RW (1995) The thinking ape: evolutionary origins of intelligence. Oxford University Press, Oxford

Darwin C (1872) The expression of the emotions in man and animals John. Murray, London, England

Descartes R (1637) Discourse on the method of rightly conducting the reason and seeking the truth in the sciences, by René Descartes. In: Charles WE (ed) The Harvard classics, vol 34. P.F. Collier & Son, New York

Fouts RS, Mills ST (1998) Next of kin: my conversations with chimpanzees. Harper Paperbacks, New York

Gardner RA, Gardner BT, Van Cantfort TE (1989) Teaching sign language to chimpanzees. State University of New York Press, Albany

Gillespie-Lynch K, Greenfield PM, Lyn H, Savage-Rumbaugh S (2011) The role of dialogue in the ontogeny and phylogeny of early word combinations. First Lang 31(4):442–460

Greenfield PM, Lyn H (2007) Symbol combination in *Pan*: language, action, and culture. In: Washburn DA (ed) Primate perspectives on behavior and cognition. American Psychological Association, Washington, pp 255–267

Greenfield PM, Savage-Rumbaugh ES (1990) Grammatical combination in *Pan paniscus*: processes of learning and invention in the evolution and development of language. In: Parker ST, Gibson KR (eds) "Language" and intelligence in monkeys and apes: comparative developmental perspectives. Cambridge University Press, New York, pp 540–578

Greenfield PM, Savage-Rumbaugh ES (1991) Imitation, grammatical development, and the invention of protogrammar by an ape. In: Krasnegor NA, Rumbaugh DM, Schiefelbusch RL, Studdert-Kennedy M (eds) Biological and behavioral determinants of language development. Lawrence Erlbaum Associates, Hillsdale, pp 235–258

Greenfield PM, Savage-Rumbaugh ES (1993) Comparing communicative competence in child and chimp: the pragmatics of repetition. J Child Lang 20(1):1–26

Greenfield PM, Smith JH (1976) The structure of communication in early language development. Academic, New York

Greenfield PM, Lyn H, Savage-Rumbaugh ES (2008) Protolanguage in ontogeny and phylogeny: combining deixis and representation. Interact Stud 9(1):34–50. doi:10.1075/is.9.1.04gre

Izard CE (1971) The face of emotion. Appleton, East Norwalk

Ladygina-Kohts NN, de Waal FBM, Wekker B (2002) Infant chimpanzee and human child: a classic 1935 comparative study of ape emotions and intelligence. Oxford University Press, New York

Lamb S (1991) Internal state words: their relation to moral development and to maternal communications about moral development in the second year of life. First Lang 113(33)

Lyn H (2007) Mental representation of symbols as revealed by vocabulary errors in two bonobos (*Pan paniscus*). Anim Cogn 10(4):461–475. doi:10.1007/s10071-007-0086-3

Lyn H (2012) Apes and the evolution of language: taking stock of 40 years of research. In: Vonk J, Shackelford TK (eds) Oxford handbook of comparative evolutionary psychology. Oxford University Press, Oxford

Lyn H, Savage-Rumbaugh ES (2000) Observational word learning by two bonobos: ostensive and non-ostensive contexts. Lang Commun 20(3):255–273

Lyn H, Greenfield PM, Savage-Rumbaugh ES (2006) The development of representational play in chimpanzees and bonobos: evolutionary implications, pretense, and the role of interspecies communication. Cogn Dev 21(3):199–213. doi:10.1016/j.cogdev.2006.03.005

Lyn H, Franks B, Savage-Rumbaugh ES (2008) Precursors of morality in the use of the symbols "good" and "bad" in two bonobos (*Pan paniscus*) and a chimpanzee (*Pan troglodytes*). Lang Commun 28(3):213–224. doi:10.1016/j.langcom.2008.01.00

Lyn H, Russell JL, Hopkins WD (2010) The impact of environment on the comprehension of declarative communication in apes. Psychol Sci 21(3):360–365

Lyn H, Greenfield PM, Savage-Rumbaugh ES (2011a) Semiotic combinations in Pan: a cross-species comparison of communication in a chimpanzee and a bonobo. First Lang 31(3):300–325. doi:10.1177/0142723710391872

Lyn H, Greenfield PM, Savage-Rumbaugh S, Gillespie-Lynch K, Hopkins WD (2011b) Nonhuman primates do declare! a comparison of declarative symbol and gesture use in two children, two bonobos, and a chimpanzee. Lang Commun 31:63–74

Moll H, Tomasello M (2007) Cooperation and human cognition: the Vygotskian intelligence hypothesis. Philos Trans R Soc B 362:639–648. doi:10.1098/rstb.2006.2000

Panksepp J, Watt D (2011) What is basic about basic emotions? Lasting lessons from affective neuroscience. Emotion Rev 3(4):387–396. doi:10.1177/1754073911410741

Parr LA, Waller BM, Vick SJ, Bard KA (2007) Classifying chimpanzee facial expressions using muscle action. Emotion 7(1):172–181. doi:10.1037/1528-3542.7.1.172

Penn DC, Holyoak KJ, Povinelli DJ (2008) Darwin's mistake: explaining the discontinuity between human and nonhuman minds. Behav Brain Sci 31:109–178. doi:10.1017/S0140525X08003543

Preston SD, de Waal FBM (2002) Empathy: its ultimate and proximate bases. Behav Brain Sci 25(1):1–20. doi:10.1017/s0140525x02000018

Russell JL, Lyn H, Schaeffer JA, Hopkins WD (2011) The role of socio-communicative rearing environments on the development of social and physical cognition in apes. Dev Sci 14(6):1459–1470

Savage-Rumbaugh ES (1986) Ape language: from conditioned response to symbol (animal intelligence). Columbia University Press, New York

Savage-Rumbaugh ES, Rumbaugh DM, Smith ST, Lawson J (1980) Reference: the linguistic essential. Science 210:922–925

Savage-Rumbaugh ES, McDonald K, Sevcik RA, Hopkins WD, Rupert E (1986) Spontaneous symbol acquisition and communicative use by pygmy chimpanzees (*Pan paniscus*). J Exp Psychol Gen 115(3):211–235

Savage-Rumbaugh ES, Murphy J, Sevcik RA, Brakke KE, Williams SL, Rumbaugh DM (1993) Language comprehension in ape and child. Monogr Soc Res Child Dev 58(3–4):1–222

Savage-Rumbaugh ES, Fields WM, Spircu T (2004) The emergence of knapping and vocal expression embedded in a *Pan/Homo* culture. Biol Philos 19(4):541–575

Tomasello M (2007) If they're so good at grammar, then why don't they talk? hints from apes' and humans' use of gestures. Lang Learn Dev 3(2):133–156

Yamanashi Y, Matsuzawa T (2010) Emotional consequences when chimpanzees (Pan troglodytes) face challenges: individual differences in self-directed 25 behaviours during cognitive tasks. Anim Welfare 19(1):25–30

Chapter 7
Animal Aesthetics from the Perspective of Comparative Cognition

Shigeru Watanabe

Abstract This chapter discusses the aesthetic behavior of animals from three aspects, namely cognitive or discriminative property, pleasure or reinforcing property, and creation or motor skills. There are many examples of discrimination of aesthetic stimuli by animals. A wide range of animals, from fish to primates, successfully

S. Watanabe (✉)
Department of Psychology, Keio University, Mita 2-15-45, Minato-Ku, Tokyo 108-8345, Japan
e-mail: swat@flet.keio.ac.jp

learn discrimination of music, while preference for particular music is rather rare in animals, although songbirds prefer some musical stimuli to others. Pigeons can discriminate good pictures from bad pictures and zebra finches prefer particular styles of paintings to others. Experimental evidence thus suggests that animals also have the ability to discriminate and enjoy aesthetic stimuli. Some animals, such as chimpanzees and elephants, draw and paint, and their behavior may be maintained by self-reinforcement, although there are numerous examples of training by other reward. However, the animals do not "enjoy" their products and the products do not have a reinforcing property to other conspecifics. This constitutes the clear difference between human art and the art-like behavior of animals.

7.1 From Experimental Aesthetics to Comparative Cognition of Art

Aesthetics was for a long time a branch of philosophy, not a scientific endeavor, so it was a real breakthrough when von Fechner (1876) founded the field of experimental aesthetics. He argued for a bottom-up approach to the concept of beauty instead of the top-down approach common in the philosophical aesthetics of his era. He developed psychophysical measurement methods that are still used today in the study of sensation, and applied this experimental approach to the perception of beauty.

The second breakthrough, by Berlyne (1976), was termed behavioral aesthetics (sometimes referred to as new experimental aesthetics). He introduced four methods for the experimental investigation of aesthetics, namely, verbal judgment, psychophysics, statistical analysis, and measurement of exploratory behavior. According to Berlyne, the basic mechanism underlying the enjoyment of art is sensory reinforcement, and this behavioral approach opened the way for using animal experiments in the study of aesthetics.

The third breakthrough is the neuroscientific approach, called neuroaesthetics (Zeki 1999). We can trace back physiological interpretation of aesthetics to Allen (1877), but recent development of brain-imaging technology opened a new approach, the "brain mechanisms of aesthetics." According to Zeki, beautiful paintings extract invariant elements of an object from the variant visual world. Accordingly, using functional magnetic resonance imaging, he found that beautiful pictures activated the anterior cingulate cortex and the orbitofrontal cortex (Kawabata and Zeki 2004). Latto (1995) described an idea of an "aesthetic primitive" that has a particular effect on our visual system. For example, horizontal or vertical straight lines are detected more easily than tilted lines; therefore, vertical and horizontal stimuli should be preferred to slanted stimuli. Ramachandran and Hirstein (1999) also suggested that paintings stimulate a particular visual area by exaggeration of discriminable features. A magnetoencephalographic study also showed activation of the prefrontal cortex by beautiful paintings, including those of impressionists (Cela-Conde et al. 2004). These areas have a role in decision making, so it is still not clear whether our brain has a particular area it uses for judgment of beauty. Perhaps affective processes and decision processes are integrated in the creation of our sense of beauty (Nadel et al. 2008). Redies (2007) proposed the hypothesis that art induces resonant, or synchronized states in the visual system (Averbeck and Lee 2004; Bichot et al. 2005). It is interesting that both a visual artist (Kandinsky) and a poet (Breton) saw beauty as vibration, or a convulsive phenomenon. Neuroaesthetics is an attempt to find the neural dimension for aesthetics.

Evolutionary or Darwinian aesthetics (see Voland and Grammer 2003; Grammer et al. 2003) is yet another approach to aesthetics. The Darwinian approach argues that aesthetics is an adaptation to the environment in which humans have evolved (Thornhill 1998).

One serious problem for the experimental study of beauty is defining what "beautiful" is. This is because beauty is not an objectively or physically defined concept (a truth embodied in the popular phrase "beauty is in the eye of the beholder"). Consequently, our criterion of beauty depends on individuals, cultures, era, and so forth. Beauty from one era may not be beautiful in a different era, just as what is considered to be art in one culture might be considered as ugly in a different culture.

However, even though "beauty" is a socially constructed concept, there seem to be certain common features or properties of beauty that exist at a basic level. Eibl-Eibesfeldt (1988) classified three different levels of perceptual bias in aesthetics: (1) a basic bias that we share with the higher vertebrates, (2) our species-specific bias, and (3) our specifically cultural bias. In fact, humans seem to share some sense of beauty with other animals. We perceive most bird song as beautiful sounds and see a beautiful pattern in the tail of a peacock. Such beautiful features of animals cannot be explained by strict natural selection, because these features seem not to have any specific survival advantage or other adaptive value.

Darwin's answer to this conundrum was sexual selection. Male peacocks evolved their beautiful tails because females liked them when they were beautiful. But there is still a puzzle: Why and how did the females acquire their aesthetics? Later, some sociobiologists claimed that such beautiful features provide a signal of an underlying "fit" genotype, implicitly "sent" from male to female (in most cases). If the signal is an honest signal and the female likes such "beauty," then beauty accumulates as it evolves by sexual selection.

There are examples in nature where animals must make qualitative judgments of a conspecific or of something created by a conspecific. One example is bowerbird. Male bowerbirds make a complicated construction and decorate it to attract females (Kusmierski et al. 1997; Miller 2000). We see the bowers as beautiful constructs, but for female bowerbirds it could be an honest signal of motor skills of the males (Douchet and Montogomerie 2003; Driscoll 2006). But an honest signal for animal does not necessarily mean "beautiful" for humans. In fact "ugly" features of animals (as judged by humans) often attract females of a species. For example, longer snoods of male turkeys attract females, and the snood length correlates with fewer parasites (Buchholz 1995), however unattractive the snood may look to human eyes.

An important message emerging from this Darwinian approach is the communicative aspect of art. Art is a tool for sending a message to conspecifics (or, in some cases such as flowers, to other species). In this sense, the message does not need to be beautiful (or attractive). Similarly, modern message art, or conceptual art, is not necessarily beautiful in an ordinary sense. The function of such art is sending a message and not one of beauty. If so, message art can be considered as having a biologically relevant type of aesthetics. Another implication of the sexual selection theory of the origin of art is a feeling of pleasure in aesthetics. If the aesthetic judgments originate from sexual behavior, it is easily associated with pleasure. This aspect of pleasure in art is discussed later.

Another biological theory of aesthetics focuses on adaptation to habitat. The origin of human aesthetics is in part an innate affiliation to plant and animal habitats (Wilson 1983). Our ancestors had to find a productive and safe habitat, select suitable food, and avoid dangerous animals. According to this theory, a beautiful stimulus is beautiful because it signals a good environment for us. Human aesthetics is, therefore, a species-specific adaptation to the environment. At present, evolutionary origin of pleasure caused by art has not been clarified but evolutionary psychologists; for example, Pinker (1997) claimed that stimuli with aesthetic components had adaptive value in the past. Cheesecake is delicious because taking glucose and protein was adaptive during human evolution. Pinker argued pleasure caused by art is the same.

Comparative cognition has two goals, namely, reconstruction of the evolutionary history of human cognition and the identification of phylogenetic contingencies that produced the cognition. If there is a biological component to the sense of beauty, perhaps it is possible to trace the evolutionary origin of human aesthetics and to understand the phylogenetic contingencies (selection pressures) that produced a certain basis for human aesthetics. These two questions are similar to those of Darwinian aesthetics. However, one feature of the comparative cognitive approach is a focus on animal research rather than the theoretical analysis of the evolution of art after hominization. Another feature of the comparative cognitive approach is the use of the experimental method, fully equipped with experimental techniques such as operant conditioning, along with its theoretical framework for the study of animal behavior.

In the comparative cognition framework, we can separate three aspects of aesthetics. The first is its reinforcing property: we enjoy beauty as sensory reinforcement. Our art is a product of our species, but does its reinforcing property have an effect only on *Homo sapiens* or does it have a similar effect on other species? If so, what kinds of species share this effect with us? The second aspect is aesthetics' discriminative property. We can categorize styles of visual art, such as cubist or impressionist, and we can categorize styles of auditory art, such as classical music or techno. We can discriminate beautiful stimuli from ugly stimuli. Beauty is a category, or a concept. Do animals have such an ability to discriminate? The third question is creation or motor skill. We are not simply discriminating and enjoying art, but also creating art ourselves. In the case of paintings, humans' creative abilities have been evident at least since the cave drawings of our ancestors more than 30,000 years ago (Lewis-Williams 2002). Probably our ancestors also enjoyed music. So in addition to the reinforcing and discriminating sensitivity used in art appreciation, motor skill is required for art creation. Our hand is a well-developed organ used to produce art, but some birds (such as New Caledonian crows) can manufacture tools, and elephants can skillfully draw a painting by gripping a brush with their trunk. The creation of art is discussed in the last section of this chapter.

One crucial factor in creating art as it pertains to human aesthetics is functional autonomy. Even though art has a biological origin, it developed for its own purpose. Creation of art does not have survival value in and of itself, and at present it is not always a successful method to attract sexual mates. Art activity is mostly maintained by self-reinforcement: that is, the activity of creating art itself reinforces the activity. How has *Homo sapiens* acquired this functional autonomy? Do animals also create art for its own sake, independent of other reward?

7.2 Discriminative and Reinforcing Properties of Auditory Art

Music is one of the characteristics of a species-specific behavior of humans. Because almost all cultures have their own music, music can be considered as being species specific to humans, like our language. To investigate the evolution of the culture of music, comparative studies, including those on nonhuman animals, have been conducted (for a review see Hauser and McDermott 2003). Here reinforcing and discriminative properties of music for animals are reviewed.

7.2.1 *Reinforcing Property of Music for Animals*

There are several similarities between music produced by humans and sounds produced by other animals (Gray et al. 2001). Some whale songs and bird songs share similar rhythms, phrases, and lengths with human music (Mercado et al. 2005; Hahnloser et al. 2002). Hearing music often causes a pleasurable experience in humans (i.e., it has reinforcing properties for us). In other words, music is a sensory reinforcement. Artificial and nonbiologically relevant sensory stimuli also have sensory reinforcement effects on animals. Music has some direct effect on physiological responses of animals—lowering blood pressure in rats (Sutoo and Akiyama 2004) and elephants (Wells and Irwin 2008), for example. Sometimes musical stimuli are used as a tool of environmental enrichment (Wells et al. 2002). But the reinforcing properties of music on nonhuman animals have not been well investigated. Most of the published results failed to demonstrate any reinforcing effect of music, even in primates.

7.2.1.1 Primates

Chimpanzees given a "jukebox" that could play different music such as Pavarotti's singing, Indian flute music, etc., did not show any musical preference (Howell et al. 2003), and Gorillas did not show an interest in music (Wells et al. 2006).

This is not to say that animals are unable to discriminate music or are completely tone-deaf. Two species of monkeys have been tested experimentally: the common marmoset and the cotton-top tamarin (McDermott and Hauser 2004, 2007). They were placed in a V-shaped maze in which entering each arm produced a different auditory stimulus. They preferred slow lullabies (of a Russian folk song) to a techno beat (*Nobody gets out alive*). This preference might be caused by preference for slow tempos, because the authors observed a preference for 60 clicks/min over 400 clicks/min in these animals. When tested with a choice between music—the lullaby or a string concerto by Mozart—and silence, both species preferred silence to music. Furthermore, their choice showed no preference between consonance and dissonance. The authors concluded that the monkeys did not have a musical preference.

As pointed out by Lamont (2005), the setting of this experiment is a kind of forced choice. Avoiding one stimulus produced an apparent preference for the other. Using electrophysiological techniques, Fishman et al. (2001) obtained dissonance-specific phase-locked oscillation in the primary auditory cortex of rhesus monkeys, and a similar greater oscillation by dissonance than consonance in Herschel's gyrus in human patients. Thus, humans and monkeys seem to process the perception of dissonance differently at some point after the primary auditory cortex.

However, adults and infants may show different patterns. In one experiment, an infant chimpanzee showed a preference for consonance. Sugimoto et al. (2010) trained an infant chimpanzee to pull a string to hear musical stimuli played by piano or marimba, and compared pulling responses for consonant music and dissonant music. (To create dissonant music, original music was modified to have dissonant intervals.) They found that the infant chimpanzee preferred the consonant version. In addition to the species difference, there are several procedural differences between the McDermott and Hauser (2007) and Sugimoto et al. experiments, for example, the age of the subjects. Infants may be more sensitive to musical stimuli.

7.2.1.2 Rodents

The reinforcing property of music has not been well examined with rodents. There was one report that rats will press a lever to produce a consonant sound more often than a lever to produce a dissonant sound (Zentner and Kagan 1998). The author's group trained rats on a concurrent chain schedule in which the terminal links were associated with different music: Bach or Stravinsky (Otsuka et al. 2009). As shown in Fig. 7.1, there were two levers in an experimental chamber, and a VI (variable interval) schedule was effective on either lever (initial link). That is, lever pressing was reinforced at variable intervals, but the reinforcement is not food reward but moving to the terminal link. When the VI schedule was completed on one of the two levers, the terminal link started. In the terminal link, the lever selected in the initial link remained in the chamber and the other lever was withdrawn. Fixed interval (FI) of 7 s was put into effect on the remaining lever and the rat obtained a food reward after 7 s. The schedule in the terminal link for two levers was identical except for music associated with the lever. One of the two music stimuli was presented during the 7 s, depending on the selection of the levers. The rats did not show a strong preference for either style of music, although one subject showed a weak preference

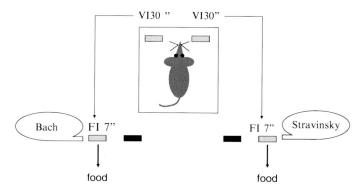

Fig. 7.1 Procedure of concurrent chained schedule. Preference of music is measured by choice of levers in the initial link (concurrent VI–VI schedule)

for Bach and another subject preferred Stravinsky. The validity of the concurrent chain procedure was examined as a method of preference measurement with conspecific vocalization evoked by an aversive experience. During responding in one terminal link the vocalization was presented, whereas white noise was presented during the other terminal link. Most of the rats preferred white noise to the conspecific vocalization. Thus, the association of auditory stimuli in terminal links in the concurrent chain schedule is sensitive in detecting relative preference (or aversion) for auditory stimuli in rats.

These experiments suggest no musical preference in rats, but it must be pointed out that humans and rats have different ranges of auditory sensation. Rats cannot hear sound below 500 Hz (Fay 1988); thus, rats cannot hear parts of human music. One recent experiment suggested that mice also sing ultrasonic songs (Holy and Guo 2005). Thus, ultrasonic music may have reinforcing effects for mice.

7.2.1.3 Birds

There are a few reports on musical reinforcement in birds. McAdie et al. (1993) trained hens to peck two keys associated with food reinforcement, then replayed a piece of *The Theme of Local Hero* contingent upon pecking one key. The music presentation did not affect the behavior. The author's group applied the concurrent chain procedure (similar to the one used in their rat experiment) to pigeons (Watanabe et al. 2009). Music by Bach and Stravinsky was used, similar to that used in the rat experiment. In the first condition works by Bach and Stravinsky were used; in the second and third conditions, one of the two music pieces and white noise were used. One subject preferred Bach and another subject preferred Stravinsky in the first condition, but their choice in the initial link was less than 60%. Overall, these results demonstrated no reinforcing effects of music for pigeons.

These experimental studies demonstrated that several species of nonhuman animals do not seem to have preferences for musical stimuli. However, the species examined

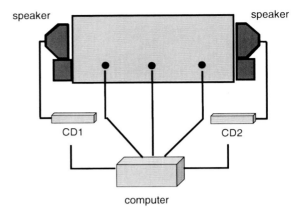

Fig. 7.2 Apparatus to measure music preference of Java sparrows. Staying at each perch is detected by photosensor and different music is played back according to the position

above are not auditory-specific animals. Species that have advanced auditory cognition and production, such as songbirds or dolphins, may be different. Stevenson (1969) investigated chaffinches' preference of conspecific song by measuring staying time on three perches. On the end perches a tape recorder played either a conspecific song or white noise. The birds visited the perch connected to the conspecific song more often. Similar preferences for the conspecific song were confirmed in several species (Zebra finch: ten Cate 1991; white-crowned sparrow: Dobson and Petrinovich 1973). Because similarities between human music and animal songs have been pointed out before (Gray et al. 2001, see also Marler and Slabbekoorn 2004), it is reasonable to hypothesize that songbirds may be sensitive to the reinforcing property of auditory stimuli such as those contained in human music.

Java sparrows show a preference for musical style (Watanabe and Nemoto 1998). A test chamber had three perches; the position of the bird was detected by a photosensor. In the first phase, the end perches connected to Bach and Schoenberg. The center perch was not connected to any music. Figure 7.2 illustrates the apparatus. Music used in the second phase was composed by Vivaldi and Carter. The third and fourth phases used white noise and the same music used in phase 1. Two of four birds stayed longer on the perch associated with Bach in phase 1 and Vivaldi in phase 2. One bird showed a preference for Bach over the white noise, but the remaining bird did not show any musical preference. Thus, a musical stimulus can have a reinforcing property for Java sparrows even though some birds did not show a preference. Such individual differences have not been reported in preference experiments of conspecific song. Thus, a reinforcing property of conspecific song is rather universal for the species, while musical stimuli show significant interindividual variation.

7.2.1.4 Fish

Among fish, goldfish (*Carassius auratus*) are regarded as "hearing specialists," because they have evolved a structure that enhances auditory signals detected by the inner ear. The Weberian ossicles connect the swim bladder to the otolith organ in this species, which enhances auditory sensitivity because the inner ear can detect the pressure component of sounds received by the swim bladder. Goldfish have a lower auditory threshold than oscars (*Astronotus ocellatus*), another hearing-generalist species (Kenyon et al. 1998).

The author's group investigated whether goldfish prefer a particular type of music (Shinozuka and Watanabe unpublished). In this experiment, the music was presented depending on the subject's position in an aquarium, and the time spent in each area was measured. An experimental aquarium equipped with underwater speakers was used. The speakers were placed on each narrow side of the aquarium. The stimulus music was identical to that used in our rat and pigeon experiments (Bach and Stravinsky). For example, Bach might be presented when the subject was in the right area, and Stravinsky would be presented when the subject was in the left area. Four of the six subjects did not have any preferences. One subject displayed a significant preference for silence. The last subject displayed avoidance of Bach's music. The results suggest that in general, musical stimuli do not have a reinforcing property for goldfish. At first, it had been predicted that goldfish would generally avoid the musical areas, because it is hypothesized that hearing specialists evolve their hearing ability in order to avoid predators in quiet habitats (Ladich 1999). In fact, compared with nonhearing specialists, their hearing sensitivity is strongly affected by ambient noise (Amoser and Ladich 2005). As a result of the present experiment, however, only one fish displayed significant avoidance of the music areas. When a strong water-stirring noise was used instead of music, the fish preferred a silent area to the noisy one; this suggests that the procedure to measure auditory preference is a valid one.

Because hens (McAdie et al. 1993) and pigeons (Watanabe et al. 2009) showed no musical preference, it is reasonable to hypothesize that there are no reinforcing properties of music for non-songbirds. The reinforcing effects of music have not been thoroughly examined and at present, humans and songbirds (Java sparrows) are exceptional species by virtue of showing preferences for particular types of music (Gess 2007).

7.2.2 *Discriminative Stimulus Properties of Music*

The discriminative stimulus property of music has been reported in various nonhuman animals. Music discrimination is a kind of category discrimination for humans. For example, we can discriminate music by Bach from music by Stravinsky or, even more easily, discriminate classical music from modern music. Such discrimination does not depend on particular pieces or musical instruments. Bach's works for organ and those for violin can be classified as one category: namely, Bach's music. Therefore, music discrimination is based on complex sensory processing rather than lower-level processing of sounds.

7.2.2.1 Primates

Octave-generalization is a kind of perceptual transposition and indicates categorical discrimination of music. Although D'Amato and Salmon (1982) failed to demonstrate octave-generalization, Wright et al. (2000) reported octave generalization in rhesus monkeys. They used a delayed matching to sample procedure, in which a center speaker played sample tones and two side speakers provided two choice tones. The animals showed octave-generalization for the song *Happy Birthday*.

7.2.2.2 Rodents

In mammals, rats have discriminated *Frère Jacques* from its reversed sequence (Poli and Previde 1991), and *Yesterday* by the Beatles from *Die Zauberfloet* by Mozart (Okaichi and Okaichi 2001). The rats were even able to discriminate *Yesterday* played by The Beatles and that by different players with different instruments, suggesting higher discrimination ability of this species.

The author's group trained rats to discriminate music by Bach from that by Stravinsky using operant conditioning (Otsuka et al. 2009). Rats were trained by using Bach's *Toccata and Fugue for organ* and Stravinsky's *The Rite of Spring* for orchestra. For half of the animals, responses made during Bach's music were reinforced on the VI-10 schedule, while responses made during Stravinsky's music were extinguished. For the other half, the reinforcement contingency was reversed. Four rats successfully learned the discrimination. The fastest rat attained 85% correct response across two consecutive days by 18 sessions, and the slowest by 59 sessions. However, two rats did not reach the criterion after 60 sessions of training. Long–Evans rats reached a 65% discrimination ratio after 25 sessions of discrimination training between *Frère Jacques* and its reversed sequence (Poli and Previde 1991). Although it is difficult to directly compare these results due to procedural differences among the experiments, Wistar rats seem to have considerable music discrimination ability.

Rats have higher auditory information processing capability,comparable to songbirds and humans,even though rats have different audio-sensing ability to those other species. According to Fay (1988), the lowest threshold audiogram is between 1,200 and 1,300 Hz in the human, around 1,000 Hz in pigeons, and around 10,000 Hz in rats. Thus, rats are sensitive to much higher frequencies than are pigeons and humans. Pigeons cannot hear sound below 200 Hz, and rats cannot hear sound below 500 Hz. Because the audiogram of rats differs so much from that of humans, it is probably untenable to say that rats and humans hear exactly the same auditory stimuli even if the stimuli are physically the same.

After the discrimination between Bach's *Toccata and Fugue for organ* and Stravinsky's *The Rite of Spring*, two additional music stimuli [Bach's *Prelude in E flat minor*, Chorale from *The Easter Cantata* (vocal and piano), and Stravinsky's *The Firebird Suite for the orchestra*] were used in the generalization test. The four rats maintained their discrimination. Note that the "*Rite of Spring*" is orchestral music, whereas *Toccata and Fugue* is organ music; hence the rats might simply have discriminated the sound of an orchestra from that of an organ. But the transfer test showed that the rats maintained their discrimination between Bach and Stravinsky when the vocal and orchestral music were presented. Likewise, *The Easter Cantata* contains piano sounds in addition to the song; the rats might discriminate the orchestra from the organ or piano. However, there was no difference in performance between Bach S+ rats (the organ S+) and Stravinsky S+ rats (the orchestra S+) in the generalization test. These results suggest that the rats did not discriminate these various stimuli based solely on the instruments used in the music. Instead, it seems that rats not only can discriminate musical stimuli but can also learn musical categories (as can the other animals described above).

7.2.2.3 Birds

There are many experiments demonstrating successful discrimination of birdsong by songbirds (for example, zebra finch by Cynx 1993; budgerigars by Okanoya and Dooling 1991; great tits by Weary 1989; song sparrows by Stoddard et al. 1992). Porter and Neuringer (1984) demonstrated that pigeons could be successfully trained to discriminate between classical (e.g., Bach's) and modern (e.g., Stravinsky's) music. In the probe trials of that experiment, the pigeons transferred discriminative responses to novel music, and their responses were almost similar to those of humans. Likewise, Java sparrows are able to discriminate the music of Bach from that of Schoenberg (Watanabe and Sato 1999). The birds were trained to move from a rest perch to a response perch when particular music was played. Five birds learned the discrimination (taking 33–52 sessions to reach the criterion of 80% correct responses), but two did not reach the criterion within 60 sessions. Pigeons trained on a music discrimination task showed less than an 80% discrimination ratio even after 60 sessions of training (Porter and Neuringer 1984). Although we have to consider differences in discriminative stimuli, training procedures, and so forth, it seems quite likely that the music discrimination abilities of songbirds may be higher than those of pigeons.

In the generalization tests, the author's group used different music by Bach and Schoenberg, and also conducted a test with Vivaldi versus Carter. The Java sparrows showed generalization not only to new Bach and new Schoenberg, but also to the Vivaldi and Carter music, suggesting that they have the ability to discern differences between, and to correctly categorize, classical music and modern music. These results are consistent with results in which pigeons showed generalization from Bach to Buxtehude and Scarlatti, and from Stravinsky to Carter and Piston (Porter and Neuringer 1984). Interestingly, the pigeons did not show generalization from Bach to Vivaldi. In this sense, Java sparrows have a more human-like music categorization.

These experiments suggest category-like discrimination of music in birds but did not clarify which cues are used for the discrimination. One possible cue could be consonance and dissonance. In modern music theory, musical intervals such as unisons, octaves, fifths, and fourths are classified as perfect consonances; major and minor thirds and sixths are classified as imperfect consonances; and major and minor seconds and sevenths are classified as dissonances. The physical characteristic of consonance is the matching of the maximum number of upper harmonic frequencies (which are integer multiples of the fundamentals). Such physical features, however, do not completely determine the difference between consonance and dissonance. Consonance is static, and produces a pleasant feeling, whereas dissonance is dynamic and intense, and produces an unpleasant feeling.

Java sparrows were trained on discrimination between consonance and dissonance in a setting similar to the one used in the music discrimination experiments (Watanabe et al. 2005). Four consonant sounds and four dissonant sounds produced by a programmable sound generator were used for the discriminative training. The fastest bird reached the criterion in 13 sessions and the slowest in 25 sessions. The generalization test used four new consonant stimuli and four new dissonant stimuli. The birds maintained their discrimination for these new sounds. In the second test, four novel dissonances consisting of tones with different intervals were presented. The subjects trained to perch for consonance performed poorly, whereas those trained to perch for dissonance performed perfectly. These results suggest that stimulus generalization by Java sparrows after being trained for consonance versus dissonance discrimination differs from human perception of consonance and dissonance. Pigeons, on the other hand, could learn to discriminate a C major chord from minor, sus4, flat five, or augmented chords, but two of five birds failed to reach the criterion within 50 sessions (Brooks and Cook 2010). The three learners failed to transfer from the C chord to the D chord. This result also suggests difference in chord discrimination between humans and birds.

Rhythm is an important factor in music, and also an important factor in vocal learning (Patel 2006). Starlings (Hulse and Kline 1993; Hulse et al. 1984, 1995) could discriminate the rhythm, and Snowball; the dancing parrot, could synchronize a movement to a musical rhythm (Patel et al. 2009). Pigeons naturally show a rhythmic movement of the head (bobbing) when they walk, so it would be reasonable to hypothesize that they could learn good rhythm discrimination. However, they did not learn to discriminate rhythmic from arrhythmic sound (Hagmann and Cook 2010), even though they could discriminate fast and slow piano tempos. Thus, both Java sparrows and pigeons can discriminate music, but they might use different cues for this discrimination.

7.2.2.4 Fish

In goldfish, the psychophysical properties of hearing have been studied extensively by Fay. He used the classical respiratory conditioning method and the generalization paradigm, and demonstrated that goldfish could discriminate between pure tones at various frequencies (Fay 1970). He also showed that goldfish could recognize complex sounds. For example, goldfish showed a generalization from pure tones to amplitude-modulated sounds and complex pure tones (Fay 1992) that included the training stimulus frequency. Goldfish also showed a generalized response for spectrally and temporally complex sounds (Fay 1995). This suggests that, like humans, goldfish respond to complex sounds in a "pitch-like" and "timbre-like" manner. Furthermore, when goldfish were trained with the presentation of such a mixture of stimuli, they showed generalized responses to sounds that contained a particular repetition rate and spectral profile. Thus, goldfish could analyze the perceptual correlation found in a sound source (Fay 1998). This suggests that goldfish have evolved a specific feature for hearing and can recognize complex auditory stimuli similar to that of modern higher vertebrates.

The author's group trained goldfish on music discrimination (Shinozuka and Watanabe unpublished). The subjects were trained to discriminate between two pieces of music (*Toccata and Fugue* by Bach and *The Rite of Spring* by Stravinsky, the same works used in the rat experiments; Otsuka et al. 2009). The operandum consisted of a small red bead connected to a stainless-steel rod by a nylon string. The reinforcer was a commercial goldfish pellet delivered by a universal feeder. A flat panel speaker unit was placed under water. The training was conducted under the multi VI-10 EXT schedule. That is, the bead-pulling behavior was reinforce on VI-10 during presentation of one type of music but extinguished during presentation of the other music. All four fish met the criterion (above 75% in three successive sessions). The number of sessions needed to achieve the criterion was 79–196. The discriminative stimuli and training were similar to those used in the previous experiment with rats. In comparison with the rats, the goldfish required more sessions to learn the discrimination. Chase (2001) showed that carp could discriminate between two genres of music (classical music and blues). Thus, fish have an ability to discriminate complex auditory stimuli such as music.

However, after the discriminative training, generalization tests with three pieces of music similar to those used in the rat experiment (Otsuka et al. 2009) were carried out. In contrast to the rats and pigeons (Porter and Neuringer 1984), there was no generalized response to novel music; all the fish responded almost randomly. One reason for this might be the musical instruments used as stimuli. For training stimuli, *Toccata and Fugue* was played by pipe organ, whereas *The Rite of Spring* was played by an orchestra. For the test stimuli, both pieces were played by an orchestra. Thus, if the subjects learned discrimination based on differences in timbre, it might have been difficult for them to discriminate between the test stimuli. In this case, the Bach S+ group would be expected to decrease their total responses in the test session, because both stimuli were played by an orchestra (as in the S− stimulus). For the same reason, the Stravinsky S+ group should also have increased their total responses in the test session. However, the results did not display this tendency. Therefore, the difference in timbre might not be the main cue for discrimination. However, Chase (2001) successfully trained carp on musical discrimination, and reported generalization to new music in that species. It is possible that discriminative behavior differs between these species, even though goldfish and carp are closely related. There are several procedural differences between the two experiments in addition to the species difference. Although a more detailed analysis will

Table 7.1 Discriminative and reinforcing properties of music

Aspects	Species	Comments	Results	Authors
Discrimination	Monkey	Octave generalization	P	Wright et al. (2000)
		Short melodies	P	D'Amato and Salmon (1982)
	Elephant	Short melodies	P	von Reinhert (1957)
	Cow	Approaching	P	Uetake et al. (1997)
	Rat	Music	P	Okaichi and Okaichi (2001)
			P	Poli and Previde (1991)
			P	Otsuka et al. (2009)
		Short melodies	P	D'Amato and Salmon (1982)
	Pigeons	Music	P	Porter and Neuringer (1984)
	Starling	Chord	P	Hulse et al. (1995)
	Java sparrow	Music	P	Watanabe and Sato (1999)
		Consonance	P	Watanabe et al. (2005)
	Carp	Music	P	Chase (2001)
	Goldfish	Chord	P	Fay (1992)
		Music	P	Shinozuka and Watanabe (unpublished)
Reinforcement	Tamarin	Preference	N	McDermott and Hauser (2004)
			N	McDermott and Hauser (2007)
	Marmoset	Preference	N	Sugimoto et al. (2010)
	Chimpanzee	Preference for consonance	P	
	Java sparrow	Preference	P	Watanabe and Nemoto (1998)
	Hen	Pecking	N	McAdie et al. (1993)
	Pigeon	Pecking	N	Watanabe et al. (2009)

P and N mean positive and negative results, respectively. Comments give a short description of musical stimuli

be necessary in future studies, it can be said that the goldfish might have acquired discrimination by learning the training stimuli by rote, while the carp might have been able to discriminate the concepts. Goldfish are domesticated, whereas carp are usually wild. Thus domestication may affect discriminative behavior of goldfish.

7.2.3 Conclusion

Table 7.1 summarizes the results of the experiments described in this section. A variety of species, from fish to songbirds, demonstrated musical discrimination abilities. Hence, an ability to discriminate complex auditory stimuli may be widespread among vertebrates, regardless of their different audiograms. Animals may be able to discriminate complex auditory stimuli by training if there is a psychophysical difference between the species. One peculiarity of goldfish is a lack of generalization to other music. This probably indicates poor cognitive processing in this species.

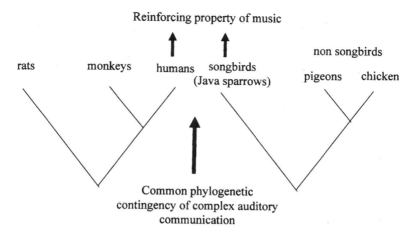

Fig. 7.3 Humans and songbirds are exceptional species showing musical preference

On the other hand, the reinforcing property depends on the species. Music is made by humans for humans, so it is adjusted to the limits and characteristics of human auditory perception. It is interesting that the primates, creatures close to humans, have not shown preferences for music (except for an infant chimpanzee). Because identical music stimuli were used in the two experiments with rats (Otsuka et al. 2009), the difference between reinforcing and discriminative properties is clear. Different species have different hearing ranges; thus, music can be a distorted or filtered stimulus for particular animals. To understand discriminative and reinforcing properties of music, we have to take account of such differences in auditory ability.

In avian studies, it is reasonable to suggest that there are no reinforcing properties of music for non-songbirds. The reinforcing effects of music have not been thoroughly examined, and at present, humans and songbirds (Java sparrows) are exceptional species showing preference for a particular music. One common characteristic of humans and songbirds is that both have well developed vocal communication, although human language is extraordinary in that it has a seemingly unlimited capacity for sending meaningful messages. Both humans and songbirds have to learn their communication system after birth. During song learning, birds first produce sub-songs, then modify them to complete, full song. Their song-producing behavior during this period is maintained by self-reinforcement. Similarly, human infants first produce babbling and acquire their native language through experience, primarily through self-reinforcement. Thus, self-reinforcement appears to play a crucial role in the acquisition of language. The reinforcing effects of complex auditory stimuli, such as music, may be involved in such self-reinforcing behavior in both humans and songbirds. One would predict that other species with well-developed auditory communication systems, such as elephants or dolphins, should show a reinforcing property of music (Fig. 7.3).

7.3 Discriminative and Reinforcing Properties of Visual Art

Just as there is a similarity between animal song and music, there is a similarity between beauty in natural objects and beauty in visual artworks. Philosophers of aesthetics have pointed out common features in nature and art (Immanuel Kant, for example). Although art is not a mere mimic of natural beauty, one of the origins of beauty in visual art should be beauty in the natural landscape (flowers, etc.). Therefore it is plausible to hypothesize that nonhuman animals share some sense of beauty with us.

7.3.1 Reinforcing Property of Visual Art

Just as an auditory stimulus can provide sensory reinforcement, so can a visual stimulus do so. Biologically relevant or natural visual stimuli have reinforcing value (e.g., an image of conspecific for a Java sparrow: Watanabe 2002; and of macaques: Fujita and Watanabe 1995; Schwartz and Rosenblum 1980). But some biologically irrelevant sensory stimuli are also reinforcing (e.g., switching on a lamp for rats: Berlyne 1969; looking at an electric toy train for chimpanzees: Butler 1953, etc.). Rensch (1957, 1958) examined preferences for visual patterns in several species. He described a preference for regular, symmetrical patterns in capuchin monkeys (confirmed later by Anderson et al. 2005). Crows and meerkats also showed a preference for regular patterns, but fish did not. Humphrey (1972) trained monkeys to press a button to see movies. The monkeys preferred a Walt Disney film with a continuous story to a looped film. Wilson and Goldman-Rakic (1994) measured the gaze of rhesus monkeys viewing faces, pictures obtained from color magazines, and color patterns, and found that the monkeys spent more time looking at the faces and the pictures. There is, however, no experimental work on the reinforcing properties of paintings in animals.

Ikkatai and Watanabe (2010) examined preferences for paintings in Java sparrows. In a long experimental chamber (108 cm × 17.5 cm × 25.5 cm), three TV monitors displayed three different styles of paintings (Japanese, impressionist, and Western modern paintings) and gray-scale patterns. Their staying time on a perch in front of each monitor was measured as an index of their preference. Considerable individual differences were observed, but in summary, five of seven birds preferred the modern to the impressionists, three preferred the Japanese to the modern, two showed reversed preferences, and six did not show differential preference between the impressionist and Japanese categories. These results suggest that something differentiated the modern paintings from the others, although the direction of the preference depended on individuals. On the other hand, the reinforcing properties of the impressionists and Japanese paintings were approximately the same. It is an interesting observation, because the Japanese paintings influenced the impressionists. This experiment did not clarify mechanisms of reward as a function of painting, but did demonstrate differential behavior directed at different styles of paintings.

7.3.2 Discriminative Stimulus Property of Complex Visual Stimuli

Behaviorally, a concept is defined as a generalization within a stimulus class and discrimination between classes. Herrnstein and Loveland (1964) were the first to find evidence for a complex visual concept in pigeons. Since that time, a long list of natural and artificial concepts formed by pigeons has been created (e.g., Herrnstein et al. 1989; Wasserman et al. 1988; Watanabe 1988; Watanabe et al. 1995). One good example is a concept of symmetry. For example, pigeons showed generalization to novel symmetric patterns after discrimination training between symmetric and asymmetric patterns (Delius and Habers 1978).

There is, however, still a lot of discussion about similarity and dissimilarity between human formation of concepts and those acquired by animals. For example, in one study, pigeons learned to discriminate many different patterns of triangles from three randomly arranged lines but, testing with distorted triangle patterns such as a partial triangle or a triangle pattern made of dots, it was shown that pigeons were under a different stimulus control from that of humans (Watanabe 1991). Of course, the human definition of a triangle is a rule that can be described verbally. On the other hand, pigeons have to establish nonverbal definition-like rules through behavioral experience alone. Comparing the pigeons' patterns of responding after training with multiple exemplars and after training with a single exemplar suggests that exposure to multiple exemplars may be essential to the formation of an artificial geometrical concept based on a definition-like rule (Watanabe 1991). Thus, a basic procedure to form a visual concept in pigeons is training with multiple exemplars, followed by testing with stimuli never seen during the discriminative training.

7.3.3 Discriminative Property of Painting Style

Painting style also consists of a visual concept. When we see paintings by Picasso and Monet, we can say which is by Picasso and which is by Monet with some accuracy, even if we have never seen these particular paintings. There could be some crucial cues evident to specialists, but it may be difficult for ordinary people to describe cues used in their discrimination. Color, subject, brush style, and so forth, may constitute the cues for such discrimination. Such experience suggests that the concept of a painting style is a polymorphous concept in which several different cues are combined.

The author's group has reported discrimination of paintings in birds (pigeons by Watanabe et al. 1995; Watanabe 2001b; Java sparrows by Ikkatai and Watanabe 2010). In their earliest study (Watanabe et al. 1995), eight pigeons were trained on discrimination with ten Picasso and ten Monet paintings. The birds required 6–24 sessions to reach the criterion. They were then tested with novel paintings by Picasso and Monet, never seen during the discriminative training, and paintings by Renoir, Cezanne, Braque, Matisse, and Delacroix. The pigeons showed generalization not only from trained Monet (or Picasso) to new Monet (or Picasso) but also from Monet to Renoir and Cezanne, and from Picasso to Braque and Matisse. Therefore they formed "concepts" of impressionism and cubism. But if the birds are unable to discriminate within a given category, their discriminative behavior is just confusion rather than a concept. Humans can discriminate each painting of Monet, and also can discriminate the impressionist from the cubist. To clarify this point, a new group of pigeons was trained on "pseudo-concept" discrimination, in which ten Picasso and ten Monet paintings were shuffled and divided into two stimulus groups. Both groups contained Monet and Picasso. The birds were able to learn the discrimination; this suggested that the birds could not only discriminate each painting, but also could discriminate on the basis of a concept of the painting style.

A cartoon is a kind of visual art in which some aspects of the object are neglected and others exaggerated. In this sense, a cartoon shares some aspects with both cubist and abstract paintings. Cerella (1980) taught pigeons to recognize the cartoon character Charlie Brown, and then tested them with scrambled pictures of Charlie Brown. In the scrambled pictures, head, trunk, and legs were connected but randomly arranged. The birds did not show any decrease in responding to the scrambled versions. This result was confirmed using the Japanese cartoon character Sazae-san (Watanabe 2001a). On the other hand, scrambling photographs of the head of a pigeon disturbed responding after discrimination of the individual pigeons (Watanabe and Ito 1991). Photographs have a strong relation to reality, but a cartoon does not. Watanabe (2001a) examined two types of objects, human or pigeon, and two types of medium, photograph or cartoon. Scrambling depressed responding to photos of pigeons the most, and that to cartoons of humans the least. An analysis of variance revealed that both object and medium have significant effects on the response rates. Therefore, the effects of scrambling depend on type of object and type of media. The suppression of response occurs more in real and in familiar objects.

Successful discrimination of paintings in pigeons might be based on their fine visual discrimination of features such as textures (Cook 1992). Greene (1983) also reported that pigeons could detect slight differences between landscape photographs taken successively. Watanabe (2010) trained pigeons to discriminate between watercolor paintings and pastel paintings of school children. The pigeons had to discriminate the paintings based on the difference in painting medium. They succeeded in learning this task. Moreover, in generalization tests with new paintings, the pigeons clearly discriminated watercolor from pastel. However, gray-scale test stimuli showed lower discrimination performance. Thus, color is an important cue for watercolor versus pastel discrimination. Because the mosaic processing of the paintings affected the discrimination, pattern cues also played a role in the discrimination.

7.3.4 Discrimination of "Beauty"

If a curator collects impressionists' paintings, the collection may contain Monet, Renoir, and so forth, but would not contain Picasso or Braque. There is a stylistic similarity among the paintings of impressionists. When another curator wants to collect "beautiful" paintings, the collection may contain not only Monet or Renoir but also Picasso, Braque, or Hokusai. There must be a rule or a collection of rules by which we decide which paintings are more aesthetically pleasing than others, but it is difficult to clarify them. Understanding the concept of beauty as the summation of discrimination of independent exemplars of beautiful pictures is theoretically possible, but difficult to examine empirically.

Criteria of "beauty" may depend on age, culture, and individual preferences. However, we may have a common sense of beauty for relatively less sophisticated paintings, such as those drawn by schoolchildren. There should be some common perceptual lower-level features in these "beautiful" paintings. If so, nonhuman animals probably could learn the human concept of "beauty" as a discriminative stimulus class based on perceptual similarity.

To examine this, the author used drawings made by children as stimuli in a set of experiments. To select good and bad paintings, evaluations of these paintings by an elementary school art teacher and by ordinary adults were used;that is, this was the basic subjective evaluation of naïve pictures by ordinary people rather than by professional artists. Pigeons were trained to discriminate ten "good" paintings from ten "bad" ones. The birds learned the discrimination in about 23 sessions. They then received a generalization test with novel paintings, and showed different levels of responding to these new paintings, again as judged "good" or "bad" by ordinary people. Thus, it would be fair to say that they had acquired the category of good versus bad. When the paintings were presented in gray-scale, good/bad discrimination was considerably disturbed, suggesting that the color was important for discrimination of beauty. When the paintings were presented after mosaic mixing, discrimination was disturbed depending on the level of processing, suggesting that the spatial pattern was important for discrimination of beauty. Thus, both color and spatial pattern play important roles in good/bad discrimination (Fig. 7.4).

7.3.5 *Strategy of Discrimination of Aesthetic Stimuli*

Sometimes both humans and birds show similar visual discrimination, but may use different cues. Humans and pigeons may use different strategies for discrimination even though these strategies result in similar discriminative performance. One difference should be local versus global cues. However, global and local are somewhat arbitrary concepts, and different researchers use the classification in different ways. For example, Aust and Huber (2001) considered that the global cues include general brightness and size. In this section, global cues refer to the configuration of elemental cues, with each local cue being one of the elements. Navon (1977) created hierarchical stimuli consisting of a larger letter configured from smaller component shapes (imagine a large H pattern consisting of small M characters, or a large M consisting of small H characters) and reported that the global cues are easier for humans to detect than are the local or elemental cues. By contrast, Cavoto and Cook

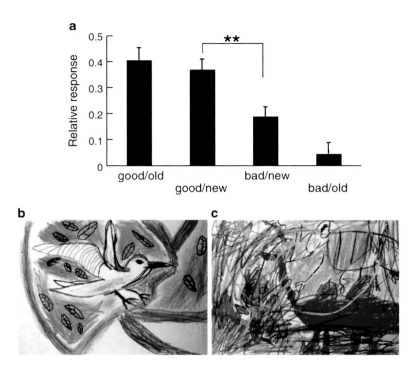

Fig. 7.4 Pigeons can discriminate "good" and "bad" paintings, and maintained their discrimination even for new paintings never shown during the discriminative training (from Watanabe 2010). (**a**) Generalization test. (**b**) Example of good painting. (**c**) Example of bad painting. **$P<0.05$

(2001) found that pigeons learned discrimination of a local-relevant cue discrimination faster than a global-relevant cue discrimination. This fact, that pigeons are adept at fine visual discrimination of small differences, also is consistent with research on how pigeons perceive stimuli by attending to local features of those stimuli (Greene 1983). On the other hand, using discrimination of artificial stimuli involving spatial location and visual features, Legge et al. (2009) reported that pigeons' preference for global or local cues differed from stimulus to stimulus. In some cases, pigeons used both global and local cues for discrimination of artificial two-dimensional objects (Lazareva et al. 2006b).

As mentioned earlier, both humans and pigeons effectively discriminate cartoons of people, but tests with unusual arrangements of the parts of the cartoons revealed differences in the discrimination behavior of the two species (Cerella 1980; Watanabe and Ito 1991). Pigeons used global arrangement for conspecific face discrimination, but did not attend to such features in human cartoon discrimination. Lazareva et al. (2006a) also reported that feature scrambling disturbed pigeons' discrimination of people and flowers, but not of cars and chairs. Scrambling tests after discrimination of simple line drawings revealed that pigeons attended to both local elements (such as geon) and their global arrangement (Kirkpatrick-Steger et al. 1996, 1998). Thus, stimulus control by elements or configuration depends on the nature of the stimuli used for the discrimination tasks. Pigeons are not, in general, local cue users.

Aust and Huber (2001) analyzed pigeons' discrimination of human images using scrambled pictures. The images were divided into small square elements, which were randomly distributed over the whole area of the image. Although scrambling at this extreme level resulted in a reduction of discrimination, pigeons' discrimination behavior was still robust. Matsusaka et al. (2004) trained pigeons on discrimination of black and white cartoon drawings and found decrement of responding depending on degree of scrambling. The decrement was steeper than that reported by Aust and Huber (2001). The training stimuli of the cartoon discrimination were different pictures of the same character, so that the same arrangement of the parts, particularly that of body parts of the characters, appeared on the training stimuli repeatedly. On the other hand, the training stimuli used in the present experiment by Aust and Huber had more variety. In addition to difference in color or black and white, common arrangement of the parts in the cartoon character enhanced stimulus control based on the arrangement. The author used a similar division and random rearrangement technique to analyze the role of elemental cues for painting discrimination in pigeons (Watanabe 2011). If pigeons attend only to local elemental cues, transfer from the original painting to a scrambled one, and from a scrambled painting to the original, would be the same. One group of pigeons was trained on discrimination between Japanese paintings and Western paintings and tested with their scrambled images, whereas the other group was trained on discrimination between scrambled Japanese paintings and scrambled Western paintings and then tested with the original paintings. Technically speaking, in the narrow sense Japanese paintings use colored glue, and the diameter of the molecules of the glue color is larger than the glue color used in other countries. More broadly, different features characterize Japanese paintings; in contrast to Western paintings, they are not representative like photographs, and they do not use shadows, outlines, or dark colors. Although Japanese paintings (Ukiyo-e) influenced the impressionists, it is easy to discriminate paintings by impressionists from Japanese paintings. If pigeons were to learn local elemental cues through the discriminative training of Japanese versus Western stimuli, then the pigeons should show bidirectional transfer: from the original to the scrambled painting and vice versa. In this experiment, pigeons learned the discrimination within 12–36 sessions, and they maintained the discrimination for novel Japanese and Western paintings in a generalization test. Moreover, as shown in the left graphs in Fig. 7.5, pigeons showed bidirectional transfer in the original-to-scrambled and scrambled-to-original tests outlined here. This suggests that the birds used local cues in both discrimination tasks.

The author examined similar bidirectional transfers after discrimination of the concept of "good" and "bad" paintings. The pigeons successfully learned the original good/bad discrimination and showed generalization to novel paintings, but could not maintain their discrimination when tested with the scrambled paintings. The pigeons also learned to discriminate scrambled good/bad paintings, but they did not discriminate the original paintings (right graph in Fig. 7.5). There was no transfer between the original and scrambled stimuli. This suggests that the pigeons used different cues for the two different discrimination tasks.

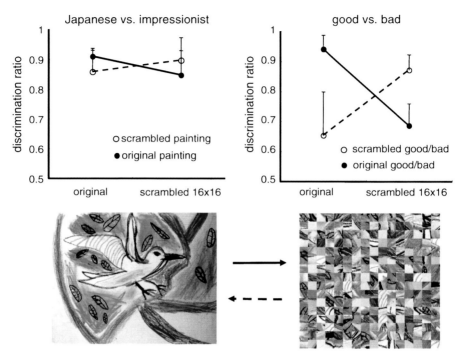

Fig. 7.5 Transfer between the original pictures to their scrambled images (Watanabe 2011). *Upper left*: discrimination of painting styles. *Upper right*: good versus bad discrimination. Pigeons use different strategies for different discriminations. *Lower*: examples of the original picture and the scrambled images

Recently, statistical aesthetics has developed a mathematical model to identify artists' works. This sparse-coding model is based on local features in a random square taken from paintings, in which a few elements can encode the square. Hughes et al. (2010) applied this sparse-coding analysis to distinguish drawings by Pieter Brugel the Elder from imitations. A similar computerized analysis was applied to identify Van Gogh paintings (Johnson et al. 2008). Thus, an analysis based on local features rather than configuration structure may, in an absolutely mechanical sense, provide enough information to distinguish artistic styles. Bidirectional transfer between the original paintings and their scrambled images in pigeons suggests that pigeons might use similar processing to discriminate Japanese from Western paintings. On the other hand, the sparse-coding model has never been applied to the detection of "good" paintings. If the neural mechanism to detect beautiful paintings is the sparse synchronous activation of neurons in the visual system, then the configured features must be the crucial aspects used to pick out the beautiful paintings. The present results from pigeons are consistent with results of these mathematical models.

7.3.6 Conclusion

Experimental works suggest that discrimination of painting styles is not an ability unique to humans. It shows that such discrimination can be (or usually is) based on rather lower-level, bottom-up processing. The pigeons probably use a different strategy for each different discrimination task, and each strategy may differ from the one used by humans (even though both species accomplished the discrimination task). On the other hand, the stimulus property of "beauty" has not been well examined. More research with animals and humans is needed to draw any conclusion, but it is plausible to assume that there is some common type of "beauty" that is widespread across some range of vertebrates. Neural activity evoked by particular types of visual stimuli may be similar among these species. Phylogenetic contingency shaped such preference, and comparative study of beauty will clarify the ultimate cause of sense of beauty.

7.4 Creation of Art

Welsch (2004) described two stages in the evolution of aesthetic phenomena in nature. The first stage is a nonaesthetic type of beauty observed in the forms and colors of lower animals such as corals or sea anemones. This kind of "beauty" is the result of a chemical process (e.g., color) or the minute structure of tissue (e.g., texture). Even if these shapes look beautiful to human eyes, the beauty is a by-product of nonaesthetic processes. The second stage is the proto-aesthetic stage observed in the forms and colors of flowers and fruits: They have to attract animals (for example, insects) for pollination. Such second-stage aesthetics function so as to send signals to other species.

However, neither type of aesthetic pertains to creating a "work" of art outside of the organism itself. An essential aspect of human aesthetics is the creation of art, and creating art has been considered to be an activity unique to humans. Human drawings have been documented since the cave drawings of our ancestors. The oldest musical instruments were flutes made of bones that are 32,000 years old (Balter 2004), although there could be even older ones made of perishable materials and *Homo sapiens* might sing their song earlier than this period. We not only sing songs, but create music with instruments. Such behavior is rare in animals, but one observation is drumming by a stick in palm cockatoos (Wood 1984, 1987), which make a drumstick, bring it to a tree hollow, and drum. The particular sound can be heard 100 m away. It is still unknown whether they use this particular sound for sexual display or territorial defense, or for some other purpose. This case of tool making is one that is not used for foraging.

To examine the creation of art, we can separate three aspects. The first is motor skills to produce art. Particularly special motor skills are required to make representational art. Most animal art is mere scratching, even if some of it looks like modern human art. "Animal scratching like" modern art has, however, history of realism in the past. The realisms or imitation of real objects on canvas is an essential prerequisite of aesthetics. Do animals draw something that represents an object apart from this? The second is functional autonomy. The evolutionary origin of human art could be an adaptation to the environment or be due to sexual selection, but at present, artists create art for art's sake. Reinforcement is the art itself; in other words, art creation is maintained by self-reinforcement. Do animals make art for art's sake? If their paintings result from training with other rewards, their activity is different from ours even if the animal could produce human-like paintings. Finally, art must have reinforcing value for the artist and other persons. Nonprofessional artists often keep their products and love to see or show the products to friends. Do animals enjoy their products in such a way also? The second point is a representative or symbolic feature.

7.4.1 Motor Skills

Discriminative and reinforcing properties are prerequisites of art creation, but motor skills are also required. Although there have been some studies of the drawing ability of non-human animals (see Zeller 2007), most animal paintings are made for exhibition (e.g., in the gift shop of a zoo) rather than to investigate a topic of scientific interest. From a Russian web site (www.animalart.ru), one can view paintings created by several different species: chimpanzees, orangutans, elephants, dolphins, whales, ravens, dogs, horses, and pigs. A beautiful art book of cat paintings has also been published (Bush and Silver 1995), but the purpose of that book is to present art-like performance by cats from the viewpoint of human artists, and not the scientific analysis of cat behavior.

Some species, particularly some birds, show fine motor skills that enable them to make complex nests and to produce complicated songs. The question is whether these animals can apply such motor skills to produce artistic products for the sake of art.

7.4.2 Does Animal Art Symbolize Something Outside?

The author bought one painting by an elephant through an Internet auction: a picture of flowers (see Fig. 7.6). Friends easily understand the painting, and nobody notices that it was made by an elephant. Later, the author found quite similar elephant paintings on YouTube (one can also visit www.elephantart.com). Not enough information is available to decide whether the elephants were trained to produce such art, but training could have an important role because there are several fixed patterns of painting.

Fig. 7.6 Painting by an elephant

Painting constitutes mapping or transcription of three-dimensional objects onto a two-dimensional canvas (Watanabe, 2011). There is a correspondence between objects in a scene and their representation in a painting. Some paintings by elephants look like representational paintings: flowers, other elephants, and so forth. The difficulty lies in deciding whether these paintings are really transcriptions of external objects or repertoires of painting behavior induced by different stimuli. Through training, animals may learn to draw something when object A is presented while drawing a different something when object B is presented. In this scenario, different drawings result from two repertoires controlled by two different discriminative stimuli (A and B). This type of behavior is called conditional discrimination, having an "IF-THEN" logical structure. To constitute representational transcription, on the other hand, we have to have a general correspondence between many objects and many paintings.

Sometimes animals indicate behaviors that are based on general rules. One interesting observation concerned the spontaneous imitation of the sound of trucks by elephants (Poole et al. 2005). This suggests that elephants have the ability to transfer sensory information to motor activity. There is no particular reward to maintain such mimicking behavior; according to the authors, the function of the mimicry is unknown. Dolphins also draw paintings with a brush. Levy (1992) reported dolphins that drew shapes (circle or T, etc.) that had also been drawn by a trainer on a different canvas; this demonstrates that they could transcribe the drawing. Dolphins also have an advanced ability to mimic human movement without any particular reward (Herman 2002); thus, mimic drawing could be explained in terms of their more general mimicking ability. Representational painting is a kind of matching-to-sample behavior in which the sample is an external object and the choice to paint is a product of motor skills. Animals with well-developed motor skills and cognitive abilities may be trained in such a task, but if there is no functional autonomy, their seemingly human-art-like transcription behavior differs significantly from human art.

As already mentioned, there have been many studies of paintings by chimpanzees, but no report of representational painting. Saito et al. (2010) compared chimpanzees' painting with those of human infants, and conclude that that there is a lack of imitation of model drawing in chimpanzees. They observed some marking on the models but no completion of incomplete facial outlines by chimpanzees. The chimpanzees were not trained to draw in this experiment, so it is unknown whether they can paint, as do elephants, by behavioral shaping.

A direct answer to the question of whether a drawing is representational would be to ask animals about their intention. We often ask children, "What is this?" or "What you want to draw?" during their painting behavior. Moja, a chimpanzee taught sign language, named one of her drawings as "bird" (described in Gucwa and Ehmann 1985). Unfortunately, such a communicative approach using signing in a face-to-face situation might be contaminated with unconscious signaling by the experimenter. Thus, the best way to demonstrate representational painting in animals is to demonstrate general drawing behavior of the matching-to-sample type when the sample is a novel object. Unfortunately, we do not have enough data to address this question. In conclusion, we do not have evidence of representational painting in animals.

Another symbolization of human art is the symbolization of inner states, or the desires and needs of the painters. For example, ancient native paintings are considered to represent the desires of the artists, such as the desire for a good hunt. Scratches made by animals may represent some aspects of their innate state, such as calmness, or aggression, but not desires such as "I want good food." Moreover, other conspecifics do not understand the messages in art. In other words, animal art does not have a narrative function.

7.4.3 *Functional Autonomy of Animal Art*

Chimpanzee is a species well known as a drawing animal (see, e.g., Lenain 1997). Morris (1962) documented 32 cases of drawings and paintings by primates. The oldest report of chimpanzee painting is *The ape and the child* by Kellogg and Kellogg (1933) or *Infant ape and human child* by Kohts (1935). It is interesting that both of these early researches concern comparison of ape with child. Chimpanzees spontaneously painted if they were given pen or brush and maintained their painting behavior without food reinforcement (Boysen et al. 1987; Tanaka et al. 2003). Schiller (1951) wrote an episode that Alpha (his chimpanzee) once tried to draw on a leaf when suitable paper was not available. Such observation, even though episodic, supports self-reinforcement in chimpanzee drawing. In addition to sensory feedback of motor movement, visual feedback might have reinforcing value, because drawing behavior on a touch-screen decreased when no trace of the drawing appeared on the screen (Tanaka et al. 2003). Morris (1962) stated that food reward disturbed, not enhanced, painting behavior of chimpanzees; however, it was also possible to train chimpanzees to trace model lines on a touch-screen by food reinforcement (Iversen and Matsuzawa 1997). Most of the painting experiments with primates are carried out in a face-to-face situation, therefore there could be social reinforcement. According to Morris (1962), Congo (a chimpanzee) had a criterion of "completion" of drawing, and it was hard to let him continue to draw after his completion. Congo seems to have had the goal of drawing but the goal may have been lack of space to draw or other physical limitations rather than "aesthetic goal." The primates did not enjoy their products, because they often tore papers after drawing, suggesting no reinforcing value to them of their products. It is still unknown whether the primates paint to create aesthetic product that has any reinforcing property.

The oldest book on elephant paintings was written by Gucwa and Ehmann (1985). In their account Siri, an Asian elephant, drew a lot of pictures, even though her trainer never trained her to do so nor rewarded her for such behavior. Elephants have good motor skills in using sticks or stones, and captive ones often scratch the floor or ground with these materials. That such behavior occurs spontaneously without training suggests functional autonomy or self-reinforcement. Captive animals usually have a limited environment and often invent new behaviors. Thus, drawing might be a by-product of captivity.

Another interesting example of animal art was observed in a dancing parrot named Snowball (Patel et al. 2009). Complex dance functions as sexual display in many different species ranging from insects to fish, birds, and mammals. But in Snowball's case, the dance did not function as innate sexual display; the parrot was dancing to human music. Patel et al. (2009) experimentally analyzed this parrot's synchronization of its body movements to music. When the tempo changed, the bird spontaneously adjusted its rhythmic movement to fit the new rhythm. Such adjustment is observed in animals that have complex vocal learning skills, such as songbirds, cetaceans, and pinnpeds. According to Patel et al., the owner of Snowball obtained him at a bird show, and his exact history of music is unknown. However, he started rhythmic bobbing movement to music soon after the owner obtained him. As far as understood, there was no special reward for him to maintain this dancing behavior. Bonobos and chimpanzees do not seem to have such synchronization ability (Kugler and Savage-Rumbaugh 2002). Thus, the ability to synchronize body movements to music seems to have evolved in several evolutionary lines independently.

The animal arts described here are human art-like behaviors of animals that take place in artificial settings. However, some animals show art-like performance in nature. The best-documented example is bird song. The function of bird song is sexual display. Thus, male birds need not continue singing after a mate has been secured. Kaplan (2009) described blackbirds and willow warblers that develop their song long after mating, and suggested that they sang for their own sake. Such improvement of the song may, however, have benefit in the next mating reason or serve a function other than sexual display. Improvement of song during acquisition could be maintained by self-reinforcement. Young birds start singing subsong, then gradually shape it into final crystallized song by matching it to a template that they heard in the sensory phase of song learning. This process is not maintained by sexual reward, but rather by the matching itself.

Another example is animal architecture. Von Frisch presented many beautiful products of animals in his interesting book, *Tiere als Baumeister* (*Animal as Architect*) (1974). For example, male bowerbirds constructed complex bowers and decorated them with many colorful materials to attract females (Madden 2008). A complex bower could be an honest signal of good sensorimotor skills, and has often been used as an example of art evolved by sexual selection (Miller 2000). Male bowerbirds arrange and rearrange decorations in their bowers to satisfy their criteria (Rogers and Kaplan 2006), but there is no direct evidence of functional autonomy in building a beautiful bower after mating. Again, functional autonomy in animal architecture is an open question.

Perhaps nonhuman animals have functional autonomy or self-reinforcement in their art-like behavior, but the reinforcing value of their products for conspecifics is not clear except for the art-like behavior as sexual display.

7.4.4 Do Animals Enjoy their Product?

Humans not only create artistic products for art but also enjoy them. The art product has a reinforcing property for conspecifics. Some animal beauty is an honest signal that has reinforcing properties for conspecifics but the reinforcing property of animal art for animals is doubtful, even if the animals create something for its own sake. Here, animal art is defined as human-like art (for example animal paintings) made by animals. Chimpanzees often tear up their paintings after completing them, and do not keep them and enjoy them. This is the big difference between animals' creative art behavior and human art. As pointed out earlier, human art is socially constructed. Society should have a consensus about art. This consensus gives artistic products a value within that society, hence the products have reinforcing properties for members of the society.

7.5 Conclusion

Figure 7.7 illustrates relationships among the aspects of animal aesthetics discussed in this chapter. Animals have to discriminate aesthetic stimuli, that is, the cognitive aspect of aesthetics. The aesthetic stimuli also have to have reinforcing property, that is, the aspect of pleasure of aesthetic stimuli. The reinforcing property has not been well examined in visual stimuli, but it is plausible to assume that different

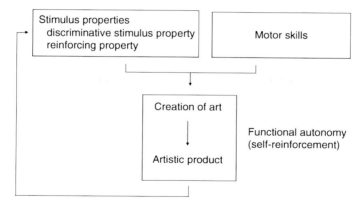

Fig. 7.7 Stimulus must have discriminative and reinforcing properties and subject has to have motor skills to create art. Functional autonomy or self-reinforcement is crucial for the creation of art

complex stimuli have different values of sensory reinforcement. In the case of auditory stimulus, songbirds, at least, are sensitive to the reinforcing property of music. The reinforcing property must be examined with careful arrangement of the stimuli, to adjust to species-specific sensory abilities.

To create art, motor skills are prerequisite as stimulus property. Furthermore, the motor skills must be maintained by self-reinforcement or functional autonomy. This aspect is still unclear in animals. However, their "creating" behavior in nature suggests the possibility of art creation maintained by self-reinforcement in animals. Finally, the art products must have reinforcing value for others. If the motor skills to create art-like product is maintained by self-reinforcement, sensory feedback from the product may have some role. But the products after the creation probably have no value for animals that created the products nor for other conspecifics. Such research suggests that we can find a precursor of aesthetics in nonhuman animals, but our species has the most advanced aesthetics in the animal kingdom particularly in aspects of creation and the reinforcing value of the art products.

Finally, an issue skipped in this chapter is the relationship between art and religion. Of course no animals, with the exception of *Homo sapiens*, have a religion. Thus, it is impossible to discuss this topic in the context of animal aesthetics. Human art has not only a biological basis but also a cultural basis. Some aspects of human art cannot be understood without understanding our cultural evolution. Finding the phylogenetic origin of unique human behavior is important for understanding human beings. However, in order to fully understand humans it is also important to single out the emergence of unique human behaviors.

References

Allen G (1877) Physiological aesthetics. Henry S King & Co, London
Amoser S, Ladich F (2005) Are hearing sensitivities of freshwater fish adapted to the ambient noise in their habitats? J Exp Biol 208:3533–3542

Anderson JR, Kuwaharam H, Kuroshimma H, Leighty KA, Fujita K (2005) Are monkeys aesthetists? Rensch (1957) revisited. J Exp Psychol Anim Behav Process 31:71–78

Aust U, Huber L (2001) The role of item- and category specific information in the discrimination of people versus nonpeople images by pigeons. Anim Learn Behav 29:107–119

Averbeck BB, Lee D (2004) Coding and transmission of information by neural ensembles. Trends Neurosci 27:225–230

Balter M (2004) Seeking the key to music. Science 306:1120–1122

Berlyne DE (1969) The reward value of light increment under supranormal and subnormal arousal. Can J Psychol 23:11–23

Berlyne DE (1976) Aesthetics and psychobiology. Appleton, New York

Bichot NP, Rossi AF, Desimone R (2005) Parallel and serial neural mechanisms for visual search in macaque area V4. Science 308:529–534

Boysen ST, Berntson GG, Prentice J (1987) Simian scribbles: a reappraisal of drawing in the chimpanzee (*Pan troglodytes*). J Comp Psychol 101:82–89

Brooks DI, Cook RG (2010) Chord discrimination by pigeons. Music Percept 27:183–196

Buchholz R (1995) Female choice, parasites load and male ornamentation in wild turkeys. Anim Behav 50:929–943

Bush H, Silver B (1995) Why cats paint. Benedikt taschen Verlag GmbH, Koeln

Butler RA (1953) Discrimination by rhesus monkeys to visual exploration motivation. J Comp Physiol Psychol 50:177–179

Cavoto KK, Cook RG (2001) Cognitive precedence for local information in hierarchical stimulus processing by pigeons. J Exp Psychol Anim Behav Process 27:3–16

Cela-Conde CJ, Marty G, Maestu F, Ortiz T, Munar E, Fernandez A, Roca M, Rossello J, Quesney F (2004) Sex-related similarities and differences in the neural correlates of beauty. Proc Natl Acad Sci U S A 101:6321–6325

Cerella J (1980) The pigeon's analysis of picture. Pattern Recognit 12:1–6

Chase AR (2001) Music discrimination by carp (*Cyprinus carpio*). Anim Learn Behav 29:336–353

Cook RG (1992) The visual perception and processing of texture by pigeons. In: Honig WK, Fetterman JG (eds) Cognitive aspects of stimulus control. Lawrence Erlbaum Associates, Hillsdale, pp 279–299

Cynx J (1993) Conspecific song perception in zebra finches (*Taeniopygia guttata*). J Comp Psychol 107:395–402

D'Amato MR, Salmon DP (1982) Tune discrimination in monkeys (*Cebus apella*) and in rats. Anim Learn Behav 10:126–134

Delius JD, Habers G (1978) Symmetry: can pigeons conceptualize it? Behav Neural Biol 22:336–342

Dobson CW, Petrinovich L (1973) Song as reinforcer in the white-crowned sparrow. Behav Biol 9:719–729

Douchet SM, Montogomerie R (2003) Multiple sexual ornaments in satin bowerbirds; ultraviolet plumage and bowers signal different aspects of male quality. Behav Ecol 14:503–509

Driscoll C (2006) The bowerbirds and the bees: Miller on art, altruism, and sexual selection. Philos Psychol 19:507–526

Eibl-Eibesfeldt I (1988) The biological foundation of aesthetics. In: Rentschler I, Herzberger B, Epstein D (eds) Beauty and the brain. Biological aspects of aesthetics. Birkhaeuser, Basel, pp 29–68

Fay RR (1970) Auditory frequency generalization in the goldfish. J Exp Anal Behav 14:353–360

Fay RR (1988) Hearing in vertebrates: a psychophysics data book. Hill-Fay Associate, Chicago

Fay RR (1992) Analytic listening in goldfish. Hear Res 59:101–107

Fay RR (1995) Perception of spectrally and temporally complex sounds by the goldfish. Hear Res 89:146–154

Fay RR (1998) Auditory stream segregation in goldfish (*Carassius auratus*). Hear Res 120:69–76

Fishman Y, Volkov IO, Noh MD, Garell PC, Bakken HAJ, Howard MA, Steinschneider M (2001) Consonance and dissonance of musical chord: neural correlates in auditory cortex of monkeys and humans. J Neurophysiol 86:2761–2788

Fujita K, Watanabe K (1995) Visual preference for closely related species by Sulawesi macaques. Am J Primatol 37:253–261

Gess A (2007) Birds like music, too. Science 317:1864

Grammer K, Fink B, Moller AP, Thornhill R (2003) Darwinian aesthetics: sexual selection and biology of beauty. Biol Rev 78:385–407

Gray PM, Krause B, Atema J, Payne R, Krumhansl C, Baptista L (2001) The music of nature and the nature of music. Science 291:50–54

Greene SL (1983) Feature memorization in pigeon concept formation. In: Commons ML, Herrnstein RJ, Wagner AR (eds) Quantitative analysis of behavior: discriminative processes. Ballinger Publishing Company, Cambridge, pp 209–230

Gucwa D, Ehmann J (1985) To whom it may concern. An investigation of the art of elephants. W.W. Norton & Company, New York

Hagmann CE, Cook RG (2010) Testing meter, rhythm, and tempo discriminations in pigeons. Behav Process 85:99–110

Hahnloser RHR, Kozhevnikov AA, Fee MS (2002) An ultra-sparse code underlies the generation of neural sequences in a songbird. Nature 419:65–70

Hauser MD, McDermott H (2003) The evolution of the music faculty: a comparative perspective. Nat Neurosci 6:663–668

Herman LM (2002) Vocal, social, and self-imitation by bottlenosed dolphins. In: Dautenhahn K, Nehaniv CL (eds) Imitation in animals and artifacts. MIT Press, Cambridge, pp 63–108

Herrnstein RJ, Loveland DH (1964) Complex visual concept in the pigeon. Science 146:549–551

Herrnstein RJ, Vaughan W Jr, Mumford DB, Kosslyn SM (1989) Teaching pigeons on abstract rule: insideness. Percept Psychophys 46:56–64

Holy TE, Guo Z (2005) Ultrasonic songs of male mice. PLoS Biol 3:2177–2186

Howell S, Schwandt M, Fritz J, Roeder E, Nelson C (2003) A stereo music system as environment enrichment for captive chimpanzees. Lab Anim 32:31–36

Hughes JM, Graham DJ, Rockmore DN (2010) Quantification of artistic style through sparse coding analysis in the drawings of Pieter Brugel the Elder. Proc Natl Acad Sci U S A 107:1279–1283

Hulse S, Kline C (1993) The perception of time relations in auditory tempo discrimination. Anim Learn Behav 21:281–288

Hulse SH, Humpal J, Cynx J (1984) Discrimination and generalization of rhythmic and arrhythmic sound patterns by European starlings (*Sturnus vulgaris*). Music Percept 1:442–464

Hulse SH, Bernard DJ, Braaten RF (1995) Auditory discrimination of chord-based spectral structures by European starlings (*Sturnus vulgaris*). J Exp Psychol Gen 124:409–423

Humphrey NK (1972) 'Interest' and 'pleasure': two determinants of monkey's visual preference. Perception 1:395–416

Ikkatai Y, Watanabe S (2010) Discriminative and reinforcing properties of paintings in Java sparrows (*Padda oryzivora*). Anim Cogn 14:227–234

Iversen I, Matsuzawa T (1997) Model-guided line drawing in the chimpanzee (*Pan groglodytes*). Jpn Psychol Res 39:154–181

Johnson CH, Hendriks E, Bereznhoy IJ, Brevdo E, Hughes SM, Daubechies I, Li J, Postma W, Wang JZ (2008) Image processing for artist identification. IEEE Signal Process Mag 25:37–48

Kaplan G (2009) Animals and music: between cultural definitions and sensory evidence. Sign Syst Stud 37:75–101

Kawabata H, Zeki S (2004) Neural correlates of beauty. J Neurophysiol 91:1699–1705

Kellogg WN, Kellogg LA (1933) The ape and child. McGrraw Hill, New York

Kenyon TN, Ladich F, Yan HY (1998) A comparative study of hearing ability in fishes: the auditory brainstem response approach. J Comp Physiol A 182:307–318

Kirkpatrick-Steger K, Wasserman EA, Biederman I (1996) Effects of spatial rearrangement of object components on picture recognition in pigeons. J Exp Anal Behav 65:465–475

Kirkpatrick-Steger K, Biederman I, Wasserman EA (1998) Effects of geon depletion, scrambling, and movement on picture recognition in pigeons. J Exp Psychol Anim Behav Process 24:34–46

Kohts N (1935) Infant ape and human child. Scientific memories of the Museum Darwininun, Moscow, cited in Morris (1962)

Kugler K, Savage-Rumbaugh S (2002) Rhythmic drumming by Kanzi an adult male bobobo (*Pan paniscus*) at the language Research Center. In: 25th meeting of the American Society of Primateologists

Kusmierski RG, Borgia G, Uy A, Crozier RH (1997) Labile evolution of display traits in bowerbirds indicates reduced effects of phylogenetic constraints. Proc R Soc Lond 264:307–313

Ladich F (1999) Did auditory sensitivity and vocalization evolve independently in otophysan fishes? Brain Behav Evol 53:288–304

Lamont AM (2005) What do monkeys' music choices mean? Trends Cogn Sci 9:359–361

Latto R (1995) The brain of the beholder. In: Gregory R, Harris J, Heard P, Rose D (eds) The artful eye. Oxford University Press, Oxford, pp 66–94

Lazareva OF, Freiburger KL, Wasserman EA (2006a) Effects of stimulus manipulation on visual categorization in pigeons. Behav Process 72:224–233

Lazareva OF, Vecera SP, Wasserman EA (2006b) Object discrimination in pigeons: effects of local and global cues. Vision Res 46:1361–1374

Legge ELG, Spetch ML, Batty ER (2009) Pigeons' (*Columba livia*) hierarchical organization of local and global cues in touch screen tasks. Behav Process 80:128–139

Lenain T (1997) Monkey painting. Reaktion Books, London

Levy BA (1992) Psychoaesthetics dolphin project. J Am Art Ther Assoc 9:193–197

Lewis-Williams D (2002) The mind in the cave: consciousness and the origin of arts. Thames & Hudson, London

Madden JR (2008) Do bowerbirds exhibit culture? Anim Cogn 11:1–12

Marler P, Slabbekoorn H (eds) (2004) Nature's music. Elsevier, San Diego

Matsusaka A, Inoue S, Jitsumori M (2004) Pigeon's recognition of cartoon: effects of fragmentation, scrambling, and deletion of elements. Behav Process 65:25–34

McAdie TM, Foster TM, Temple W, Matthews LR (1993) A method for measuring the aversiveness of sounds to domestic hens. Appl Anim Behav Sci 37:223–238

McDermott J, Hauser MD (2004) Are consonant intervals music to their ear? Spontaneous acoustic preferences in non-human primates. Cognition 94:B11–B21

McDermott J, Hauser MD (2007) Nonhuman primates prefer slow tempos but dislike music overall. Cognition 104:654–668

Mercado E, Herman LM, Pack AA (2005) Song copying by humpback whales: themes and variations. Anim Cogn 8:93–102

Miller GF (2000) The mating mind. How sexual selection choice shaped the evolution of human nature. Bull Psychol Art 2:20–25

Morris D (1962) The biology of art. Methuen, London

Nadel M, Munar E, Capo MA, Rosello J, Cela-Conde CJ (2008) Towards a framework for the study of the neural correlates of aesthetic preference. Spat Vis 21:379–396

Navon D (1977) Forest before trees: the precedence of global features in visual perception. Cogn Psychol 9:353–383

Okaichi Y, Okaichi H (2001) Music discrimination by rats. Jap J Anim Psychol 51:29–34

Okanoya K, Dooling RJ (1991) Perception of distance calls by budgerigars (*Melopsittacus undulates*) and zebra finches (*Poephila guttata*). J Comp Psychol 105:60–72

Otsuka Y, Yanagi J, Watanabe S (2009) Discrimination and reinforcing stimulus properties of music for rats. Behav Process 80:121–127

Patel A, Iversen J, Bregman M, Schulz I (2009) Experimental evidence for synchronization to a musical beat in a nonhuman animal. Curr Biol 19:827–830

Petel A (2006) Musical rhythms, linguistic rhythm and human evolution. Music Percept 11:409–464

Pinker S (1997) How the mind work. Allen Lane Penguin Press, London

Poli M, Previde EP (1991) Discrimination of musical stimuli by rats (*Rattus norvegicus*). Int J Comp Psychol 5:7–18

Poole J, Tyack P, Stoeger-Horwatch A, Watwood S (2005) Elephants are capable of vocal learning. Nature 434:455–456

Porter D, Neuringer A (1984) Musical discrimination by pigeons. J Exp Psychol Anim Behav Process 10:138–148

Ramachandran VS, Hirstein W (1999) The science of art: a neurological theory of aesthetic experience. J Conscious Stud 6:15–51
Redies C (2007) A universal model of esthetic perception based on the sensory coding of natural stimuli. Spat Vis 21:97–117
Rensch B (1957) Asthetische Faktoren bei Farb- und Formbevorzugungen von Affen. Z Tierpsychol 14:71–99
Rensch B (1958) Die Wirksomkeit aesthetischer Factoren bei Wirbeltieren. Z Tierpsychol 15:447–461
Rogers LJl, Kaplan G (2006) Elephants that paint, birds that make music: do animals have an aesthetic sense? In: Read CA (ed) Cerebrum. Dana Press, New York, pp 1–14
Saito A, Hayashi M, Takeshita H, Matsuzawa T (2010) Drawing behavior of chimpanzees ad human children: the origin of the representational drawing. In: Proceedings of the 3rd international workshop on Kansei, pp 111–114
Schiller P (1951) Figural preferences in the drawings of a chimpanzee. J Comp Physiol Psychol 44:101–111
Schwartz GG, Rosenblum LA (1980) Novelty, arousal, and nasal marking in the squirrel monkey. Behav Neural Biol 28:116–122
Shinozuka K, Watanabe S (unpublished) Reinforcing and discriminative stimulus properties of music in goldfish (submitted for publication)
Stevenson JG (1969) Song as a reinforcer. In: Hind RA (ed) Bird vocalization. Cambridge University Press, Cambridge, pp 49–60
Stoddard PK, Becher MD, Losesche P, Campbell SE (1992) Memory does not constrain individual recognition in a bird with song repertoires. Behaviour 122:274–287
Sugimoto T, Kobayashi H, Noritomo N, Kiriyama Y, Takeshita H, Nakamura T, Hashiba K (2010) Preference for consonance music over dissonance music by an infant chimpanzee. Primates 51:7–12
Sutoo D, Akiyama K (2004) Music improves dopaminergic neurotransmission: demonstration based on the effect of music on blood pressure regulation. Brain Res 1016:255–262
Tanaka M, Tomonaga M, Matsuzawa T (2003) Finger drawing by infant chimpanzees (*Pan troglodytes*). Anim Cogn 6:245–251
ten Cate C (1991) Behavior-contingent exposure to taped song and zebra finch song learning. Anim Behav 42:857–859
Thornhill R (1998) Darwinian aesthetics. In: Crawford C, Krebs DL (eds) Handbook of evolutionary psychology. Erlbaum, Mahwah, pp 543–572
Uetake K, Hurnik JF, Johnson L (1997) Effect of music on voluntary approach of daily cows to an automatic milking system. Appl Anim Behav Sci 53:175–182
Voland E, Grammer K (eds) (2003) Evolutionary aesthetics. Springer, Berlin
von Fechner GT (1876) Vorschule der Aesthetik. Breitkopf & Hartel, Leipzig
von Frisch K (1974) Tieres als Baumeister. Ullstein, Frankfurt
von Reinhert J (1957) Akustische Dressurversuche an einem Indischen Elefanten. Z Tierpsychol 14:100–126
Wasserman EA, Kiedinger RE, Bhatt RS (1988) Conceptual behavior in pigeons: categories, subcategories, and pseudo categories. J Exp Psychol Anim Behav Process 14:235–246
Watanabe S (1988) Failure of visual prototype learning in the pigeon. Anim Learn Behav 16:147–152
Watanabe S (1991) Effects of ectostriatal lesions on natural concept, pseudoconcept and artificial pattern discrimination in pigeons. Vis Neurosci 6:497–506
Watanabe S (2001a) Discrimination of cartoon and photographs in pigeons: effects of scrambling of elements. Behav Process 53:3–9
Watanabe S (2001b) Van Gogh, Chagall and pigeons. Anim Cogn 4:147–151
Watanabe S (2002) Preference for mirror images and video image in Java sparrows (*Padda oryzivora*). Behav Process 60:35–39
Watanabe S (2010) Pigeons can discriminate "good" and "bad" paintings by children. Anim Cogn 13:75–85

Watanabe S (2011) Discrimination of painting style and beauty: pigeons use different strategies for different tasks. Anim Cogn 14(6):797–808

Watanabe S, Ito Y (1991) Individual recognition in pigeon. Bird Behav 36:20–29

Watanabe S, Nemoto M (1998) Reinforcing property of music in Java sparrows (*Padda oryzivora*). Behav Process 43:211–218

Watanabe S, Sato K (1999) Discriminative stimulus properties of music in Java sparrows. Behav Process 47:53–58

Watanabe S, Wakita M, Sakamoto J (1995) Discrimination of Monet and Picasso in pigeons. J Exp Anal Behav 63:165–174

Watanabe S, Uozumi M, Tanaka K (2005) Discrimination of consonance and dissonance in Java sparrows. Behav Process 70:203–208

Watanabe S, Suzuki T, Yamazaki Y (2009) Reinforcing property of music for non-human animals: analysis with pigeons. Philosophy 121:1–21

Weary DM (1989) Categorical perception of bird song: how do great tits (*Parus major*) perceive temporal variation in their song? J Comp Psychol 103:320–325

Wells DH, Irwin RM (2008) Auditory stimulation as enrichment for zoo-housed Asian elephants (*Elephantus maximus*). Anim Welf 17:335–340

Wells DL, Graham L, Hepper PG (2002) The influence of auditory stimulation on the behavior of dogs housed in a rescue shelter. Anim Welf 11:385–393

Wells DL, Coleman D, Challis MG (2006) A note on the effect of auditory stimulation on the behavior and welfare of zoo-housed gorilla. Appl Anim Behav Sci 100:327–332

Welsch W (2004) Animal aesthetics. Contemp Aesthet:2

Wilson EO (1983) Biophilia. Harvard University Press, Cambridge

Wilson FA, Goldman-Rakic PS (1994) Viewing preferences of rhesus monkeys related to memory for complex pictures, colours and faces. Behav Brain Res 60:79–89

Wood GA (1984) Tool use by the palm cockatoo Probosciger aterrimus during display. Corella 8:94–95

Wood GA (1987) Further field observations of the palm cockatoo Probosciger aterrrimus in the Cape York peninsula, Queensland. Corella 12:48–52

Wright AA, Rivera JJ, Hulse SH, Shyan M, Neiworth JJ (2000) Music perception and octave generalization in rhesus monkeys. J Exp Psychol Gen 129:291–307

Zeki S (1999) Inner vision: an exploration of art and the brain. Oxford University Press, London

Zeller A (2007) "What's in a picture?" A comparison of drawings by ape and children. Semiotica 166:181–214

Zentner MR, Kagan J (1998) Infants' perception of consonance and dissonance in music. Infant Behav Dev 21:483–492

Part II
Emotion in Humans

Chapter 8
The Unique Human Capacity for Emotional Awareness: Psychological, Neuroanatomical, Comparative and Evolutionary Perspectives

Horst Dieter Steklis and Richard D. Lane

Abstract We propose that the ability to be consciously aware of one's own and others' emotions is a unique human capacity. Emotion may be divided into implicit (visceromotor and somatomotor) and explicit (conscious feeling and reflective awareness) components. Based on a brain model of implicit and explicit emotional

---

H.D. Steklis (✉)
Departments of Psychology and Family Studies and Human Development,
University of Arizona, Tucson, AZ 85721, USA
e-mail: steklis@email.arizona.edu

R.D. Lane
Departments of Psychiatry, Psychology and Neuroscience, University of Arizona,
1501 N. Campbell Ave., Tucson, AZ 85724-5002, USA
e-mail: lane@email.arizona.edu

processing and an evolutionary perspective on the frontal lobe and human cognition, it is proposed that human emotion has a visceromotor and somatomotor foundation that is likely shared with other mammals. What may be unique to human cognition is the ability to engage in shared intensions and collaborative activities, which includes a species-unique motivation to share emotions, experiences and activities with other persons. The capacity for shared emotional experiences (knowing that you and the other person are experiencing the same feelings) likely requires mediation by a mentalizing network that includes the medial prefrontal cortex. We review evidence that humans appear to be unique in being able to reflect upon their own emotions, to generate a complex range of differentiated experiences, to appreciate complexity in the emotional experiences of other people, and to intentionally and knowingly share emotional experiences with other people. However, individual differences in this capacity among people are considerable. Although emotional awareness promotes self-regulation and adapative social behavior, impairments in this function can lead to social isolation, mental illness and adverse physical health consequences. We propose that the unique human ability to be emotionally aware requires mediation by the medial prefrontal cortex (BA10), a structure that is far more developed in humans than in other species, but also requires advantageous ontogenetic experiences. Such experiences include a particular type of mentalizing (accurate cognitive empathy) by parents and other caretakers, mediated in part by this same medial prefrontal cortical structure, in order for the full function of this capacity to be realized.

8.1 Introduction

On January 8, 2011 the nation and the world were stunned by a tragic shooting incident in Tucson, Arizona. While holding a public meeting with her constituents in front of a grocery store, Congresswoman Gabriele Giffords was shot in the head at close range by a 22 year-old man. Six people including a federal judge and a 9 year-old girl were killed, and 13 others were wounded by the gunman. The city, the nation and the world were shaken and horrified. Within days the President of the United States came to Tucson to visit the surviving victims and their families, and gave a soothing and uplifting televised speech that did much to help people deal with their grief, come to terms with what happened and begin to find a way forward.

While there are many issues raised by this incident, including many areas of public policy, there are two aspects of the tragedy that are so commonplace from the standpoint of human behavior that we tend to take them for granted. Yet, what may be considered obvious elements of the tragedy may in fact represent uniquely human capacities.

The first element is that across the city, the nation and the world, people shared their shock, horror and grief with one another. This was accomplished in many ways from local private conversations to memorial services and funerals to television commentary and internet postings to the televised speech by the President. One might say that the emotions that were aroused by the tragedy were too difficult to bear alone and needed to be shared with others. More generally, the need to share experiences is thought to be a uniquely human characteristic (Tomasello et al. 2005), and this incident, while unusual in its specific details, nevertheless illustrates typical human experience that is observed daily in all known cultures around the world.

A second feature of this tragedy that stands out is that virtually everyone struggled to understand what could have motivated this 22 year-old man to do what he did. What was going on in his mind? What were his motivations? Why would he do this? Many speculate that he was out of touch with reality and suffered from a major psychiatric disorder. The need to understand the mind of the person responsible for such an act is also considered a universal human characteristic (Frith and Frith 1999). We naturally want to understand the mental states that can explain behavior, particularly when that behavior is emotionally upsetting to us.

These two features of human behavior, the need to share experiences and the capacity for mentalizing, are critical to our argument that emotional awareness is a uniquely human characteristic. We will propose that mentalizing is a basic cognitive capacity that involves understanding the thoughts and feelings of others as well as oneself, and that other animals, including non-human primates, are not nearly as developed as human beings in having this capacity (Premack and Woodruff 1978). Moreover, these phenomena are inter-related in that: (1) knowing what someone else is experiencing requires introspection, (2) knowing what one is experiencing oneself requires input from others about one's own mental states (Gergely and Watson 1996), either currently or in the past, and (3) the sharing of personal experiences may be the operational expression of this mentalizing function.

We will begin by describing the basic human capacity for emotional awareness. We next explain variability across people in the capacity for emotional awareness as a cognitive skill that undergoes development similar to other cognitive skills. Empirical evidence demonstrating the existence of such individual differences and their clinical consequences will be reviewed. Next, we review evidence that the capacity for emotional awareness is limited in non-human primates and other intelligent animals relative to that of humans. We conclude by discussing the implications of these observations from an evolutionary perspective.

8.2 Theory of Levels of Emotional Awareness

Lane and Schwartz (1987) proposed that an individual's ability to recognize and describe emotion in oneself and others, called emotional awareness, is a cognitive skill that undergoes a developmental process similar to that which Piaget described for cognition in general. A fundamental tenet of this model is that individual differences in emotional awareness reflect variations in the degree of differentiation and integration of the schemata (implicit programs or sets of rules) used to process emotional information, whether that information comes from the external world or the internal world through introspection.

Emotional awareness is considered to be a separate line of cognitive development that may proceed somewhat independently from other cognitive domains

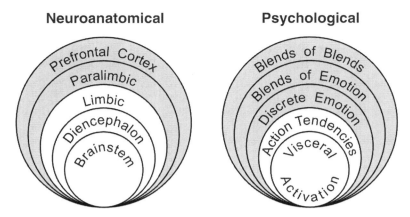

Fig. 8.1 Parallels in the hierarchical organization of emotional experience and its neural substrates. The shell structure is intended to convey that each succeeding level adds to and modulates lower levels but does not replace them. Although each model contains five levels, a one-to-one correspondence between each level in the psychological and neuroanatomical models is not intended. Lower levels with white background correspond to implicit processes. Higher levels with gray background correspond to explicit processes

(Lane and Pollerman 2002). The concept that development can proceed at different rates in different domains of knowledge is known as horizontal decalage (Fischer 1980). In principle, it is entirely possible that a developmental arrest can occur in one domain while development in other domains of intelligence continues unabated, or vice versa.

The model posits five "levels of emotional awareness" that share the structural characteristics of Piaget's stages of cognitive development (Piaget 1937). The five levels of emotional awareness in ascending order are awareness of physical sensations, action tendencies, single emotions, blends of emotions, and blends of emotional experience (the capacity to appreciate complexity in the experiences of self and other).

The five levels therefore describe the cognitive organization of emotional experience. The levels are hierarchically related in that functioning at each level adds to and modifies the function of previous levels but does not eliminate them. For example, blends of emotions (Level 4 experiences), compared to action tendencies (Level 2 experiences), should be associated with more differentiated representations of somatic sensations (Level 1). The feelings associated with a given emotional response can be thought of as a construction consisting of each of the levels of awareness up to and including the highest level attained. The trait level of function is the level at which a given individual typically functions. The nested hierarchy of emotional awareness is depicted schematically in the right half of Fig. 8.1.

Modern conceptions of cognitive development have refined Piaget's views but are still consistent with the model proposed here. Karmiloff-Smith (1992), for example, holds that the development of knowledge proceeds through a process called "representational redescription." Cognitive development from this perspective consists of the transformation of knowledge from implicit (procedural, sensorimotor) patterns to explicit (conscious thought) representations through use of language or other semiotic mode. This transformation renders thought more flexible, adaptable and creative. This viewpoint is consistent with the theory that the way language is used to describe emotion modifies what one knows about emotion and how emotion is experienced consciously (Werner and Kaplan 1963).

The five levels of emotional awareness can therefore be mapped onto the distinction between implicit and explicit processes (Lane 2000). In cognitive science the distinction between implicit and explicit roughly corresponds to the distinction between unconscious and conscious processes (Kihlstrom et al. 2000). Levels 1 (bodily sensations) and 2 (action tendencies) phenomena, viewed in isolation, would not necessarily be considered indicators of emotion, but are critical components of emotional responses. The peripheral physiological arousal and action tendencies associated with emotion are implicit in the sense that they occur automatically and do not require conscious processing in order to be executed efficiently. If one focuses conscious attention on a somatic sensation or action tendency in isolation, emotion is implicit in the sense that the quality of experience needed to call it an emotion requires processing at higher levels. Levels 3–5 consist of conscious emotional experiences at different levels of complexity and by definition are explicit. Level 3 involves the conscious experience of a single emotion, e.g., happy. Level 4 involves the conscious experience of two or more emotions simultaneously, e.g., happy and sad. Level 5 involves awareness of Level 4 experiences of one's own and simultaneous awareness of Level 4 experiences of another person, e.g., I feel happy and sad, and you feel proud and grateful. The levels of emotional awareness framework therefore puts implicit and explicit processes on the same continuum, and at the same time distinguishes between types of implicit (Level 1 vs. Level 2) and explicit (Level 3 vs. Level 4 vs. Level 5) processes.

8.3 Normative and Clinical Observations with the Levels of Emotional Awareness Scale

One can measure an individual's level of emotional awareness using the Levels of Emotional Awareness Scale (LEAS), a paper-and pencil performance measure that asks subjects to describe in an open-ended manner how they and another person would feel in scenarios described in two to four sentences (Lane et al. 1990). Scoring of the LEAS consists of the degree of differentiation of words used to describe the experiences of self and other following the five-level hierarchy just described. The LEAS does not rely on the respondent's own ratings of their ability to put emotions into words or their ability to differentiate between emotions and somatic sensations.

The LEAS correlates moderately positively with two cognitive-developmental measures (Lane et al. 1990), the Sentence Completion Test of Ego Development by Loevinger (Loevinger and Wessler 1970; Loevinger et al. 1970) and the cognitive complexity of descriptions of parents by Blatt and colleagues (Blatt et al. 1979). These results support the claim that the LEAS is measuring a cognitive-developmental continuum and that the LEAS is not identical to these other measures. Two independent studies of undergraduates, involving 63 and 55 subjects, respectively, revealed that greater emotional awareness is associated with greater self-reported impulse control ($r = 0.35$, $p < 0.01$ and $r = 0.30$, $p < 0.05$, respectively) (unpublished observations by Dr. Lisa Feldman-Barrett). Given that the five levels are hierarchically related and that action tendencies represent Level 2 function, this finding is consistent with the theory that functioning at higher levels of emotional awareness (Levels 3–5) modulates function at lower levels (actions and action tendencies at Level 2) (Lane 2000). Several additional findings indicate that higher scores on the LEAS are associated with more differentiated (i.e., more accurate) emotion information processing. Scores on the LEAS are positively correlated with the understanding emotions section of the Mayer Salovey Caruso Emotional Intelligence Test (Lane et al. 1996), the perception of emotions in stories of the Multifactor Emotional Intelligence Scale (Ciarrochi et al. 2003), the Range and Differentiation of Emotional Experience Scale (Kang and Shaver 2004), the accuracy of facial emotion recognition (Lane et al. 1996, 2000) and self-reported empathy (Ciarrochi et al. 2003). It has also recently been observed that higher levels of emotional awareness are associated with more differentiated somatic symptoms (Lane et al. 2011). The latter observation is consistent with both theory and empirical evidence that higher levels of emotional awareness are associated with greater differentiation of function at lower levels.

Clinically, it has been shown that patients with borderline personality disorder score lower on the LEAS than age-matched control subjects (Levine et al. 1997) and that individuals with the "disorganized attachment style" have lower LEAS scores than those with the "organized attachment style" (Subic-Wrana et al. 2007). Patients with irritable bowel syndrome (IBS) do not on average have lower LEAS scores than healthy controls, but among those with IBS lower scores on the LEAS are associated with greater pain (Lackner 2005). Patients on a psychosomatic inpatient ward with somatoform disorders had lower LEAS scores than patients with disorders involving psychological distress such as depression. This same study showed that somatoform patients showed significant increases in LEAS scores after three months of multi-modal inpatient treatment that integrated body-based techniques with intensive group and individual psychotherapy (Subic-Wrana et al. 2005). These findings support the theory that impairments in emotional awareness can occur developmentally, that lower emotional awareness is associated with a greater tendency to experience emotional distress as bodily symptoms and that emotional awareness can improve with therapeutic interventions that facilitate the transition from implicit to explicit processing.

The LEAS has also yielded useful findings in a variety of other clinical settings. Patients with essential hypertension had lower LEAS scores than those with hypertension secondary to other medical conditions such as renal disease (Consoli et al. 2010). Patients with eating disorders (anorexia and bulimia) were observed to have lower LEAS scores than matched controls (Bydlowski et al. 2005), consistent with

Hilde Bruch's classic observation that eating disorders are associated with an impairment in interoceptive awareness of one's own emotions (Bruch 1973). Patients with PTSD have lower LEAS scores than matched controls, and LEAS scores were inversely correlated with the severity of PTSD symptoms, particularly symptoms involving dissociation (Frewen et al. 2008). Patients with morbid obesity were observed to have lower LEAS scores than controls, and it was also observed that among the obese patients the higher the LEAS scores the greater their social anxiety (Consoli 2005). The latter finding indicates that greater emotional awareness is associated with a greater awareness of the negative emotional responses that morbid obesity elicits from others. A related finding is that individuals with generalized anxiety disorder have higher LEAS scores than matched controls (Novick-Kline et al. 2005), indicating that emotional awareness can be a double-edged sword. In contrast, patients with depression were found to have decreased awareness of the emotions of others (Donges et al. 2005; Berthoz et al. 2000), consistent with the pathological introspective focus that can occur with depression. Together these findings indicate that the LEAS can detect variations in emotional awareness that have meaningful clinical correlations.

8.4 A Model of the Neural Substrates of Implicit and Explicit Emotional Processes

The distinction between implicit (non-conscious) and explicit (conscious) processes is foundational in cognitive neuroscience because their neural substrates are dissociable (Gazzaniga et al. 2002). The distinction was first applied to memory. Explicit memory for facts and events requires participation of medial temporal lobe structures (such as the hippocampus) and diencephalon, whereas implicit memory requires structures such as the striatum (skills and habits), neocortex (priming), amygdala and cerebellum (classical conditioning) and reflex pathways (non-associative learning). Implicit processes have also been demonstrated in a variety of other cognitive domains, including attention, perception, and problem solving (Gazzaniga et al. 2002). This body of research has led to a growing recognition that consciousness is the tip of the cognitive iceberg in the sense that the vast majority of cognitive processing occurs outside of conscious awareness (Gazzaniga 1998).

Antonio Damasio's distinction between primary emotion and feeling, and their dissociable neural substrates (Damasio 1994), paved the way for the application of the implicit–explicit distinction to emotion. Primary emotion is the phylogenetically older behavioral and physiological expression of an emotional response. Primary emotion occurs automatically and without the necessity of conscious processing. Feeling, on the other hand, involves the conscious experience of that emotional state. According to Damasio, primary emotion and feeling are separable, both conceptually and neuroanatomically (Damasio 1994). While primary emotion is necessary for successful adaptation to environmental challenges and the physiological adjustments needed to meet those challenges, a conscious feeling state enables

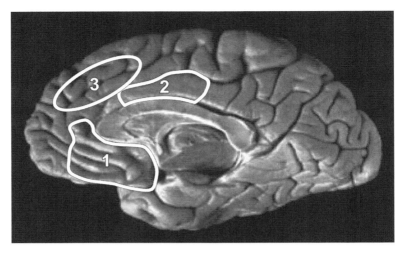

Fig. 8.2 Structures on the medial surface of the frontal lobe that participate in (*1*) background feelings, (*2*) attention to feelings and (*3*) reflective awareness of feelings

previous emotional experiences to be consciously recalled or current experiences to be accessed and used for decision-making and navigation of the social world.

Although the neural substrates of each level of emotional awareness are not yet known, a rudimentary neuroanatomical model that distinguishes between implicit and explicit processes can be formulated (Lane 2000) (see left half of Fig. 8.1). The details of this model have been elaborated elsewhere and will not be repeated here (Lane 2000, 2008). This model distinguishes between the neural substrates of implicit emotion on the one hand and three different aspects of the conscious experience of emotion: background feelings, focal attention to feelings, and reflective awareness. As described above, implicit emotion includes those sensory and motor aspects of emotional responses that precede the emergence of an emotional feeling state. Background feelings are bodily states on the periphery of awareness that color conscious experience but are not noticed as such unless attended to. Focal attention to feelings refers to the condition in which one's own subjective emotional state is the object of directed attention. Reflective awareness involves thinking about the contents of conscious emotional experience, typically after it has been the object of focal attention. The structures in the medial prefrontal cortex preferentially associated with the three types of explicit emotional experiences are depicted in Fig. 8.2.

Parallels between the corresponding neuroanatomical and psychological models are schematically depicted in Fig. 8.1. Both models are hierarchical and show a similar architecture of concentric shells. The concentric architecture means that each new level subsumes and modulates previous levels. Although both the psychological and neuroanatomical models designate five levels, there is no intention to suggest a direct correspondence between a given level in one model and that of the other model. In general, implicit functions at Levels 1 and 2 in the psychological model correspond to Levels 1–3 in the neuroanatomical model, and explicit functions at Levels 3–5 in the psychological model correspond to Levels 4 and 5 in the neuroanatomical model.

8.5 Reflective Awareness

Within the field of consciousness research a distinction is made between phenomenal and reflective awareness. Phenomenal awareness refers to the actual content of consciousness (focal attention to feelings and background feelings) whereas reflective awareness involves attending to or performing a cognitive operation upon the contents of conscious experience (Farthing 1992). Reflective awareness, or metacognition, requires the creation of a representation of experience, and this representation will affect how future emotional information is interpreted and experienced. As indicated by the theory of emotional awareness described above, the capacity for reflection is an essential function that drives the development of emotional awareness forward.

One of us (Lane et al. 1997) examined the role of medial prefrontal cortex in reflective awareness in a PET study of the pattern of neural activation associated with attending to one's own emotional experiences. To confirm that subjects were allocating their attention as instructed, we had them indicate on a keypad how each emotion-evoking picture made them feel. In essence, we were examining an aspect of conscious experience involving commentary on that experience (Weiskrantz 2000). By having subjects attend to and label their experience in this study, we were examining reflective awareness.

We studied ten healthy men as they viewed twelve picture sets, each consisting of pleasant, unpleasant and neutral pictures from the International Affective Picture System (Lang et al. 2001). Pictures were presented for 500 ms every 3.0 s. Twelve PET-derived measures of rCBF were obtained in each subject, one for each picture set. During half the scans subjects attended to their emotional experience (indicating on a keypad whether the picture evoked a pleasant, unpleasant or neutral feeling); during the other half they attended to spatial location (indicating whether the scene depicted was indoors, outdoors, or indeterminate). Across subjects, picture sets were counterbalanced across the two attention conditions.

During attention to subjective emotional responses increased neural activity was elicited in rostral ACC (BA32) and medial prefrontal cortex (BA10; coordinates: 0, 50, 16), right temporal pole, insula and ventral ACC. Under the same stimulus conditions, when subjects attended to spatial aspects of the picture sets, activation was observed in parieto-occipital cortex bilaterally, a region known to participate in the evaluation of spatial relationships. Activation of dorsomedial prefrontal cortex during attention to and labeling of emotional experience and activation of parieto-occipital areas during the evaluation of spatial relationships have been replicated in other independent laboratories (Gusnard et al. 2001; Ochsner et al. 2004) using fMRI.

Frith and Frith (1999) hypothesize that the ability to mentalize, i.e. the ability to understand the mental state of others or oneself, evolved from the action system for the purpose of identifying the intentions of others and anticipating their future actions. We know that emotional states may fundamentally consist of action tendencies (Frijda 1986), which can be construed as equivalent to the intentions of the self. This dovetails beautifully with the Piagetian perspective on cognitive developmental in that all mental representations at the conceptual level are fundamentally derived from action schemes. It is therefore reasonable to consider, as a first approximation, that the paracingulate cortex, which includes the dorsomedial prefrontal cortex, is a substructure within the prefrontal cortex that participates in establishing the representations of mental states (including emotional states) of both self and other.

This conclusion is supported by several findings. Johnson and colleagues (Johnson et al. 2002) demonstrated activation of this region during self-reflective thought (considering whether statements or attributes pertained to the self). Vogeley and associates (Vogeley et al. 2001) observed activation of this region during the process of evaluating what someone else is thinking (theory of mind), as have others previously (Happe et al. 1996). The relevance of this area to emotional experience is supported by a meta-analysis of 162 functional imaging studies of emotion (Kober et al. 2008), which showed that this area of medial prefrontal cortex was an important component of one of the six significant clusters of activation during imaging of emotional states. Clearly this area is involved in the representation of both thought and emotion.

Ochsner and colleagues (Ochsner et al. 2004) observed that a common area of medial prefrontal cortex was activated when attending either to the emotional experiences of the self or to another, and also observed that neighboring subareas were uniquely activated by attending to self or other. In a comprehensive review of functional neuroimaging studies of reasoning and episodic memory, Christoff and Gabrieli (2000) concluded that medial prefrontal cortex BA10 might be specialized for the explicit processing of internal states. Although the functions of BA10 may be broader and involve integrating the outcomes of two or more separate cognitive operations in the pursuit of a higher behavioral goal (Ramnani and Owen 2004), such a conclusion does not exclude the possibility that BA10 plays a key role in reflective awareness of emotion.

8.6 The Comparative Approach: A Caution

We have argued and presented evidence for a neural model of implicit and explicit emotional processing—or emotional awareness—in humans. In the remaining sections, we turn to comparative and evolutionary considerations in exploring which aspects of this capacity are uniquely human and in reflecting on their potential adaptive value. Specifically, we discuss behavioral and neurobiological research on nonhuman primates—hereafter simply referred to as "primates"—that bears on emotional awareness. We conclude that the ability to mentally represent complex emotional states—particularly ones relying on language—in self and others is not shared with other primates and may thus be uniquely human. Because our argument rests on comparative evidence from several primate species, we need first to explain what such a comparative approach to the problem entails, and to expose some prominent assumptions and their pitfalls inherent in the comparative study of behavior and cognition.

The comparative study of behavior, including the expression of emotions in other animals, has a long and venerable history, arguably brought to wide public attention by Darwin's 1872 landmark publication "On the Expression of the Emotions in Man and Animals." It pointed our attention to substantial continuities in the forms and functions of facial and other bodily expressions of emotions in humans and other primates, and thus launched the next century of comparative emotions research (e.g., Ekman 1973).

Darwin championed evolutionary continuity over discontinuity, proposing that there were few if any differences between the mental capacities of humans and the "higher mammals," including "similar passions, affections, and emotions" (Bolhuis and Wynne 2009). In retrospect, Darwin's generous view of mental continuity may well be "one of his few significant mistakes" (Wynne and Bolhuis 2008), because it fostered the widely held view still common today that any mental differences are ones of degree rather than kind. This in turn has invited easy anthropomorphism and unnecessarily rich mentalistic interpretations of nonhuman animals' behavior, as evident in the work of both contemporary researchers (Bolhuis and Wynne 2009) and the lay public. As we will see shortly, the issue of continuity vs. discontinuity is also evident in research on the awareness and representational capacities of nonhuman primates.

No one denies that there are important evolutionary continuities in the behavior and mental capacities of animals. It is, however, an overly ardent focus on evolutionary continuities that can produce a blind spot for important discontinuities. In their recent review of comparative cognition studies, Penn et al. (2008) aptly state: "The profound biological continuity between human and nonhuman animals masks an equally profound functional discontinuity between the human and nonhuman mind" (p. 110). Understandably, it is tempting for scientist and layman alike to think that the degree of evolutionary or genetic relatedness of two species, say human and chimpanzee, should reliably correspond to degree of overall phenotypic resemblance. Along these lines, we might propose, for example, that the emotions, temperament or intelligence of chimpanzees are far closer to our own than those of our genetically more distant primate relatives. Even more distantly related species, like birds, by the same reasoning, we would expect to show hardly any resemblance to us in these mental traits. Consequently, it may come as a shock or surprise when assumed "human-like" tempered chimpanzees co-habiting with humans unpredictably turn violent, inflicting serious physical injury on their life-long human companions. This is unsurprising, however, to those working with chimps in the wild, where emotional impulsivity and violent temper are a regular feature of chimpanzee social struggle for status (Peterson and Wrangham 1996). Indeed, chimpanzee personality differs from human personality in the addition of a factor or dimension called "dominance" that includes items like "not fearful," "not cautious," "not timid" and "bullying" (King and Figueredo 1997). Equally surprising may be the mounting findings from several avian species, particularly corvids, of apparently sophisticated meta-cognitive or representational abilities heretofore believed possible only in apes and humans (see Emery and Clayton 2004). This is especially so if we believe that such sophisticated cognitive abilities can only be generated by the evolution of a neocortex, which birds lack.

The main point we wish to make is that the common assumption, that more closely related species will also be most similar in their cognitive abilities, is a poorly founded and often misguided assumption (Bolhuis and Wynne 2009). There are at least two reasons for this. One is that similar selection pressures in distantly related species can lead to the evolution of convergent cognitive and behavioral adaptations. Thus, the cognitive and behavioral skills (e.g., self-recognition, vocal mimicry, sophisticated tool use) of crows and other corvids, some of which are equal to or exceed those of monkeys and apes, are produced by a forebrain that lacks a laminated cortex (Kirsch et al. 2008). Here, similar functional capacities and solutions to ecological problems rely on dissimilar neural equipment.

But even in closely related species, shared genetics and anatomy is no guarantee of functional continuity or equivalence. Dissimilar ecologies encountered by the two species can generate strong selection pressures for functional divergence or innovation in cognition and behavior over relatively short periods of evolutionary time. Natural selection and artificial selection, for example, jointly produced the present breeds of domestic dogs from their wolf ancestors, beginning as early as 100,000 years before the present, with strong artificial selection beginning about 10,000 years ago. As a result, domestic dogs are more sensitive to human attentional states and bodily cues (e.g., looking, pointing, turning, see Udell et al. 2010 for review) than chimpanzees (Hare 2007). Well-known, too, are the results from a silver fox domestication project (see Trut et al. 2004), where strong artificial selection over 30 generations produced a population of "tame" foxes from wild ones. The taming appeared to be the result of altered developmental schedules, along with neurochemical and hormonal changes. Hence, strong selection pressures can relatively quickly alter neural functioning, temperament, and behavior.

Because humans and chimpanzees share more than 98% of DNA, we may be led to think that this near genetic identity in DNA sequence must greatly limit the amount of cognitive or behavioral differentiation. However, significant change can be achieved by altering the "epigenome"—the mechanisms that regulate DNA expression—without necessarily altering gene structure or sequence. The DNA base pair similarity, therefore, while a useful index of phylogenetic affinity, can be a poor indicator of DNA expression or functional similarity. Humans and chimpanzees, since their divergence from a common ancestor 6–8 million years ago, have built and occupied substantially different ecological and social worlds, and we should expect therefore that the minds of ape and human have been adaptively shaped in response to these different environments. Indeed, as we will note again later, because of this ecological divergence, what is an ecologically valid cognitive task for a human is not necessarily an ecologically valid one for a chimpanzee and vice versa. For example, Pinker (2010) has proposed that humans have evolved special psychological faculties, including language, causal reasoning and social cooperation, as key to constructing a "cognitive niche." Because chimpanzee psychology appears to have been shaped more by social competition than cooperation, cognitive experiments that rely on their social cooperation for this reason may have less success than ones using a socially competitive design (Hare and Tomasello 2004; and see below).

Here, the more general point is that while chimpanzees and other group-living animals also engage in social cooperation (e.g., Harcourt and de Waal 1992), human social cooperative relationships depend on perhaps special empathic (see below) and enhanced mind reading capacities (i.e., as key elements of a cognitive niche) that are responsible for the greater complexity of human cultural and socio-political arrangements. Given that humans began to construct this cognitive niche after the split from a common ancestor with African apes, it is not unreasonable to expect that chimpanzees lack these specializations.

Some may still argue that because the human cognitive niche evolved gradually over millions of years, coming to full fruition only recently in a dramatically different post-Pleistocene environment, human language and other cognitive abilities evolved gradually as well and thus differ only in degree from fundamentally similar capacities in apes. In other words, brain expansion during human evolution produced greater intellect, enhanced language capacity and more complex representational capacities by degrees but did not produce capacities of a different kind. This view may derive support from the related assumption held by many evolutionary psychologists that there has been insufficient time in the post-Pleistocene era for significant further cognitive specializations to evolve; that, de facto, humans navigate the modern world with a "Stone-Age" mind.

It is difficult, however, to sustain this argument, given our earlier discussion about the relatively rapid effects of strong selection on neuro-cognitive features. Moreover, recent studies of the human and chimpanzee genomes have provided evidence of significant recent positive selection for adaptive substitutions that are linked with post-Pleistocene rapid cultural evolution (Hawks et al. 2007), i.e., within the past 10,000 years. Apparently, human evolution proceeded at a relatively slower rate in the distant past, but then accelerated recently with human geographic expansion and cultural diversification. It appears likely that there was recent positive selection for key genetic changes in the human lineage, *pari passu* with increased cultural diversification and complexity. Several genetic differences between humans and chimpanzees, for example, are linked to the production and perception of language in humans or to aspects of brain development, supporting the view that the human cognitive niche depended on the evolution of specialized, likely unique abilities (see Pinker 2010). The genetic evidence thus puts into serious doubt the assumption of many evolutionary psychologists (i.e., those holding the "Stone-Age" mind view) that there has been no human evolutionary change in the last 10,000 or more years. More importantly, it alerts us to the likelihood of significant discontinuities between humans and apes in cognition and behavior.

8.7 Emotions and Emotional Awareness in Nonhuman Primates

We now turn to a consideration of the evidence from naturalistic behavior observations, experimental studies of behavior and cognition, and comparative neurobiology as it bears on the question of the presence of emotional awareness in primates. By emotional awareness, first and foremost, we mean a kind of self-awareness, a mental state that we identify in ourselves as being conscious of our internal feelings or affects. As we have suggested earlier, language provides for the symbolic mental representation of a wide range of differentiated and blended emotions. Awareness of complex emotions, in turn, allows us to regulate or control emotions in contextually appropriate ways. Importantly, our capacity for emotional awareness appears to be linked to our ability to recognize, understand, and experience the emotions of others, commonly referred to as empathy. Human social relationships, cooperative and otherwise, and social bonds critically depend on the accurate decoding of emotions and the sharing of emotional experience. Consequently, we will examine evidence for emotional regulation and empathic capacities.

For the time being, we are leaving open the question of whether or not emotional self-awareness is distinct from self-awareness of mental content generally. While in humans self-awareness appears to be domain-general, this does not preclude the possibility that in other species emotional awareness is a domain-specific capacity. Thus other primates may have varying degrees of awareness in the emotional domain, depending upon the particular adaptive benefits of this capacity. However, this may not be coupled to other kinds of awareness as it is in humans, particularly the awareness of one's own and others' thoughts, goals or beliefs, known as "theory of mind" (hereafter ToM). In humans, for example, deficits in emotional self-awareness correlate with deficits in ToM (e.g., Subic-Wrana et al. 2010). Indeed, Gallup (1982) has argued that mindreading (i.e., ToM) presupposes introspection. Bundled together, mindreading and introspection facilitate emotional self-regulation and make possible the mental simulation of one's own or another's future mental states. The human brain evidently evolved to become "a great anticipation machine" (Dennett 1996), but to get to this point no doubt required the concomitant evolution of language, as we have noted before. It is easy to imagine that this kind of mental simulation would also be of adaptive benefit to other social primates or indeed any other social species, which is why much research has addressed ToM in primates and other species. We will, therefore, also consider whether other primates show evidence of ToM, as its presence may be construed as indirect support for introspection and emotional self-awareness.

8.8 Naturalistic Behavior Observations

Let us first consider what qualifies as evidence in support of emotional awareness in the naturally occurring behaviors of nonhuman primates. For present purposes, by "naturally occurring behavior" we simply mean behaviors that are shown spontaneously in social interactions with conspecifics, such as in social groups in the wild or in captivity. Behaviors shown in interactions or test situations with humans, while not "unnatural," will be treated separately (see next section), because they may reflect the effects of close interactions with humans and their cultural environment (i.e., enculturation effects) or language training.

One obvious difficulty is that providing conclusive evidence for emotional awareness as an internal or subjective state in another animal—or in another person—may be too high a goal. At best, we will be able to make inferences about similar qualities of an internal state on the basis of similarity in behavior, and, where available, similarity in physiological or neurological mechanisms to such internal states in humans. Clearly, similarity in the behavioral expression and physiological substrates of an emotion, therefore, while not proof of similarity in the subjective experience of the emotion, nevertheless, constitutes stronger evidence than behavioral or physiological similarity alone.

The unavoidable uncertainty about an animal's subjective experience—excepting those species that use language to report on their internal states—may well be responsible for the relative shortage of studies on animal emotion (Aureli and Whiten 2003). Many animal behaviorists avoid making inferences about subjective states and anthropomorphic interpretations of these by using descriptions or categories of behavior in place of descriptors for internal emotional states (e.g., flight instead of fear). However, Aureli and Whiten (2003) suggest that researchers set aside the problem of how animals feel and treat emotional states, or any other "state of mind," as "intervening variables" that have utility because they simplify complex causal linkages between independent and dependent variables. For example in a rat, "thirst" posited as an intervening variable, they suggest, allows us to more economically model how various factors (e.g., fluid deprivation, consuming dry food) influence drinking behavior (e.g., rate of bar pressing, amount of water consumed), even though we have no clue about how a rat feels or experiences "thirst." We might call this the "as if" approach, in that in this example the rat behaves "as if" it is thirsty, although it may have no internal experience nor awareness of its mental state.

While the "as if" approach has utility and is a good compromise solution, it is not always the most parsimonious. We suggest that if an animal behaves, for example, as if it is angry and the physiological manifestations (e.g., increased blood pressure) and neural processing and regulatory mechanisms (e.g., brain regions and pathways, neurochemistry) are similar to those known to be necessary and sufficient for a reported emotional experience (i.e., in humans and other language-using species), then the most parsimonious conclusion is that the animal shares the emotional experience. This is a fairly conservative approach to inferring mental states in other animals, and, as we will see shortly, many researchers are happy to infer subjective states based on functional behavioral evidence and phylogenetic proximity alone. With this in mind, we will now look at some examples of the natural behaviors of primates that may be indicators of emotional awareness.

Old World monkeys, like baboons and macaques, and the great apes have natural repertoires of facial expressions, vocalizations and other body signals (e.g., cowering, strutting, piloerection) for a variety of emotions, and there is resemblance between some elements of these emotional displays—especially in African apes—and those of humans. This is not to say that these multi-modal signals exclusively serve the expression of emotions, as indeed was once thought to be the case for nonhuman primate vocalizations. Much research, for example, on primate vocalizations has established that these also convey information about the caller's identity, dominance rank, kinship, or quantity of food, and that fellow group members use this information for strategic social action (e.g., see review by Owren et al. 2003). Similarly, a series of elegant studies has shown that chimpanzees understand the emotional content of their species-typical facial expressions of emotion (see review by Parr and Mastripieri 2003, and further discussion below).

It is vital, however, to note that the fact that primates recognize, understand, and act on vocal and other expressions of emotions does not by itself provide evidence for emotional awareness. For example, children can identify facial and vocal displays of emotion and recognize how these are connected to actions before they are able to reflect on or talk about these emotional experiences (Thompson and Lagattuta 2008). De Waal (1996) describes several instances of "spontaneous deception" in a captive colony of chimpanzees that he believes require awareness of the chimpanzee's own and other's emotions. In one case—described as a "dramatic instance of self correction"—a male used his fingers repeatedly to push his lips over his teeth to conceal a bared-teeth display (a fear or submissive signal) from another male challenger. In another case, an adult female appears to feign a mood shift from aggression to conciliation by giving soft pants (a "friendly" call) and reaching out a hand toward a younger female she had been pursuing aggressively, only to suddenly grab and bite the younger female who had come close. Such cases suggest that chimpanzees understand the effect of their emotional displays on others and thus use emotional self-awareness to regulate their emotional expressions for deceptive purposes. Alternative interpretations, however, cannot be ruled out. If chimpanzees are aware of their emotions and can change their emotional displays at will, then why not simply inhibit or abort the bared-teeth display without the more cumbersome manual manipulation of the lips? Perhaps the action of the fingers was a part of the nervous or anxious behavior rather than an intentional aid to change the emotional expression. Indeed, the richer interpretation of the male's behavior seems inconsistent with the second example, in that the latter supposedly illustrates a highly effective instrumental alteration of the female's own emotional displays. Unfortunately, in this case, too, we can argue more simply that the changes in emotional displays were not willfully deceptive but rather indicative of actual changes in mood (i.e., from aggressive, to conciliatory, to aggressive) in response to subtle cues received from the other female.

It is not self-evident that triggering or inhibiting an emotional display in context appropriate ways requires conscious awareness of the emotional state. Subordinate mountain gorilla males inhibit their copulation vocalizations while copulating out of sight but within audible distance of a dominant male (Steklis, personal observation). But is it necessarily evidence of willful deception, when an important part of primate socialization is the learning of when and how to use context-appropriate emotional displays? Learning and demonstrating such socially appropriate behavior, while surely enhanced by emotional awareness, does not necessarily entail it.

In children emotional self-awareness emerges together with general self-awareness, and they begin to see themselves and others as "psychological beings with internal mental lives" (Thompson and Lagattuta 2008). As a result, by age 3–4, children have fully developed self-conscious emotions, such as pride, guilt, and shame, that are based on enhanced self-awareness and mental perspective taking (or ToM), and they are capable of emotional regulation, false belief attribution, and deception (Thompson and Lagattuta 2008). This deeper level of emotional understanding of self and other has been aptly called cognitive empathy as distinct from emotional empathy, which can become dissociated in some clinical conditions, such as autism (Smith 2006). Cognitive empathy involves mentalizing—the capacity to ponder (or model) the contents of one's own or another's mind. Emotional empathy, by contrast, need not involve mentalizing but is more like what has been called "raw empathic arousal" (Thompson and Lagattuta 2008) or "emotional contagion" (de Waal 1996), that in very young children may trigger helping behavior. Prior to the development of a well-developed concept of self, however, childrens' empathic responses can also lead to emotional over-arousal, distress and confusion, underscoring the importance of self-awareness to cognitive empathy and emotional regulation.

While emotional empathy (or contagion) is likely widespread among social species because it provides a basis for social cooperation and altruistic behavior (de Waal 1996), there is debate about the presence or degree of development of cognitive empathy in nonhuman species. Many primates, for example, show consoling behavior in response to the distress of a group member (de Waal 2008), which is quite different from acts of reconciliation after conflicts in that the party doing the reconciling was an uninvolved bystander, suggesting that an understanding of both the social context and the distressed individual's emotional state (i.e., cognitive empathy) are involved. Indeed, de Waal (2008) argues that such evidence from naturalistic behavior points to cognitive empathy as shared among humans and apes. A recent critical review of these behaviors in primates (Koski and Sterck 2010), however, rightly contends that it is premature to conclude from these observations alone that either reconciling or consoling behaviors involve cognitive empathy. Thus both behaviors may be in response to the actor's own emotional discomfort (e.g., a state of anxiety) rather than based on the inference of another's emotional state (Koski and Sterck 2010). In groups of ravens bystanders similarly console conflict victims, but they appear to do so either to alleviate the victim's distress or to reduce the likelihood of renewed aggression (Fraser and Bugnyar 2010). By the same behavioral criteria, therefore, both apes and ravens show identical levels of emotional understanding and empathy.

It is of interest that while de Waal (2008) believes apes are capable of cognitive empathy, he is reticent to attribute the self-conscious emotions of shame or guilt to any of our nonhuman brethren, despite appearances to the contrary. Since there is much evidence to show that humans perceive and respond to unconsciously presented emotional displays (see review by Tamietto and de Gelder 2010), basic empathic responses among primates do not by themselves constitute evidence of conscious awareness of emotions nor of the kinds or levels of empathy involved. Further, given that self-conscious emotions such as guilt and shame in humans depend on the full development of cognitive empathy, the apparent absence of these emotions in other animals suggests that they also lack the full development of cognitive empathy.

At this point we want to be clear that although the natural behavior of primates doesn't allow us to decide the degree or complexity of cognition involved in their empathic responses, this does not rule out their employing sophisticated representational capacities in other domains. For example, much animal behavior (e.g., foraging, tool use) is consistent with the notion that animals have a mental map of physical space (e.g., as in food-caching birds) or a plan for the future use of a tool. Chimpanzees in the Tai Forest of West Africa (Boesch 1991) and wild capuchin monkeys in Brazil (Fragaszy et al. 2010) carefully select stone hammers and anvils to crack otherwise hard to open nuts, transporting the stones to the nut cracking sites. Likewise, many primates and other social species no doubt are able at least to some degree to anticipate how others are likely to respond to their own actions, allowing for the strategic deployment of behavior or even manipulation of other's behavior (i.e., "Machiavellian intelligence," see Byrne and Whiten 1988). Gilbert and Wilson (2007) have argued that many animals have evolved a capacity for "prospecting"—"to predict the hedonic consequences of events they've experienced before." However, they suggest that humans far outstrip other animals by mentally simulating future events and their accompanying emotions (i.e., "prefeeling" the future) they have never experienced before. In children this level of prospection follows the development of language around 3–4 years of age, again suggesting that language provides a necessary support for this level of representational capacity. Further, as we will see shortly, human prospection and the representation of the mental content of self and of other minds (i.e., ToM) may well depend on the evolution of specialized neural circuitry. Insofar as a representation of another's mental content is vital to pedagogy, a lack of this representational skill in other animals is consistent with the lack of solid evidence for teaching (i.e., true pedagogy) in non-human animals (Byrne and Bates 2010).

Throughout the preceding section we have tried to point out the difficulty of drawing firm conclusions about the mental representational abilities of primates based on naturalistic observations of behavior. Thus, while such observations of behavior are not incompatible with rich mentalistic explanations, including conscious awareness of one's own and another's emotional states, simpler explanations often suffice (see also review by Barrett et al. 2007). Experimental studies of behavior and cognition, through the use of appropriate controls, reduce the number of alternative explanations and allow us to home in on the underlying cognitive abilities. But, as we will see in the next section, even well controlled studies do not completely eliminate alternative explanations, and hence controversy abounds.

8.9 Experimental Studies

Being consciously aware of one's own or another's emotions is a kind of "metacognition"—a part of the "cognitive executive…that monitors and controls the ongoing cognitive processing… (an) awareness of the processes of mind" (Smith 2007,

p. 64). As reviewed earlier, emotional self-awareness has been experimentally well investigated in humans. There are relatively few experimental investigations of this capacity in other primates, and we begin this section with a discussion of these studies. Because in humans explicit emotional self-awareness is dependent on a mental representation of the self that is linked to ToM abilities (see Focquaert and Platek 2007), it is possible, though it does not follow directly, that in other primates introspective capacities are similarly linked to ToM. We will, therefore, next discuss evidence for self-recognition and ToM in primates.

The series of studies by Parr and colleagues (reviewed in Parr and Mastripieri 2003) are unique in their explicit focus on emotional awareness in chimpanzees. In a novel procedure, called "matching-to-meaning," they demonstrated that without prior training chimpanzees readily matched the emotional content (or meaning) of video clips to pictures of species-typical facial expressions of emotions and that the emotional video content elicited physiological responses similar to those accompanying human experience of emotions (Parr 2001). For example, subjects viewed video showing a familiar veterinary procedure of conspecifics being injected with a needle and correctly matched this negative emotional content to facial expressions with negative emotional valence. This shows that chimpanzees are able to categorize positive and negative emotions and associate these with their own facial expressions of emotions, and that they may subjectively experience these classes of emotion in the manner humans do. It does not show that the chimpanzees are consciously aware of making these categorical distinctions. Parr (2001) appropriately concludes that while providing evidence of basic emotional awareness, further studies are needed to demonstrate whether chimpanzees have a conscious (i.e., explicit) awareness or understanding of different kinds of emotions in the way that humans do.

In humans explicit emotional awareness and understanding developmentally follow on the heels of an emerging concept of self. The recognition of one's own image in a mirror is generally taken as evidence of at least a basic recognition of or distinction of self from others in both humans and other animals (Gallup 1979). Mirror recognition appears in human children around age 2 and slowly develops thereafter into a full understanding of themselves as psychological beings distinct from others. As a result, the mirror recognition test has been used in other species as evidence for the existence of at least a rudimentary kind of conscious self-awareness. Because the mirror recognition test is relatively simple, it has been easily adapted for use with a diversity of mammals and birds.

The most common form of the test is the "mark test," in which after a period of familiarization with a mirror a mark is surreptitiously placed on a part of the head or body that is only visible to the subject by viewing itself in the mirror. Self-directed behavior toward the mark (e.g., touching the mark) is taken as evidence of self-recognition. As may be expected, many more species fail than pass this mirror recognition test (Byrne and Bates 2010), but there have been some surprises. At first, it was expected that species phylogenetically closest to us—the great apes—or those with relatively large brains—such as dolphins and elephants—would pass the test, which they do, while dogs, monkeys, and gibbons consistently fail the mark test. This is the case even after adjustments to the mark test are made in monkeys who may avoid looking into a mirror because staring directly into a face is perceived as a threat (Macellini et al. 2010). The surprise, however, is that magpies—a corvine bird—also pass the mark test (Prior et al. 2008). Although, as we have noted before, corvids and other birds lack a neocortex, they have brain to body weight ratios well within the range of apes and other mammals who pass the mark test (Prior et al. 2008). Convergent selection pressures on brains and cognition, perhaps in response to the cognitive demands of social life, may therefore have produced the capacity to understand that a mirror image is of one's own body in some mammals and birds.

The question remains, however, whether this level of understanding of self in other species is equivalent to self-awareness in humans. Prior et al. (2008), for example, do not claim that magpies have "a level of self-consciousness...typical of humans" (p. 1647). Rather, they argue more conservatively that while magpie mirror self-recognition is fully comparable to what has been demonstrated for apes it can only be concluded that they understand that the body in the mirror belongs to them, which is likely a prerequisite to full self-recognition.

Indeed, since in humans self-awareness and hence the response to self in a mirror is a developmental process consisting of stages instead of an all or none phenomenon (Rochat 2003), it seems useful to consider that other species reach different stages of self-awareness as indicated by their response to mirrors. Thus, monkeys who generally fail the mirror mark test (see Anderson and Gallup 1999), nevertheless have a limited kind of mirror recognition (or stage of self-awareness) in that they don't necessarily treat their body image as if it belonged to a stranger. For example, they readily match the movement of a body part in the mirror to their own respective body part (i.e., visual-kinesthetic matching) and use a mirror for visually guided behavior. Unlike apes, however, monkeys do not seem to have an internal representation of their body and hence a corresponding expectation of their body's appearance, such that when a mark is placed on it they don't regard it as an alteration to their body (Macellini et al. 2010). In terms of Rochat's (2003) proposed five stages of self-awareness, monkeys reach stage one (i.e., perception of body movement in the mirror as one's own) while apes and other species that pass the mark test reach stage three (i.e., full identification of the self-experienced body and identity with that reflected in a mirror). However, it is not clear from the mirror mark test alone that apes reach stages four (i.e., temporal stability of the self) or five (i.e., third person perspective of self). Stage five appears in children at age 4–5, along with self-conscious emotions and ToM competence, including understanding of false beliefs, capacities that apes seem to lack (see below).

This conservative interpretation of the mirror mark test results at this point seems most sensible to us in that it also fits well with the observation that autistic children show mirror self-recognition in spite of their ToM deficits and failure to understand the more complex self-conscious emotions like embarrassment and shame (see also fuller discussion in Focquaert and Platek 2007). As mentioned earlier, the latter emotions depend on a more fully developed self-awareness that entails a representation of how the "me" is seen by others, that is, a mental model of a third person's view (or cognitive empathy), which is precisely what is impaired in autism. Hence, mirror self-recognition as evidenced in the mirror mark test while surely a prerequisite for ToM is not itself a direct measure of level of self-awareness or ToM.

Let's turn now to the question of ToM capacity in primates. There is a substantial literature on ToM abilities in primates—primarily the great apes—and many recent reviews have well captured the novel and diverse experimental approaches to the problem as well as the salient disagreements over the interpretation of results (e.g., Santos et al. 2007; Focquaert and Platek 2007; Penn and Povinelli 2007; Call and Tomasello 2008; Byrne and Bates 2010; Fitch et al. 2010; Rosati et al. 2010). In consequence, we will not review this literature again, but instead focus on a few key experiments with great apes around which the differences in interpretation hinge.

In evaluating the experimental evidence, it needs to be stressed that ToM is not an all-or-nothing capacity, but rather that there are several levels of ToM that require varying complexity of metacognitive skill—from understanding the sensory perceptions of others to understanding others' intentions, desires and goals to modeling what others know or believe, including awareness of false beliefs. As well, ToM capacities, like other complex cognitive skills (e.g., language), are founded and dependent on simpler cognitive skills, such as attending to or following a conspecific's gaze, present in a variety of mammals and birds (see Fitch et al. 2010, for review). Adjusting one's position so as to follow another's gaze—referred to as "geometric gaze following"—is a more complex representational (i.e., low level ToM) capacity in that it implies an understanding of another's visual perspective, which may be present in great apes and corvids (Fitch et al. 2010). In children, gaze following and understanding others' intentions and goals are stepping stones toward and predictive of later emerging, more elaborate ToM capacities (Wellman and Brandone 2009). As is the case for levels of self-awareness discussed earlier, primates and humans may differ not in the absence or presence of ToM capacities but rather in achieving different levels of ToM.

That being said, the present debate is squarely focused on whether or not particular experimental tasks used with chimpanzees or other great apes—on whom the majority of research has focused—in fact require apes to make inferences about others' mental states or content at all or whether reliance on the subject's own representational abilities to reason about past and present observable behaviors is sufficient (e.g., see Heyes 1998; Penn and Povinelli 2007; versus Call and Tomasello 2008). In short, the debate is about whether or not any of these experiments to date can be said to unequivocally demonstrate ToM in primates.

To appreciate this key difference of interpretation, we need to look more closely at the experimental evidence. The key experiment—critiqued at length by Penn and Povinelli (2007)—was a food retrieval competition task involving pairs of chimpanzees, one subordinate and one dominant (Hare et al. 2001). The two chimpanzees were in separate chambers opposite each other, connected by a central area where food items were hidden behind one of two cloth bags that occluded the food from view of the dominant. The pair could see each other or the central area only when the opaque door on their respective chamber was partially raised. The subordinate's door was always raised to allow viewing of where the food was hidden and whether or not the dominant was watching as well. Several procedural variants were used, but the basic design included an informed condition (the dominant could see where the food was hidden) and an uninformed condition (the dominant's door was closed).

The subordinate was released into the central area before the dominant was released, allowing the subordinate to approach and retrieve the food first. A total of 33 dyads (formed from nine subordinate and three dominant chimpanzees) were tested. Results showed that subordinates approached the hidden food more often and retrieved more food when the dominant was uninformed than when informed, which Hare and colleagues interpreted as strong evidence for ToM (i.e., the subordinate's actions were based on an understanding of the dominant's knowledge of where the food was hidden).

Penn and Povinelli (2007) make a persuasive case that these results and ones from other similar experiments (e.g., with corvids) can more parsimoniously be explained by lower level associative behavioral learning rules (e.g., don't approach food when dominant has oriented to it) that do not require the subject to make any inferences about another's mental content. To Penn and Povinelli, positing that the subordinate acts on the basis of an inference about the dominant's knowledge of where the food is does no additional explanatory work beyond the subordinate's learned adjustments to the dominant's behavior, and it is therefore a "causally superfluous" account. In their view, "chimpanzees, corvids and all other non-human animals only form representations and reason about observable features, relations and states of affairs from their own cognitive perspective" (p. 737).

As further support for their position, Penn and Povinelli (2007) refer to negative results from an "opaque visor experiment" which they believe is a more direct test of ToM in apes. In this clever experiment "highly enculturated" chimps were trained wearing two mirrored, different colored visors, one of which was see-through, the other opaque. Subsequently, the subjects were given the opportunity to use normal begging gestures to request food from experimenters who wore either the see-through or opaque visor. The chimps requested food equally often from experimenters wearing opaque or see-through visors, i.e., begging behavior was not less frequent when the experimenters wore the opaque visor. These findings indicate that they did not generalize their own previous visual experience wearing the two kinds of visors to the visual experiences of the visor-wearing experimenters. The authors note that an analogous test with appropriately modified blindfolds in 18-month-old children produced positive results. Along the same lines, we suggest that a further test of Penn and Povinelli's position concerning the lack of ToM in apes would be the demonstration that humans with ToM impairments (e.g., as in autism spectrum disorder) could nonetheless perform tasks analogous (and appropriately modified) to the ToM tasks used on chimps.

In response to Povinelli and colleagues, Call and Tomasello (2008) concede that for any particular experiment it is impossible to rule out a "surface level" behavioral rules explanation. However, they argue that such an explanation becomes less viable and indeed less parsimonious when applied to the large number and varied experimental procedures where positive results on levels of ToM have been obtained. For example, in one of the conditions of the food competition experiments (Hare et al. 2001), in some of the trials the experimenters switched the dominant pair member after she had viewed the baiting with another dominant who had not observed the baiting. Subordinate subjects observing the switch, immediately adjusted their responses consistent with the substitute's lack of knowledge about the food's location: They obtained more food on trials when the dominant was switched compared to trials when the dominant remained the same. As predicted, this result is consistent with the subordinate chimps using a flexible cognitive strategy to adapt to the novel circumstance. In addition, there are several experiments reviewed by Call and Tomasello (2008) in which chimpanzees act out what they understand another intends to do, where a behavioral rules explanation is impossible. Hence, taken collectively, the results lead Call and Tomasello to conclude that "chimps like humans understand that others see, hear and know things" (p. 190).

Despite Penn and Povinelli's compelling critique, we concur with Call and Tomasello's conclusion, as have other reviewers of the same sets of studies (e.g., Carruthers 2008; Fitch et al. 2010). One reason in addition to the ones supplied by Call and Tomasello (2008) for why we concur with their overall conclusion about ape ToM capacities, is the recent further demonstration of ToM in a large sample of both bonobos ($n=34$) and chimpanzees ($n=106$) that are semi-free ranging in African sanctuaries for orphaned apes (Herrmann et al. 2010). Bonobos are close genetic (con-generic) relatives of chimpanzees, who show distinct differences from chimpanzees in temperament and socio-ecology that were predicted to be related to species differences in task performance. None of the apes had been previously exposed to any of the tests and their participation was voluntary. All apes were given a series of tasks spanning the physical and social knowledge domains. The ToM tasks required subjects to understand an actor's goals and perceptions. In comparing mean performance for the ToM tasks, the authors found a predicted higher performance for bonobos. Thus, bonobos who are more cautious and socially tolerant, which in other circumstances allows them to outperform chimps on a cooperative task (Hare et al. 2007), also outperformed chimps on the ToM tasks. This interesting result recalls the earlier failure of chimps on ToM tasks requiring chimpanzee-human cooperation as compared to their success on competitively-structured ToM tasks (Hare et al. 2001). It also underscores our earlier caution with regard to ape–human continuity in cognition, namely, that significant cognitive differences can evolve in closely related species as a result of divergent socio-ecological pressures.

Even if we grant our ape cousins rudimentary ToM capacities—what Call and Tomasello (2008) describe as having a "perception–goal psychology"—apes do not appear to achieve "a full-fledged, human-like belief–desire psychology" (Call and Tomasello 2008). As noted before, a diagnostic test of this level of ToM is the attribution of false beliefs to another. Some argue that the capacity to attribute false thoughts is a litmus test for ToM generally because it excludes the possibility of a subject reasoning from behavioral contingencies alone (Penn and Povinelli 2007; Carruthers 2008). Moreover, it might be argued that because autistic children are unable to pass a false belief test due to their general incapacity for understanding their own thoughts and those of others (Happe 2003), apes unable to pass a false belief test therefore also lack ToM capacity. It is possible, however, that in apes—or in other nonhuman species with thought attribution capacities—the level of meta-cognition attained may be insufficient to allow false belief attribution.

One condition in the set of chimpanzee food retrieval competition experiments (Hare et al. 2001) tested the attribution of false beliefs. In this condition, the dominant pair member was "misinformed" (i.e., had a false belief) about the food's location: The dominant was allowed to observe the initial food placement, after which the food was switched to the other location while the dominant's compartment door was closed, so that the dominant could not see the changed food location. If the subordinate, whose door was partially open throughout the trial, understood that the dominant held a false belief about the food's location, then it should retrieve more food in the misinformed than in the uninformed condition (i.e., in which the dominant did not observe the food being hidden and thus was ignorant of its location). However, subordinate subjects did not discriminate between the misinformed and the uninformed conditions, leading the experimenters to conclude that while chimps understand a competitor's state of ignorance, they cannot distinguish this mental state from one of false belief.

There is additional negative experimental evidence for false belief attribution in chimpanzees and orangutans using altogether different experimental paradigms. Kaminski et al. (2008) compared the ability of chimpanzees and young children to distinguish between knowledge ignorance and false belief attribution. Six-year old children, but neither 3-year olds nor chimpanzees, passed the test. Similarly, Gomez (1998, 2004), testing a female orangutan on a human–ape cooperative food retrieval task, concluded that the orangutan only understood a human's state of knowledge ignorance. It is of interest that, as in chimpanzees (Hare 2001), success on this task was achieved only after a competitive component was added to the procedure. This lends support to the idea that in apes, and perhaps other nonhuman species, the demonstrated level of ToM is a domain-specific cognitive adaptation to competitive social contexts (Hare 2001; Tomasello and Herrmann 2010).

We have previously alluded to the possibility that language training in apes facilitates higher levels of ToM and self-awareness generally, especially since in children a concept of self and ToM are linked to language development. As others have also noted, language-trained apes because of their close contact with humans over many years are highly human-enculturated, and as a consequence of the enculturation alone exhibit behaviors and cognitions not shown in non-enculturated apes (Tomasello 1999; Gomez 2004; Byrnit 2005). For example, chimpanzees, hand-reared for at least 1 year looked more at a human face and eyes for clues to attention and were generally more adept at getting the attention of an inattentive human in cooperative tasks than non-hand-reared chimpanzees (Gomez 2004).

Using a modified version of the orangutan ToM experiment mentioned earlier, Whiten (2000) showed that three language-trained chimpanzees performed at a higher level (i.e., in attributing a state of knowledge ignorance) than the non-language trained orangutan. Hence, enculturation and language learning may enhance ToM capacities in apes, though this needs to be explored more systematically, especially as it has not been shown that this language-facilitated level of ToM also includes understanding of false beliefs.

Of course, linguistic capacity in an ape potentially also can serve as a direct means to assess self-awareness, in that one can ask the subject about it's internal states or its understanding of its reflection in a mirror. The gorilla female "Koko" was raised with human companions and taught ASL from age one (in 1972), acquiring over 1,000 gesture signs and a rich understanding of spoken English over the next three decades (e.g., Patterson and Gordon 2001). Similar to what has been well documented for the linguistic achievements of chimpanzees and bonobos (Rumbaugh et al. 2003), Koko uses her sign vocabulary in intentional, meaningful "conversations" with her human companions/experimenters and on occasion with gorilla companions. From the beginning, Koko had access to mirrors, which by age 3.5 years she used for self-directed behaviors, and she "passed" the mirror mark test when given for the first time at age 19 (Patterson and Cohn 1994). Of particular interest here is that when in the context of the mirror mark test Koko was asked "who are you? "and "who is that?," she responded with a variety of multi-sign phrases that consistently contained one or more of the signs "Koko," "me" or "gorilla." Moreover, examination of her daily multi-sign productions near the time of the mirror test showed that she used these three signs infrequently and context-appropriately in relation to herself or in distinguishing herself from others (Patterson and Cohn 1994). Koko also acquired an extensive vocabulary for describing internal states, including emotions (e.g., mad, sad, thirsty, hurt, sorry), which she used appropriately when she was queried about her feelings in a manner comparable to tests used with children (Patterson 1980). These and related observations of Koko (e.g., use of self-conscious emotion words like "embarrassment" or "shame," engaging in self-talk and symbolic play) and of other language-trained apes strongly suggest that they are capable of generally higher levels of self-awareness, including emotional awareness, than has been demonstrated in experiments with non-language trained apes. Additional systematic, well-controlled research, however, is needed on both language-trained and non-language-trained apes to more firmly establish and define the extent of these metacognitive capacities and to assess the specific influence of language training and/or enculturation.

In this section, we have ranged over a fairly wide array of experimental work on ToM and related self-awareness capacities in great apes in an effort to define how these may differ from the metacognitions of humans. Before moving on to an examination of the underlying neural machinery, we will present some general conclusions and implications that can be derived from the evidence so far.

First, we agree with the general conclusion reached by Call and Tomasello (2008) and others, namely, that the experimental evidence indicates that chimpanzees and other great apes achieve a level of ToM that is bound to knowledge, perception, desire and goal attribution but does not reach belief attribution. To this we would add the caveat that at least in chimpanzees and orangutans ToM may be domain-specific to, or facilitated by, competitive social contexts. A direct implication of this conclusion is that chimpanzees (and perhaps all apes) do not have full cognitive empathy as we previously suggested, and perhaps they also lack the fully differentiated or blended emotions of humans. This is likely as the existence of self-conscious emotions has not been convincingly demonstrated in apes. Recall that these emotions require a higher level of self- and other-awareness than has been shown for apes.

This conclusion about ape emotional awareness is consistent with Koski and Sterck's (2010) assessment of chimpanzee empathic capacities. They have proposed a model of chimpanzee empathy based on the gradual development of empathy alongside cognitive capacities in children—from emotional contagion in neonates, to egocentric, to veridical, and to full cognitive empathy by age 3–4 years. Based on their assessment of chimpanzee cognitive capacities, they hypothesize that chimpanzees attain the level of veridical empathy. This level allows apes to respond appropriately to another's needs so long as these are similar to its own needs in the same context. Compared to full cognitive empathy, veridical empathy does not entail an understanding of another's unique circumstances or beliefs that, for example, may produce conflicting emotions, or emotions contradicted by overt behavior. In their view, the chimpanzee level of empathy exceeds the level of emotional contagion typical of young infants, but it does not appear to reach the level of cognitive empathy of 4-year old children. As a result, "humans may be more tuned into others' needs than are chimpanzees," while chimpanzees are "less empathic and prosocial in general than even young children are" (Koski and Sterck 2010).

Second, it seems clear to us that in comparing humans to apes what is truly different about human ToM is its link to language. Not only does this linkage catapult humans into a higher order meta-representational cognitive domain, but perhaps more importantly it also changes what we might call the motivational structure of ToM. By this we mean that humans have evolved as "epistemic engines" (Churchland and Churchland 1983)—as information seekers, manufacturers *and* information sharers. We might say that we need to read other minds so as to better populate them with our own thoughts. The human drive to share new information is basic to the (evolved) function of language and requires ToM to facilitate human collaborative relationships. Language, in turn, facilitates the full conception of a self along with autobiographical memory (Parker 1998) and the conscious experience of well-differentiated emotions. This suite of "social-cognitive skills" represents a qualitative rather than a quantitative difference between apes and humans (Herrmann et al. 2007), and it allows humans to engage in complex forms of social competition and cooperation that are the basis for cultural diversification, adaptation and evolution (Tomasello and Herrmann 2010). By contrast, apes are generally neither self-reflective creatures, nor are they particularly interested or motivated to share their thoughts or knowledge. Even language-trained apes do not naturally display joint attentional behavior when communicating with each other or with humans, nor do they draw others' attention to interesting things that they experience, thus apparently failing to appreciate the "intersubjectivity" of language (see review by Byrnit 2005). Apparently, humans, along with some other species (e.g., dogs) have evolved the special ability to attend closely to human communicative cues and to understand the essentially cooperative nature of communication (Hare 2007).

Third, based on the behavioral evidence it is difficult at present to conclude whether or not and why human metacognitive abilities are unique. Although we have focused on ToM and related capacities in apes and humans—because that is where most of the research has been concentrated—these metacognitive abilities appear to have a wider phylogenetic distribution than we might have anticipated (e.g., in dogs and corvids). Within the order primates, ToM abilities are not restricted to great apes and humans, in that rhesus monkeys, for example, also appear to understand what a human knows about the location of a food in a competitive task (reviewed in Santos et al. 2007). Consequently, it will take much further analysis and thought to understand the selection pressures (i.e., socio-ecological conditions) that have brought about the apparent convergent evolution of these capacities in different lineages (e.g., see Fitch et al. 2010; Byrne and Bates 2010). As already noted, in some species (e.g., ravens, rhesus monkeys, chimpanzees, and orangutans) a domain-specific form of ToM may have evolved in response to shared social-competitive contexts. As will become clear in the following section, neurobiological data on self-awareness and ToM in primates and other animals greatly aid our understanding of species similarities and differences in these abilities and how they evolved.

8.10 Neural Correlates

In humans the neural correlates of emotional awareness, mindreading, and cognitive empathy are well known (Lane 2008; Adolphs 2009). Recall that mindreading or ToM and cognitive empathy are essentially the same function—i.e., both involve taking another's perspective, cognitively and emotionally. Moreover, the ability to take another's perspective is linked to the capacity to introspect, with both controlled by a shared set of brain structures, principally in the frontal and temporal lobes (see Frith and Frith 2010). We should note that many brain areas are activated during ToM tasks, and it still needs to be determined which areas are specialized for this purpose and which components together constitute a system for mentalizing (see review by Mar 2011). In this section we will focus on those brain areas that are well established in humans as vital for mentalizing and for which we have comparative anatomical and/or functional data that allow us to determine if they are uniquely human specializations. Our conclusions, in turn, will inform our evolutionary considerations in the final section of this chapter.

Determining what is unique about human brain structures or mechanisms requires not only comparative anatomical data but also comparative data on brain function simply because neuroanatomical similarity is no guarantee of functional similarity. For example, while there is much commonality in the neural circuitry of emotions among primates (Gothard and Hoffman 2010)—an expression of evolutionary conservatism—such shared, homologous components nevertheless may have evolved different functions that can only be revealed through further physiological-functional studies. Morphologically, human and ape brains differ in overall brain size and in the relative size or proportion of brain structures (Aldridge 2010). However, while there may be no new or unique neural structures that differentiate the human brain from the ape brain, we should expect that size-dependent reorganization entails functional reorganization, including the emergence of novel brain functions. Ideally, therefore, we need comparative studies of the neural correlates of social cognition in each of the apes and humans in order to draw firm conclusions about functional differences. In reality, very few such studies have been done on apes, so that for the most part our best inferences about functional differences at this stage will need to rely on the comparative neurobiological data.

It has been known for some time that in social primates frontal and temporal cortical areas and closely connected subcortical structures, especially the amygdaloid complex, constitute a neural system that is vital to the formation and maintenance of social relationships (Kling and Steklis 1976; Steklis and Kling 1985). In humans, along with overall brain and neocortical enlargement, these frontal and temporal brain regions have become significantly expanded and functionally further elaborated for complex socio-emotional processing, including the representation of emotions and thoughts about self and others (i.e., ToM). Because the prefrontal cortex is critically—though not exclusively—involved in these capacities, much comparative research has focused on this region. There has been debate about whether or not the prefrontal cortex is disproportionately enlarged and contains more white matter in humans compared to great apes (Semendeferi et al. 2002; Sherwood et al. 2005; Schoenemann et al. 2005). More recent and comprehensive comparative morphological work (Aldridge 2010), however, shows that, compared to all apes, the human prefrontal cortex (i.e., cortex anterior to the precentral sulcus) is morphologically different and significantly larger than in all apes, indicating that significant reorganization of cortical and subcortical components occurred in the course of human evolution. This fits well with the emerging picture of the prefrontal cortex as playing a key role in a variety of social cognitions in which humans appear to excel (e.g., self-image, mentalizing about self and others, emotion, thought and behavioral regulation, for review see Heatherton 2011).

A subarea of the prefrontal cortex—known as the "frontal pole" or Brodmann's area 10—may be specialized for these mentalizing functions. This area is disproportionately enlarged in humans compared to other apes, both absolutely and as a percent of brain volume (Semendeferi et al. 2001). Neural imaging and lesion studies in humans have clearly shown area 10 to be centrally involved in various forms of self-regulation and self-projection (reviewed in Buckner and Carroll 2006). The latter include shifting perspectives from self to other (i.e., ToM), decoupling the self from the present so as to project the self into an imagined future (i.e., prospecting), and the recollection of autobiographical memories, which as we have noted before, is critical to a sense of self that is temporally continuous. This area is a key component of the human brain's "default network," a system that is preferentially activated when attention is focused on introspection, prospection, and taking another's perspective (Buckner et al. 2008, for review). A comparative study of neural activity in homologous areas of humans and chimpanzees during resting states (Rilling et al. 2007) found an overall similar pattern of default network activation, which was cautiously interpreted as evidence supportive of chimpanzee capacity for introspection or mentalizing. There were also some notable differences, however, with regard to levels of activation in specific components of the default system. Differences included a higher level of activation in chimpanzee ventromedial prefrontal cortex, which the authors suggest indicates a more "emotion-laden" episodic memory retrieval while resting, and lower levels of activation in left hemisphere areas involved in human language processing (Rilling et al. 2007). This pattern of similarities and differences in resting state brain activation match our previous conclusions of chimpanzees attaining some level of ToM ability, but that in the absence of language their sense of self, including the abstract representation of emotional states, are perhaps qualitatively different from those of humans.

We previously suggested that one significant way in which the human self may be experientially different from that of other primates is in its reliance on autobiographical memory. Imaging studies of area 10 show a consistent functional division of labor in humans, with lateral portions of this area consistently activated during episodic memory retrieval, which includes autobiographical memory, while medial portions are active while attending to one's own emotions (Gilbert et al. 2006). We might speculate that this division of labor in human area 10 represents in part a functional specialization for autobiographical memory retrieval and hence forms the basis for a uniquely human experience of the self. There is indeed good reason to distinguish autobiographical memory from episodic memory more generally. In a recent review, Fivush (2011) argues that episodic memory consists of the "specific what, where, and when of an experience," while autobiographical memory involves "autonoetic consciousness, the awareness of self having experienced the event in the past, which involves mental time travel." Fivush suggests that episodic memory is shared with many other species because it does not necessitate "autonoetic awareness." Autobiographical memory, in contrast, builds further on episodic memory by including the self as the experiencer of the events and in linking events together so as to form a coherent life narrative. Autobiographical memory thus is dependent on a third person view of the self, combined with symbolic language for providing a running narrative of events experienced by the self. Apes and other cognitizing animals may well have episodic memory capacities, more or less developed in accordance with socio-ecological necessities, but in the absence of symbolic language nevertheless lack an autobiographical component and the resultant connected deeper sense of self.

The ape–human distinction we mentioned earlier concerning the capacity to infer mental states about intentions and desires versus ones about beliefs, may also involve human neural specializations in frontal and other areas. Recording event-related potentials in children and young adults showed the mid-frontal region (i.e., prefrontal cortex) to be active during tasks of reasoning about desires and beliefs (Liu et al. 2009a, b). Interestingly, children who failed a false belief task did not show frontal activation. Of further particular note was the demonstrated neural dissociation between desire-reasoning vs. belief-reasoning tasks, given that the latter capacity develops later in children. Results showed that reasoning about beliefs involved posterior brain areas (especially temporal–parietal junction) in addition to frontal areas, perhaps requiring a longer developmental time (Liu et al. 2009b). This demonstration of a neuro-developmental dissociation between desire and belief reasoning, of course, dovetails well with our previous conclusion concerning ape capacity for understanding desires but not beliefs. It will take more comparative study to tell us if the apes' failure to understand beliefs reflects differences in frontal and/or temporal–parietal organization and function.

Another important aspect of frontal lobe reorganization during human evolution may have involved its connectivity and interaction with the adjacent anterior insular cortex (or fronto-insular cortex, FI) and anterior cingulate cortex (ACC), both of which contribute to emotional self-awareness, empathy and social awareness (e.g., monitoring and regulating social emotions), and rapid decision-making in contexts of uncertainty (see reviews by Allman et al. 2010; Lamm and Singer 2010; Craig 2010; Medford and Critchley 2010). Of special interest are the relatively recently studied von Economo neurons (VENs) found exclusively and in great number in FI and ACC of apes and humans, but there are far more of them in the human brain. For example, compared to great apes there is a 100-fold difference in average number of VENs in the FI of adult humans (Fig. 2 in Allman et al. 2005). These large bipolar neurons have fast-conducting, long axons that connect the FI and ACC with each other and to frontal (especially area 10) and temporal lobe structures (e.g., temporal pole, amygdala) involved in socio-emotional processing (Allman et al. 2010). The location, morphology, and connectivity of VENs are consistent with their functioning to quickly relay to frontal and temporal areas and to bring into awareness the results of processing interoceptive (e.g., pain, heartbeat, gut motility) and exteroceptive (e.g., social signals of approval or disapproval) information in FI and ACC. Thus, Allman et al. (2005) suggest that VENs relay the output of FI and ACC—awareness of a motivation to act based on a "gut feeling" or "intuition" in response to external (especially social) circumstances—to frontal and temporal areas, where "fast intuitions are melded with slower, deliberative judgments." In their view, the VEN system is itself not the basis for ToM, but rather provides a critical input to the frontal-temporal structures that are the neural substrate for ToM. Human clinical and imaging studies support this view in that VEN ontogeny (i.e., migration patterns and connectivity) appears to be adversely affected in autism spectrum disorder (Allman et al. 2005), while in fronto-temporal dementia (FTD) VENs are selectively and dramatically reduced in number in FI and ACC (Allman et al. 2010; Seeley 2010).

Although there are no VENS in the temporal lobe, there is good evidence that temporal lobe structures to which they likely project do indeed play a vital part in socio-emotional processing, including mentalizing. FTD is particularly noteworthy from the present standpoint in that it selectively compromises socio-emotional functioning, while initially sparing many aspects of cognitive functioning. FTD severely impairs empathy, emotion recognition and regulation, particularly the processing of self-conscious emotions, that together can be described as a "loss of self" (Levenson and Miller 2007). The temporal pole, in particular, which is well connected to the orbito-frontal cortex and amygdala, is a component of the brain's default network and imaging studies have shown its activation in a variety of ToM tasks, including the inferring of emotional states in others (Olson et al. 2007, for review). This is consistent with cases of FTD in which the temporal pole is selectively affected, in that such patients are commonly described as severely lacking in empathy (Olson et al. 2007). Thus the temporal pole is perhaps part of a phylogenetic reorganization of frontal and temporal lobe structures that mediate enhanced emotional and cognitive perspective-taking, personality and a unique sense of self in humans.

Recently, VENs have also been discovered in the insular cortex of a diverse set of mammalian lineages (Butti and Hof 2010), and not just in large-brained ones as was previously thought, suggesting that despite their morphological similarity and anatomical location VENs play different roles in ecologically and behaviorally diverse species. One way in which species differences in functional specialization of VENs

EMOTIONAL AWARENESS IN PRIMATES

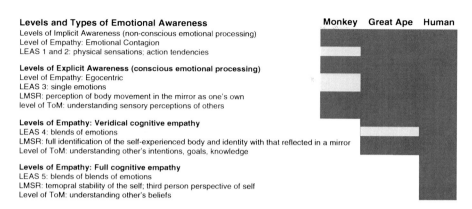

Fig. 8.3 Emotional awareness in primates

may be achieved is through differences in gene-protein expression. Compared to other neurons, human VENs preferentially express neurotransmitter receptors known to be important in social bonding and the anticipation of reward or punishment under conditions of uncertainty (Allman et al. 2005). VENs in ACC of great apes and humans also preferentially express proteins related to visceral monitoring (i.e., pain, immune, and digestive functions), but humans show a significantly larger percentage of VENs expressing these proteins (Stimpson et al. 2011). This biochemical difference may well contribute to an enhanced interoceptive awareness and thus a role of intuition and gut feelings in human decision-making compared to that of apes.

Another way to achieve functional specialization is in the maturational schedule of VENs as a component of species differences in life history. Humans and many other primates are relatively slow maturing, and human VENs mature late postnatally, increasing in number until age 4 (Allman et al. 2010). This opens the possibility of postnatal environmental influences on VEN development. Here it is of interest that one of the gorillas in Allman's sample had been reared in a human environment and, alongside Koko, had been taught to communicate with ASL. Remarkably, the FI and ACC in this gorilla showed VEN numbers approaching the lower end of the range for humans (Allman et al. 2010). It is tempting to speculate that the unusual amount of socio-cognitive stimulation induced or facilitated the proliferation of VENs.

In this section we have explored the question of whether humans have enhanced and/or specialized neural circuitry for mentalizing, especially emotional and other kinds of self-awareness (see Fig. 8.3 for a summary comparison of humans, great apes and monkeys). At present, the best evidence available points to anatomical and functional reorganization in frontal and temporal lobe regions of the human brain as key substrates for these apparently unique forms of human socio-emotional processing and metacognition. In particular, human VEN specializations in FI and ACC provide critical input to and interconnect these regions so as to form overall a specialized system for human metacognition, full cognitive empathy, and self-regulation, including the experience of social emotions like guilt and shame.

Cognitive empathy, in particular, requires emotion regulation via the executive brain, of which the disproportionately expanded frontal polar cortex of humans forms a critical part. VENs may well be at the center of the human capacity to assimilate and respond adaptively to the growing complexity of human social networks (Allman et al. 2002).

8.11 Discussion and Evolutionary Considerations

In a recent book by Paul Ekman and the Dalai Lama (2008), the authors concluded that emotional awareness is the key to psychological balance and compassion. Paul Ekman was one of the first Western scientists to study emotion objectively including seminal research on facial expression. The Dalai Lama is the current spiritual leader of the Buddhist religion. Their joint conclusion reflects an integration of Eastern and Western perspectives. They argue that awareness of one's own emotions is difficult and requires effortful skill development. They see language as essential to this ability, in that without different labels for each mental state, it is difficult to reflect on their nature or consider how one wants to enact them in future emotional episodes. They state that without words, one cannot reflect on what has or could occur. Awareness of emotion is therefore critical to the regulation of one's own emotional states and thus to adaptive social behavior. Yet, this critical aspect of emotion is not considered an inherent or core aspect of emotion.

One reason for this may be that it has not been previously established that emotional awareness is a unique human characteristic. Neural models of human emotion typically include a hierarchy of brain structures with paralimbic structures such as anterior cingulate cortex, orbitofrontal cortex, and insula at the top of the hierarchy (Davidson et al. 2000; Gazzaniga et al. 2009). Neuroanatomical models of emotion in humans are often validated by lesion, single unit and connectivity studies in non-human primates. These models do not typically aim to account for the conscious experience or awareness of emotion. Yet, if the ability to be consciously aware of one's own emotions is an important emotional function, and if non-human primates lack this capacity, then a complete neural model of human emotion cannot be adequately validated using animal models.

We believe that the foregoing review indicates that emotional awareness is a unique human characteristic and that its neural substrates should be part of the fundamental neural architecture of emotion. We agree that the medial prefrontal cortex subserves a variety of functions unrelated to emotion. For example, the paracingulate cortex serves a critical role in coordinating a network involved in mentalizing, which may, depending upon the circumstances, involve understanding what someone is thinking and not what someone is feeling (Frith and Frith 1999). However, one of us has argued previously that mentalizing is a domain-general function that applies to emotion as well as cognition (Lane 2000; Lane 2008), particularly the function discussed above known as cognitive empathy. When applied to the self, also as noted above, it involves reflecting upon one's own feeling. Although this function could be considered a "cognition-emotion interaction," or a key ingredient of the "emotion regulation" processes known as reappraisal or suppression, the central importance of emotional awareness in adaptive behavior suggests that the neural basis of it should be considered part of the fundamental neural circuitry of emotion.

Perhaps a second reason why this critical aspect of emotion is not considered an inherent or core aspect of emotion is that this ability is variable across people. The existence of such individual differences fits well with the fact that mentalizing is a domain-general function that can be used in many different ways depending upon the type of information that is processed. Among individuals who have been physically, emotionally or sexually abused, it is commonly observed that they are extremely attentive to the mental states of others, related to the anticipation of threat, but are far less adept at being aware of their own feelings (Paivio and Laurent 2001). Thus, the capacity for emotional self-reflection or awareness of the emotions of others may be underdeveloped or lacking depending upon one's developmental history. This history may result in dissociations in the awareness of the feelings of the self and others.

If this view of mentalizing as a domain-general function is correct, it provides another vantage point on the evolutionary sequence of the emergence of mentalizing and emotional awareness. According to this view, the domain-general function of mentalizing did not originate for the purpose of attending to other people first, or attending to the self first. Rather, once the mentalizing function evolved it was used in the service of both functions and provided the neural basis for what Tomasello and colleagues (2005) consider to be the unique aspect of human cognition, which is the capacity to share experiences and collaborate. In order to share experiences one needs to know what someone else is experiencing, what you yourself are experiencing, and be able to compare and contrast the two. This ability to collaborate, no doubt greatly facilitated by the advent of language, is the engine for human creativity, as expressed in the arts, the humanities, scientific discoveries and culture in general.

We agree that human evolution led to a unique kind of self-awareness (see also Focquaert and Platek 2007) that evolved pari passu with language capacity, enabling "future directed co-operation" (Brinck and Gardenfors 2003) as a key human adaptation. The model presented by Brinck and Gardenfors (2003) fits particularly well with our view of ape–human differences in cognition, in drawing attention both to the language-dependent and greater cognitive demand of "future-directed co-operation" compared to "competitive co-operation." In their model, the former kind of cooperation that apes are incapable of, the greater cognitive sophistication is required because it concerns goals (e.g., future resource allocation in times of food shortage) neither presently encountered nor perhaps ever encountered before. Hence the mental simulation of future scenarios or social strategies becomes a vital adaptive mental quality. It would also favor the refinement of human emotion regulation (via awareness), and especially the "domestication" of human emotions (see Hare 2007) in the service of cooperation, as well as the evolution of pro-social, self-conscious emotions (e.g., guilt and shame) that in humans are dependent for their effectiveness on full self-awareness. In this evolutionary view, therefore, the unique aspects of human mentalizing evolved *both* for understanding and modeling one's own and another's mind.

This dual nature of mentalizing also helps to explain the phenomenon of cognitive empathy and its role in the development of emotional awareness. Simulation theory states that the ability to know what someone else is feeling requires the ability to imagine how you yourself would feel if you were that person in that situation (Gallese and Goldman 1998). Mitchell and colleagues (2005) have generated neuroimaging evidence indicating that self-reflection may be used to infer the mental states of others when they are sufficiently similar to self. From this vantage point self-awareness is necessary or at least useful for the awareness and understanding of others. On the other hand, a child's ability to understand what it is feeling is dependent upon a parent or caretaker accurately identifying what the child is feeling and responding to the needs that generated the feeling, e.g., feeding in response to expressions of hunger (Gergely and Watson 1996). From this vantage point the parent or caretaker's capacity for cognitive empathy is necessary or at least useful for the child to become aware of its own experiences. In a similar manner in adults, a psychotherapist typically helps clients to identify feelings that the client could not adequately understand or represent to him- or herself (Greenberg 2004).

Conversely, developmental experiences of abuse are typically accompanied by failures of cognitive empathy on the part of the perpetrators (Marshall and Holtzworth-Munroe 2010) and other caregivers, which likely contributes to the limitations in awareness of one's own emotions in victims of abuse (Pollak and Sinha 2002). Relatedly, the perpetrators of abuse are themselves often victims of abuse, and their abusive behavior may reflect a deficit in the capacity for emotional awareness of both self and other. These considerations suggest that the capacity to have emotional awareness arose in humans phylogenetically, but requires a particular kind of social environment during ontogeny in order for the capacity to be adequately developed and expressed.

This leads us to reconsider the mind of the perpetrator of the shooting atrocity in Tucson. We can't help but wonder how someone could engage in such "inhuman," unfeeling behavior. Our discussion above suggests that at the time of the shooting this person's capacity for mentalizing, empathy and compassion was indeed greatly compromised. Rather than simply categorizing this behavior as mysterious and pathological, it can instead be understood at least in part as an expression of either the temporary or more permanent failure of the capacity for emotional awareness. This function itself is fragile, dependent upon a particular kind of context for its development and is quite variable across the human population. Knowing this may energize efforts to nurture its development in children and to create remedial programs to enhance emotional awareness in adults. A better understanding of its neural substrates and phylogenetic origins could potentially enhance such efforts. It is an important question for the future to determine the extent to which this capacity can be changed once an individual has reached adulthood with limitations in this function, and if it cannot, why not.

Acknowledgements We are grateful to Netzin Steklis for her contributions to many stimulating discussions on the topic of emotional awareness in primates, for her thorough reading of the manuscript, and her constructive suggestions for improvement.

References

Adolphs R (2009) The social brain: neural basis of social knowledge. Annu Rev Psychol 60:693–716

Aldridge K (2010) Patterns of differences in brain morphology in humans as compared to extant apes. J Hum Evol 60:94–105

Allman J, Hakeem A, Watson K (2002) Two phylogenetic specializations in the human brain. Neuroscientist 8:335–346

Allman JM, Watson KK, Tetreault NA, Hakeem AY (2005) Intuition and autism: a possible role for Von Economo neurons. Trends Cogn Sci 9:376–373

Allman JM, Tetreault NA, Hakeem AY, Manaye KF, Semendeferi K, Erwin JM et al (2010) The von Economo neurons in frontoinsular and anterior cingulate cortex in great apes and humans. Brain Struct Funct 214:495–517

Anderson JR, Gallup GG Jr (1999) Self-recognition in nonhuman primates: past and future challenges. In: Haug M, Whalen RE (eds) Animal models of human emotion and cognition. American Psychological Association, Washington, pp 175–194

Aureli F, Whiten A (2003) Emotions and behavioral flexibility. In: Mastripieri D (ed) Primate psychology. Harvard University Press, Cambridge, pp 289–323

Barrett L, Henzi P, Rendall D (2007) Social brains, simple minds: does social complexity really require cognitive complexity? Philos Trans R Soc B 362:561–575

Berthoz S, Ouhayoun B, Parage N (2000) Etude preliminaire des niveaux de conscience emotionnelle chez des patients deprimes et des controles (Preliminary study of the levels of emotional awareness in depressed patients and controls). Ann Med Psychol (Paris) 158:665–672

Blatt S, Wein S, Chevron E, Quinlan D (1979) Parental representations and depression in normal young adults. J Abnorm Psychol 88:388–397

Boesch C (1991) Teaching among wild chimpanzees. Anim Behav 41:530–532

Bolhuis JJ, Wynne CDL (2009) Can evolution explain how minds work? Nature 458:832–833

Brinck I, Gardenfors P (2003) Co-operation and communication in apes and humans. Mind Lang 18:484–501

Bruch H (1973) Eating disorders: obesity, anorexia nervosa and the person within. Basic Books, New York

Buckner RL, Carroll DC (2006) Self-projection and the brain. Trends Cogn Sci 11:49–57

Buckner RL, Andrews-Hanner JR, Schacter DL (2008) The brain's default network: anatomy, function, and relevance to disease. Ann N Y Acad Sci 1124:1–38

Butti C, Hof PR (2010) The insular cortex: a comparative perspective. Brain Struct Funct 214:477–493

Bydlowski S, Corcos M, Jeammet P, Paterniti S, Berthoz S, Laurier C, Chambry J, Consoli S (2005) Emotion-processing deficits in eating disorders. Int J Eat Disord 37:321–329

Byrnit J (2005) Primate theory of mind: a comparative psychological analysis. Psykologisk Institut, Aarhus Universitet, Aarhus

Byrne RW, Bates LA (2010) Primate social cognition: uniquely primate, uniquely social, or just unique? Neuron 65:815–830

Byrne RW, Whiten A (1988) Machiavellian intelligence: social expertise and the evolution of intellect in monkeys, apes and humans. Clarendon, Oxford

Call J, Tomasello M (2008) Does the chimpanzee have a theory of mind? 30 years later. Trends Cogn Sci 12:187–192

Carruthers P (2008) Meta-cognition in animals: a skeptical look. Mind Lang 23:58–89

Christoff K, Gabrieli JDE (2000) The frontopolar cortex and human cognition: evidence for a rostrocaudal hierarchical organisation within the human prefrontal cortex. Psychobiology 28:168–186

Churchland PS, Churchland PM (1983) Stalking the wild epistemic engine. Nous 17:5–18

Ciarrochi J, Caputi P, Mayer JD (2003) The distinctiveness and utility of a measure of trait emotional awareness. Pers Individ Dif 34:1477–1490

Consoli S (2005) Social anxiety is associated with higher levels of emotional awareness in obese patients waiting for gastric banding surgery. In: Symposium conducted at the American Psychosomatic Society 63rd annual meeting, Vancouver

Consoli SM, Lemogne C, Roch B, Laurent S, Plouin P-F, Lane RD (2010) Differences in emotion processing in patients with essential and secondary hypertension. Am J Hypertens 23: 515–521

Craig ADB (2010) The sentient self. Brain Struct Funct 214:563–577

Lama D, Ekman P (2008) Emotional awareness: overcoming the obstacles to psychological balance and compassion. Times Books, NY

Damasio AR (1994) Descartes' error: emotion, reason, and the human brain. G.P. Putnam's Press, New York

Davidson RJ, Putnam KM, Larson CL (2000) Dysfunction in the neural circuitry of emotion regulation: a possible prelude to violence. Science 289:591–594

Dennett DC (1996) Kinds of minds: toward an understanding of consciousness. Basic Books, New York

de Waal FBM (1996) Good natured. Harvard University Press, Cambridge

de Waal FBM (2008) Putting the altruism back into altruism: the evolution of empathy. Annu Rev Psychol 59:279–300

Donges U-S, Kersting A, Dannlowski U, Lalee-Mentzel J, Arolt V, Suslow T (2005) Reduced awareness of others' emotions in unipolar depressed patients. J Nerv Ment Dis 193:331–337

Ekman P (ed) (1973) Darwin and facial expression: a century of research in review. Academic, New York

Emery NJ, Clayton NS (2004) The mentality of crows: convergent evolution of intelligence in corvids and apes. Science 306:1903–1907

Farthing GW (1992) The psychology of consciousness. Prentice Hall, Englewood Cliffs

Fischer K (1980) A theory of cognitive development: the control and construction of heirarchies of skills. Psychol Rev 87:477–531

Fitch WT, Huber L, Bugnyar T (2010) Social cognition and the evolution of language: constructing cognitive phylogenies. Neuron 65:795–814

Fivush R (2011) The development of autobiographical memory. Annu Rev Psychol 62:559–582

Focquaert F, Platek SM (2007) Social cognition and the evolution of self-awareness. In: Platek SM, Keenan JP, Shackleford TK (eds) Evolutionary cognitive neuroscience. MIT Press, Cambridge, pp 457–498

Fraser ON, Bugnyar T (2010) Do ravens show consolation? Responses to distressed others. PLoS One 5:1–8

Frewen P, Lanius R, Lane R, Neufeld R, Densmore M (2008) Neural correlates of individual differences in levels of emotional awareness during trauma script-imagery. Psychosom Med 70:27–31

Frijda N (1986) The emotions. Cambridge University, Cambridge Press

Frith C, Frith U (1999) Interacting minds—a biological basis. Science 286:1692–1695

Frith U, Frith C (2010) The social brain: allowing humans to boldly go where no other species has been. Philos Trans R Soc B 365:165–176

Gallese V, Goldman A (1998) Mirror neurons and the simulation theory of mind-reading. Trends Cogn Sci 2:493–501

Gallup GG Jr (1979) Self-awareness in primates. Am Sci 67:417–421

Gallup GG Jr (1982) Self-awareness and the emergence of mind in primates. Am J Primatol 2:237–248

Gazzaniga M (1998) The mind's past. University of California Press, Berkeley

Gazzaniga M, Ivry RB, Mangun GR (2002) Cognitive neuroscience: the biology of the mind. Norton, New York

Gazzaniga MS, Ivry RB, Mangun GR (2009) Emotion. In: Cognitive neuroscience: the biology of the mind, vol 3. W. W. Norton & Company, New York, pp 364–387

Gergely G, Watson JS (1996) The social biofeedback theory of parental affect-mirroring. Int J Psychoanal 77:1181–1211

Gilbert DT, Wilson TD (2007) Prospection: experiencing the future. Science 317:1351–1354

Gilbert SJ, Spengler S, Simons JS, Steele JD, Lawrie SM, Frith CD et al (2006) Functional specialization within rostral prefrontal cortex (area 10): a meta-analysis. J Cogn Neurosci 18:932–948

Gomez JC (1998) Assessing theory of mind with nonverbal procedures: problems with training methods and an alternative "key" procedure. Behav Brain Sci 21:119–120

Gomez JC (2004) Apes, monkeys, children, and the growth of mind. Harvard University Press, Cambridge

Gothard KM, Hoffman KL (2010) Circuits of emotion in the primate brain. In: Platt ML, Ghazanfar AA (eds) Primate neuroethology. Oxford University Press, New York, pp 292–315

Greenberg LS (2004) Emotion-focused therapy. Clin Psychol Psychother 11:3–16

Gusnard D, Akbudak E, Shulman G, Raichle M (2001) Medial prefrontal cortex and self-referential mental activity: relation to a default mode of brain function. Proc Natl Acad Sci U S A 98: 4259–4264

Fragaszy DM, Greenberg R, Visalberghi E, Ottoni EB, Izar P, Liu Q (2010) How wild bearded capuchins select stones and nuts to minimize the number of strikes per nut cracked. Anim Behav 80:205–214

Happe F (2003) Theory of mind and the self. Ann N Y Acad Sci 1001:134–144

Happe F, Ehlers S, Fletcher P, Frith U, Johansson M, Gillberg C et al (1996) 'Theory of mind' in the brain. Evidence from a PET scan study of Asperger syndrome. Neuroreport 8:197–201

Harcourt AH, de Waal FBM (eds) (1992) Coalitons and alliances in human and nonhuman animals. Oxford University Press, New York

Hare B (2001) Can competitive paradigms increase the validity of social cognitive experiments on primates? Anim Cogn 4:269–280

Hare B (2007) From nonhuman to human mind: what changed and why? Curr Dir Psychol Sci 16:60–64

Hare B, Tomasello M (2004) Chimpanzees are more skillful in competitive than in cooperative tasks. Anim Behav 68:571–581

Hare B, Call J, Tomasello O (2001) Do chimpanzees know what conspecifics know? Anim Behav 61:139–151

Hare B, Melis AP, Woods V, Hastings S, Wrangham R (2007) Tolerance allows bonobos to outperform chimpanzees on a cooperative task. Curr Biol 17:619–623

Hawks J, Wang ET, Cochran GM, Harpending HC, Moyzis RK (2007) Recent acceleration of human adaptive evolution. Proc Natl Acad Sci 104:20753–20758

Heatherton TF (2011) Neuroscience of self and self-regulation. Annu Rev Psychol 62:363–390

Herrmann E, Call J, Hernández-Lloreda MV, Hare B, Tomasello M (2007) Humans have evolved specialized skills of social cognition: the cultural intelligence hypothesis. Science 317:1360–1366

Herrmann E, Hare B, Call J, Tomasello M (2010) Differences in the cognitive skills of bonobos and chimpanzees. PLoS One 5:1–4

Heyes CM (1998) Theory of mind in nonhuman primates. Behav Brain Sci 21:101–114

Johnson S, Baxter L, Wilder L, Pipe J, Heiserman J, Prigatano G (2002) Neural correlates of self-reflection. Brain 125:1808–1814

Kaminski J, Call J, Tomasello M (2008) Chimpanzees know what others know, but not what they believe. Cognition 109:224–234

Kang SM, Shaver PR (2004) Individual differences in emotional complexity: their psychological implications. J Pers 72:687–726

Karmiloff-Smith A (1992) Beyond modularity: a developmental perspective on cognitive science. MIT Press, Cambridge

Kihlstrom JF, Mulvaney S, Tobias BA, Tobis IP (2000) The emotional unconscious. In: Eich E, Kihlstron J, Bower G, Forgas JP, Niedenthal PM (eds) Cognition and emotion. Oxford University Press, New York, pp 30–86

King JE, Figueredo AJ (1997) The five-factor model plus dominance in chimpanzee personality. J Res Personality 31:257–271

Kirsch JA, Gunturkun O, Rose J (2008) Insight without a cortex: lessons from the avian brain. Conscious Cogn 17:475–483

Kling AS, Steklis HD (1976) A neural substrate for affiliative behavior in nonhuman primates. Brain Behav Evol 13:216–238

Kober H, Barrett LF, Joseph J, Bliss-Moreau E, Lindquist K, Wager TD (2008) Functional grouping and cortical-subcortical interactions in emotion: a meta-analysis of neuroimaging studies. Neuroimage 42:998–1031

Koski SE, Sterck EHM (2010) Empathic chimpanzees: a proposal of the levels of emotional and cognitive processing in chimpanzee empathy. Eur J Dev Psychol 7:38–66

Lackner J (2005) Is IBS a problem of emotion dysregulation? Testing the levels of emotional awareness model. In: Symposium conducted at the American Psychosomatic Society 63rd Annual Meeting, Vancouver

Lamm C, Singer T (2010) The role of anterior insular cortex in social emotions. Brain Struct Funct 214:579–591

Lane RD (2000) Neural correlates of conscious emotional experience. In: Lane R, Nadel L, Ahern G, Allen J, Kaszniak A, Rapcsak S et al (eds) Cognitive neuroscience of emotion. Oxford University Press, New York, pp 345–370

Lane RD (2008) Neural substrates of implicit and explicit emotional processes: a unifying framework for psychosomatic medicine. Psychosom Med 70:214–231

Lane RD, Pollerman B (2002) Complexity of emotion representations. In: Barrett LF, Salovey P (eds) The wisdom in feeling. Guilford, New York, pp 271–293

Lane RD, Schwartz GE (1987) Levels of emotional awareness: a cognitive-developmental theory and its application to psychopathology. Am J Psychiatry 144:133–143

Lane RD, Quinlan DM, Schwartz GE, Walker PA, Zeilin SB (1990) The levels of emotional awareness scale: a cognitive-developmental measure of emotion. J Pers Assess 55:124–134

Lane RD, Sechrest B, Riedel R, Weldon V, Kaszniak A, Schwartz G (1996) Impaired verbal and nonverbal emotion recognition in alexithymia. Psychosom Med 58:203–210

Lane R, Fink G, Chua P, Dolan R (1997) Neural activation during selective attention to subjective emotional responses. Neuroreport 8:3969–3972

Lane R, Sechrest L, Riedel R, Shapiro D, Kaszniak A (2000) Pervasive emotion recognition deficit common to alexithymia and the repressive coping style. Psychosom Med 62:492–501

Lane R, Carmichael C, Reis H (2011) Differentiation in the momentary rating of somatic symptoms covaries with trait emotional awareness in patients at risk for sudden cardiac death. Psychosom Med 73:185–192

Lang PJ, Bradley MM, Cuthbert BN (2001) International Affective Picture System (IAPS): instruction manual and affective ratings. Technical report. The University of Florida, The Center for Research in Psychophysiology

Levenson RW, Miller BL (2007) Loss of cells—loss of self: frontotemporal lobar degeneration and human emotion. Curr Dir Psychol Sci 16:289–294

Levine D, Marziali E, Hood J (1997) Emotion processing in borderline personality disorders. J Nerv Ment Dis 185:240–246

Liu D, Meltzoff AN, Wellman HM (2009a) Neural correlates of belief- and desire-reasoning. Child Dev 80:1163–1171

Liu D, Sabbagh MA, Gehring WJ, Wellman HM (2009b) Neural correlates of children's theory of mind development. Child Dev 80:318–326

Loevinger J, Wessler R (1970) Measuring ego development, vol. I: construction and use of a science completion test. Jossey-Bass, San Francisco

Loevinger J, Wessler R, Redmore C (1970) Measuring ego development, vol. II: scoring manual for women and girls. Jossey-Bass, San Francisco

Macellini S, Ferrari PF, Bonini L, Fogassi L, Paukner A (2010) A modified mark test for own-body recognition in pig-tailed macaques (Macaca nemestrina). Anim Cogn 13:631–639

Mar RA (2011) The neural bases of social cognition and story comprehension. Annu Rev Psychol 62:103–134

Marshall AD, Holtzworth-Munroe A (2010) Recognition of wives' emotional expressions: a mechanism in the relationship between psychopathology and intimate partner violence perpetration. J Fam Psychol 24:21–30

Medford N, Critchley HD (2010) Conjoint activity of anterior insular and anterior cingulate cortex: awareness and response. Brain Struct Funct 214:535–549

Mitchell JP, Banaji MR, Macrae CN (2005) General and specific contributions of the medial prefrontal cortex to knowledge about mental states. Neuroimage 28:757–762

Novick-Kline P, Turk C, Mennin D, Hoyt E, Gallagher C (2005) Level of emotional awareness as a differentiating variable between individuals with and without generalized anxiety disorder. J Anxiety Disord 19:557–572

Ochsner K, Knierim K, Ludlow D, Hanelin J, Ramachandran T, Glover G, Mackey S (2004) Reflecting upon feelings: an fMRI study of neural systems supporting the attribution of emotion to self and other. J Cogn Neurosci 16:1746–1772

Olson IR, Plotzker A, Ezzyat Y (2007) The Enigmatic temporal pole: a review of findings on social and emotional processing. Brain 130:1718–1731

Owren MJ, Rendall D, Bachorowski JA (2003) Nonlinguistic vocal communication. In: Mastripieri D (ed) Primate psychology. Harvard University Press, Cambridge, pp 359–394

Paivio SC, Laurent C (2001) Empathy and emotion regulation: reprocessing memories of childhood abuse. J Clin Psychol 57:213–226

Parker A (1998) Primate cognitive neuroscience: what are the useful questions? Behav Brain Sci 21:128

Parr LA (2001) Cognitive and physiological markers of emotional awareness in chimpanzees (*Pan troglodytes*). Anim Cogn 4:223–229

Parr LA, Mastripieri D (2003) Nonvocal communication. In: Mastripieri D (ed) Primate psychology. Harvard University Press, Cambridge, pp 324–358

Patterson FG (1980) Innovative uses of language by a gorilla: a case study. In: Nelson KE (ed) Children's language, vol 2. Gardner, New York, pp 497–561

Patterson FGP, Cohn RH (1994) Self-recognition and self-awareness in lowland gorillas. In: Parker ST, Mitchell RW, Boccia ML (eds) Self-awareness in animals and humans. Cambridge University Press, New York, pp 273–291

Patterson FGP, Gordon W (2001) Twenty-seven years of Project Koko and Michael. In: Galdikas B, Briggs N, Sheeran L, Shapiro G, Goodall J (eds) All apes great and small, vol I: African apes. Kluwer, NY, pp 165–176

Penn DC, Povinelli DJ (2007) On the lack of evidence that non-human animals possess anything remotely resembling a 'theory of mind'. Philos Trans R Soc B 362:731–744

Penn DC, Holyoak KJ, Povinelli DJ (2008) Darwin's mistake: explaining the discontinuity between human and nonhuman minds. Behav Brain Sci 31:109–178

Peterson D, Wrangham R (1996) Demonic males: apes and the origins of human violence. Harvard University Press, Cambridge

Piaget J (1937) La construction du réel chez l'enfant. Delachaux et Niestlé, Neuchâtel

Pinker S (2010) The cognitive niche: coevolution of intelligence, sociality, and language. Proc Natl Acad Sci 107((Suppl 2)):8993–8999

Pollak SD, Sinha P (2002) Effects of early experience on children's recognition of facial displays of emotion. Dev Psychol 38:784–791

Premack D, Woodruff G (1978) Does the chimpanzee have a theory of mind? Behav Brain Sci 1:515–526

Prior H, Schwarz A, Gunturkun O (2008) Mirror-induced behavior in the Magpie (Pica pica): evidence of self-recognition. PLoS Biol 6:1642–1650

Ramnani N, Owen AM (2004) Anterior prefrontal cortex: insights into function from anatomy and neuroimaging. Nat Rev Neurosci 5:184–194

Rilling JK, Barks SK, Parr LA, Preuss TM, Faber TL, Pagnoni G et al (2007) A comparison of resting-state brain activity in humans and chimpanzees. Proc Natl Acad Sci 104:17146–17151

Rochat P (2003) Five levels of self-awareness as they unfold early in life. Conscious Cogn 12:717–731

Rosati AG, Santos LR, Hare B (2010) Primate social cognition: thirty years after Premack and Woodruff. In: Platt ML, Ghazanfar AA (eds) Primate neuroethology. Oxford University Press, New York, pp 117–143

Rumbaugh DM, Beran MJ, Savage-Rumbaugh SS (2003) Language. In: Mastripieri D (ed) Primate psychology. Harvard University Press, Cambridge, pp 395–423

Santos LR, Flombaum JI, Phillips W (2007) The evolution of human mindreading: how nonhuman primates can inform social cognitive neuroscience. In: Platek SM, Keenan JP, Shackleford TK (eds) Evolutionary cognitive neuroscience. MIT Press, Cambridge, pp 433–456

Schoenemann PT, Glotzer LD, Sheehan MJ (2005) Reply to "Is prefrontal white matter enlargement a human evolutionary specialization?". Nat Neurosci 8:538

Seeley WW (2010) Anterior insula degeneration in frontotemporal dementia. Brain Struct Funct 214:465–475

Semendeferi K, Armstrong E, Schleichter A, Zilles K, Van Hoesen G (2001) Prefrontal cortex in humans and apes: a comparative study of area 10. Am J Phys Anthropol 114:224–241

Semendeferi K, Lu A, Schenker N, Damasio H (2002) Humans and great apes share a large frontal cortex. Nat Neurosci 5:272–276

Sherwood CC, Holloway RL, Semendeferi K, Hof PR (2005) Is prefrontal white matter enlargement a human evolutionary specialization? Nat Neurosci 8:537–538

Singer T (2006) The neuronal basis and ontogeny of empathy and mind reading: review of literature and implications for future research. Neurosci Biobehav Rev 30:855–863

Smith A (2006) Cognitive empathy and emotional empathy in human behavior and evolution. Psychol Rec 56:3–21

Smith JD (2007) Species of parsimony in comparative studies of cognition. In: Washburn DA (ed) Primate perspectives on behavior and cognition. American Psychological Association, Washington, pp 63–80

Steklis HD, Kling AS (1985) Neurobiology of affiliative behavior in nonhuman primates. In: Reite M, Field T (eds) The psychobiology of attachment and separation. Academic, New York, pp 93–134

Stimpson CD, Tetreault NA, Allman JM, Jacobs B, Butti C, Hof PR et al (2011) Biochemical specificity of von Economo neurons in hominoids. Am J Hum Biol 23:22–28

Subic-Wrana C, Bruder S, Thomas W, Lane R, Kohle K (2005) Emotional awareness deficits in inpatients of a psychosomatic ward: a comparison of two different measures of alexithymia. Psychosom Med 67:483–489

Subic-Wrana A, Beetz M, Paulussen J, Wiltnik J, Beutel M (2007) Relations between attachment, childhood trauma, and emotional awareness in psychosomatic inpatients. In: Symposium conducted at the American Psychosomatic Society 65th Annual Meeting, Budapest

Subic-Wrana C, Beutel M, Knebel A, Lane RD (2010) Theory of mind and emotional awareness deficits in somatoform patients. Psychosom Med 72:404–411

Tamietto M, de Gelder B (2010) Neural bases of the non-conscious perception of emotional signals. Nat Rev Neurosci 11:697–709

Thompson RA, Lagattuta KH (2008) Feeling and understanding: early emotional development. In: McCartney K, Phillips D (eds) Blackwell handbook of early childhood development. Blackwell Publishing Ltd, Oxford, pp 317–337

Tomasello M (1999) The cultural origins of human cognition. Harvard University Press, Cambridge

Tomasello M, Herrmann E (2010) Ape and human cognition: what's the difference? Curr Dir Psychol Sci 19:3–8

Tomasello M, Carpenter M, Call J, Behne T, Moll H (2005) Understanding and sharing intentions: the origins of cultural cognition. Behav Brain Sci 28:675–691

Trut LN, Plyusnina IZ, Oskina IN (2004) An experiment on fox domestication and debatable issues of evolution of the dog. Russ J Genet 40:644–655

Udell MAR, Dorey NR, Wynne CDL (2010) What did domestication do to dogs? A new account of dogs' sensitivity to human actions. Biol Rev 85:327–345

Vogeley K, Bussfeld P, Newen A, Herrmann S, Happe F, Falkai P et al (2001) Mind reading: neural mechanisms of theory of mind and self-perspective. Neuroimage 14:170–181

Weiskrantz L (2000) Blindsight: implications for the conscious experience of emotion. In: Lane R, Nadel L, Ahern G, Allen J, Kaszniak A, Rapcsak S et al (eds) Cognitive neuroscience of emotion. Oxford University Press, New York, pp 277–295

Wellman HM, Brandone AC (2009) Early intention understandings that are common to primates predict children's later theory of mind. Curr Opin Neurobiol 19:57–62

Werner H, Kaplan B (1963) Symbol formation: an organismic-development approach to language and the expression of thoughts. Wiley, New York

Whiten A (2000) Chimpanzee cognition and the question of mental re-representation. In: Sperber D (ed) Metarepresentations: a multidisciplinary perspective. Oxford University Press, New York, pp 139–170

Wynne CDL, Bolhuis JJ (2008) Minding the gap: why there is still no theory in comparative psychology. Behav Brain Sci 31:152–152

Chapter 9
The Development of Mentalizing and Emotion in Human Children

Shoji Itakura, Yusuke Moriguchi, and Tomoyo Morita

Abstract For human infants, agents—defined as other humans—are the fundamental units of their social world. Agents provide very special stimuli to infants. Researchers of object–person differentiation have proposed a set of rules that infants probably use during their interaction with people as opposed to objects.

S. Itakura (✉)
Department of Psychology, Graduate School of Letters, Kyoto University,
Yoshida-honmachi, Sakyo-Ku, Kyoto 6068501, Japan
e-mail: sitakura@bun.kyoto-u.ac.jp

Y. Moriguchi
Joetsu University of Education, Joetsu, Japan

T. Morita
National Institute for Physiological Science, Okazaki, Japan

This chapter reviews investigations into how children understand and detect both human and nonhuman agents and communicate with them, starting with a definition of mentalizing and summary of the course of its development. A study is presented on infant imitation of a robot's action and a false-belief task with robots, proposing a new research domain called "developmental cybernetics," which studies the interaction between children and robots (Itakura et al., Infancy 13:519–532, 2008). It has been predicted that in ordinary twenty-first-century households, robotics technology will be as common as refrigerators and dishwashers (Asada and Kuniyoshi, Robot intelligence, Iwanami Shoten, Tokyo, 2006). Therefore, exploring developmental cybernetics is important. Finally, two more studies from the perspective of cognitive neuroscience are presented, with a discussion on the usefulness of the neurocognitive approach in understanding the development of mentalizing, alongside two studies concerned with this issue.

9.1 Introduction

For human infants, agents—defined as other humans—are the fundamental units of their social world. Agents provide very special stimuli to infants. Researchers of object–person differentiation have proposed a set of rules that infants probably use during their interaction with people as opposed to objects. For example, Premack (1990) suggested that infants perceive people as perceptual events that are both self-propelled and goal-directed objects. In such cases, adults also perceive people as agents with intention. Spelke et al. (1995) described an infant's concept of a human as follows: "Three aspects of human interactions that are accessible in principle to young infants are contingency (humans react to one another), reciprocity (humans respond in kind to one another's actions), and communication (humans supply one another with information)." They suggested that infants interpret an object's movement with these three principles and the "principle of contact." To explain the contact principle, they used the habituation procedure and showed that infants tend to assume that an object, if it moves, was set in motion by a push from another object (or person). On the other hand, social agents do not require the application of an external force to move. They demonstrated such a perception of agency in 7-month-olds. Agents are not simply physical objects to which new properties have been added. On the contrary, they are animate entities that can move on their own, breathe, eat, drink, look, and engage in actions with objects, or interact with other agents (Gomez 2004).

From the perspective of social cognitive development, Johnson (2003) raised two questions: (1) when do children first attribute a mental state to others? and (2) to whom do they attribute it?

Until 10 years ago, a majority of perceptual and cognitive development studies were conducted without evidence for the brain (Johnson 2005). However, the recent understanding of brain function has improved significantly. Many researchers believe that the interface between cognitive and brain development should now be explored. These techniques can be applied to improve our understanding of the development of mentalizing.

In this chapter, we review investigations into how children understand and detect both human and nonhuman agents and communicate with them. We start with a definition of mentalizing and summarize the course of its development (Sect. 9.2). We then introduce a study on infant imitation of a robot's action and a false-belief task with robots, proposing a new research domain called "developmental cybernetics" (Sect. 9.3), which studies the interaction between children and robots (Itakura et al. 2008). It has been predicted that in ordinary twenty-first-century households, robotics technology will become as common as refrigerators and dishwashers (Asada and Kuniyoshi 2006). Therefore, exploring developmental cybernetics is important. Finally, we introduce two more of our studies from the perspective of cognitive neuroscience (Sect. 9.4) and discuss the usefulness of the neurocognitive approach in understanding the development of mentalizing; two studies concerned with this issue are examind.

9.2 The Development of Mentalizing

"Mentalizing," used by Frith and Frith (2003), is synonymous with the term "theory of mind." They wrote that the phrase "'theory of mind' was not to be taken literally of course, and it certainly did not imply the possession of an explicit philosophical theory about the contents of the mind (Frith and Frith 2003). They pointed out that the theory of mind implicitly assumes that the behavior of others is determined by their desires, attitudes, and beliefs. These are not states of the world but states of the mind; this consideration is crucial. However, in our definition, mentalizing attributes mental states such as goal-directedness, intention, and mind to humans and nonhuman agents. In other words, mentalizing concerns the question of how humans perceive nonhuman agents and attribute mental states to them. Thus, the development of mentalizing is the development of a mind that discovers other minds.

Frith and Frith (2003) characterize four key transitions in the development of mentalizing.

(a) From birth to 3 months. Infants only a few weeks old smile and vocalize more toward humans than objects, even human-like dolls (Legerstee 1992). Eye movements and biological motion can grab infants' attention at an extremely early age. For example, they track the movement of self-propelled objects (Crichton and Lange-Kuettner 1999). Three-month-olds also show more interest in the kinematic patterns of point-light displays of a human walking than in random movement (Bertenthal et al. 1984) .

(b) From 9 months. At this age, infants begin to engage in triadic interactions that involve the referential triangle of child, adult, and some outside entity on which they share attention. Gergely et al. (1995) defined the infant's ability to reason about goals, the "principle of rationality." Infants at this age can distinctly represent the goals of agents and the means used to reach them. The ability to represent goals and to reason (rationality) is considered an important prerequisite for the ability to represent intentions.

(c) From 18 months. This developmental watershed, which marks the end of infancy, is significant for the onset of pretend play, that is, considered an important precursor of the theory of mind. Leslie (1987) postulates that a child at this age has to maintain separate representations of real events from representations of thoughts that no longer need to refer to such events (Leslie 1987). Reliable imitation of intentional actions performed by others, regardless of whether these actions achieve their goal, also emerges at approximately 18 months, as demonstrated by Meltzoff (1995).

(d) From age five years. Children from this age reliably understand false-belief tasks (see Sect. 9.4 for details) that require the attribution of a false belief to others, after which children start to understand more difficult tasks that require the attribution of a belief about another person's beliefs: second-order tasks.

Finally, an implicit version of mentalizing, which emerges first and is concerned with desires, goals, and intentions, is usually dated around 18 months (Frith and Frith 2003). However, we believe that the ability to mentalize (such as sensitivity to social contingency, face recognition, gaze following, and biological motion) is based on social cognition in early infancy. From the perspective of developmental cognitive neuroscience, clarifying the neural mechanism of mentalizing in early infancy must be the next challenge.

9.3 Developmental Cybernetics

In this section, we introduce an exciting new research field in the development of mentalizing called "developmental cybernetics." In the future, robots will not only perform household chores but also serve as caregivers and educators to children. To date, no scientific evidence has ascertained whether children, particularly younger ones, will be amenable to receiving care, let alone learning, from robots as readily as from humans. Despite recent rapid growth in research on developmental cybernetics, it is entirely unknown as to which essential human characteristics must be built into robots to facilitate such learning.

9.3.1 Inference of Robot Intention

One of the earliest fundamental forms of learning from another human is imitation. Imitation begins at birth with neonates who copy adult behaviors within their innately endowed behavioral repertoire (for a review see Meltzoff 2005). As they grow, infants begin to imitate novel behaviors performed either live or televised by adults (Barr and Hayne 2000; Meltzoff 2005). In addition, they can re-enact an adult's novel behavior even after a long delay (Behne et al. 2005). More strikingly, Meltzoff (1995) demonstrated that when adults performed an action that appeared to fail to accomplish their intended goal (e.g., instead of pulling apart two halves of a dumbbell, the adults' hands slipped and the dumbbell stayed intact), 18-month-olds could "imitate" the unobserved but intended act (e.g., pulling the dumbbell apart) rather than the observed but unintended act (e.g., the slippage of the adult's hands off the ends of the dumbbell with the two halves not separated).

Similar results have been found in 15-month-olds (Bellagamba and Tomasello 1999; Johnson et al. 2001; Meltzoff 1999) but not in 12-month-olds (Bellagamba and Tomasello 1999). Similarly, Carpenter et al. (1998) showed that 14- to 18-month-olds were more inclined to imitate an adult's intended actions than accidental actions. These findings indicate that by their second year, infants do not blindly imitate the behavior of others but, instead, base their imitation on their understanding of the intentions and goals of others. This development is perhaps built on another developmental milestone at around 9–12 months of age when infants begin to understand that adult behavior is goal-directed and intentional (Baldwin et al. 2001; Behne et al. 2005; Luo and Baillargeon 2005; Phillips and Wellman 2005; Shimizu and Johnson 2004; Woodward 1998; Gergely and Csibra 2004).

Why are human infants so inclined to copy another person's behavior to the extent that they even "imitate" intended but unconsummated acts? Meltzoff (2005) proposed a "like me" hypothesis whose central tenet is that infants are innately endowed with the ability to see correspondence between the actions of others and those performed by their own body. With experience, infants learn to map their own and failed actions with their internal mental states. Such an innate capacity to construe others' actions as "me relevant" coupled with an acquired understanding of their own mental state allows infants to crack the problem of other minds. They use their own intentional actions as a framework for interpreting the intentional actions of others. As a result, they can selectively imitate another's intended (but not unintended) actions. The existing developmental evidence of infant imitation involving humans as models is largely consistent with the "like me" hypothesis. It is also consistent with evidence that infants do not produce the target act when a mechanical device's behavior has failed to complete an action (e.g., pulling the dumbbell apart; Meltzoff 1995). This inanimate device did not look at the human or interact with the target in a human fashion, either of which might have been sufficient to avoid triggering the "like me" interpretive framework.

What are the basic characteristics of an agent that enable infants to make "me relevant" mapping, infer the agent's goals, and thus imitate the agent's intended but unconsummated actions? One possibility is that such an agent must share human morphological characteristics. This suggestion seems reasonable given the evidence that person recognition in general and face recognition in particular begin in early infancy and develop rapidly (Johnson and Morton 1991; Quinn and Slater 2003). The ability to recognize and interpret faces can, in theory, serve as an essential enabling factor for infants to carry out such "like me" mapping and thus successfully imitate intentional and goal-directed actions.

However, Johnson et al. (1998) suggested that infant recognition of intentional agents is not necessarily isomorphic with person recognition but rather based on a set of nonarbitrary object recognition cues. Johnson et al. (1998) They showed that a novel orangutan-like object (with eyes and nose but no mouth) that appeared to be self-propelled and interacted with infants contingently led 15-month-olds to imitate its unconsummated acts. Furthermore, infants also displayed significantly more communicative behaviors toward the orangutan-like object than another physically similar but faceless, inanimate object.

The results of Johnson et al. (1998) clearly indicate that fully fledged human morphological characteristics are not necessary to engender imitation of intentional acts in infants. However, it is unclear whether infant imitation of intentional acts is engendered by the behavioral similarities between humans and the orangutan-like object (e.g., self-propelled and contingent movements) or the morphological similarities between the two (particularly their eyes). For example, the infants might have attributed intentions to the object due to its eyes. Indeed, infants at birth are already sensitive to stimuli containing eyes (Farroni et al. 2004). With increasing age, they increasingly treat objects with eyes substantially different from those without (Graham et al. 2007). Furthermore, before they could imitate the intended but unconsummated acts, they already were able to use another's eye gazes to infer mental states (Brooks and Meltzoff 2002). Thus, perhaps the presence of a pair of eyes alone is sufficient for infants to "imitate" an agent's intentional acts. Alternatively, the presence of eyes must be coupled with certain contingent and meaningful actions to ensure the imitation of intentional acts. The following study tested these possibilities.

Itakura et al. (2008) modified the paradigm of Meltzoff's (1995). Instead of using human adults, a robot named Robovie with eyes and mechanical arms served as a model (Fig. 9.1). Robovie was developed at the Advanced Telecommunications Research Institute International, Intelligence Robotics and Communication Laboratory, Japan.

As Meltzoff's (1995) human models did, the robot performed novel actions either successfully or unsuccessfully, and its behavior was videotaped and presented on a television monitor to children between 24 and 35 months old. We chose this age range because existing studies (Bellagamba and Tomasello 1999; Johnson et al. 2001; Meltzoff 1995) have shown that by 2 years of age, children can successfully imitate an adult's intended but unconsummated actions. In the Eye Contact condition, both before and after performing a novel action, the robot made eye contact with a human adult, who was also present throughout the video presentation. In the No Eye Contact condition, although the human adult was present and behaved exactly as he did in the Eye Contact condition, the robot did not make eye contact with him. Thus, in the Eye Contact and No Eye Contact conditions, eyes were present. If the presence of eyes alone is sufficient, children would correctly imitate both the successful and unsuccessful acts performed by the robot in both conditions. Otherwise, children would successfully imitate the unconsummated acts in the Eye Contact condition but not in the No Eye Contact condition.

Modeled after Meltzoff (1995), three sets of objects were used: a dumbbell, a cup and beads, and a peg with an elastic band. Robovie was controlled on the basis of the type of action trials. In the Successful Demonstration condition, Robovie pulled the dumbbell apart, put the beads into the cup, and hung the elastic band on the peg.

Fig. 9.1 (**a**) Robovie failed to pull the dumbbell apart in the unsuccessful demonstration + eye contact condition. (**b**) Robovie failed to pull the dumbbell apart in the unsuccessful demonstration + no eye contact condition

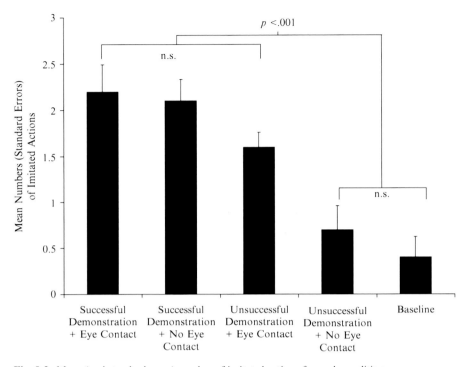

Fig. 9.2 Mean (and standard error) number of imitated actions for each condition

In the Unsuccessful Demonstration condition, Robovie tried to pull the dumbbell apart but failed. It tried to put the beads into the cup, but they dropped outside the cup. Robovie also tried to hang the elastic band on the peg, but it fell on the table.

The groups of children were divided into five conditions: Successful Demonstration + Eye Contact, Successful Demonstration + No Eye Contact, Unsuccessful Demonstration + Eye Contact, Unsuccessful Demonstration + No Eye Contact, and a baseline condition in which children were simply given one of the objects to manipulate. Children were coded as having produced the target action if they showed such behavior. The results are shown in Fig. 9.2.

The results in the Successful Demonstration conditions showed that young children imitated successful actions regardless of whether the robot made Eye Contact with a human. In the Unsuccessful Demonstration condition, however, children only completed the unobserved but intended action when the robot made Eye Contact with the human.

There are two main findings in this study. First, young children imitate a nonhuman agent's action. Second, an eye must be coupled with interactive activities with another human to complete the intentional actions. These findings will help robotic scientists design robots that not only mimic human morphologies and biomechanical movements but also convey a sense of "intentionality."

9.3.2 False Belief of a Robot

In a paper entitled "Does the chimpanzee have a 'theory of mind'?" Premack and Woodruff (1978) examined whether a chimpanzee's mind works like a human's mind. However, the paper implicitly assumes that the behavior of others is determined by their desires, attitudes, and beliefs (Frith and Frith 2003). These are not states of the world, but states of the mind. However, we have not found robust evidence of the theory of mind in any nonhuman species as yet. In contrast to such uncertainty about the theory of mind in nonhuman species, human children clearly exhibit a complex capacity to understand the minds of others at an early point in their development.

In general, from 4 or 5 years of age children understand that other humans have beliefs that may differ from their own. The most common test of children's ability to explain an action with reference to another's belief is the "false-belief task" (Wimmer and Perner 1983). In this study, a child is told about Maxi, whose mother places a piece of chocolate in a green cupboard. While Maxi is outside playing, the mother moves the chocolate from the green cupboard to a blue cupboard. Children are then asked to report Maxi's belief ("Where does Maxi think the chocolate is?"), to predict her action ("Where does Maxi look for the chocolate?"), or to explain the completed action ("Why did Maxi look for the chocolate in the green cupboard?"). The critical feature of the false-belief task is that correct answers to all three questions require the subject to concentrate on Maxi's belief rather than the actual location of the chocolate.

In the light of the previous discussion, we investigated whether young children infer a robot's mental state in a standard false-belief task (Itakura 2006). Robovie was again used, as in the study outlined in Sect. 9.3.1. The participants were 58 young children (27 boys, 31 girls, age range 54–80 months, mean 65.4 months). We chose this age range because many studies have demonstrated that children between 4 and 5 years of age pass the false-belief task. Both versions of the video stimuli were presented by a video monitor. Robovie places the doll in Box A and then leaves the room. During Robovie's absence, a man removes the doll from Box A and places it in Box B. The second condition was identical except that a human, not a robot, performed the actions.

Each subject was shown these two videos and given four questions after watching them individually. The order of presentation was counterbalanced. The following four questions were asked:

1. "Where will it/he look for a doll?" (Prediction task)
2. "Where does it/he think the doll is?" (Representation task)
3. "Which box has the doll?" (Reality task)
4. "Which box used to have the doll?"(Memory task)

The results of the experiment are shown in Fig. 9.3.

The reality and memory questions in the human and robot conditions were the same; most children answered these questions correctly. In addition, the prediction

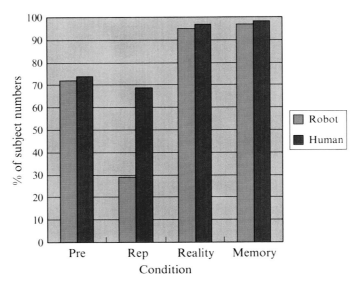

Fig. 9.3 Results of false belief task with a robot

question was the same in both conditions. However, the representation tasks in the human and robot conditions were significantly different. These results show that children attribute false beliefs, but *mental verbs*, to robots.

In this study, we provide evidence suggesting that young children discriminate between a robot and a human when the question involved such a mental verb as "think." Young children seem to have difficulty linking searching and thinking behaviors in a robot. In the light of the results reported in Sect. 9.3.1, children apparently need to be shown the robot acting as a communicative agent, so they infer that it is actually "thinking."

9.3.3 Word Learning from a Robot

It is generally assumed that young children learn words from another person. Recent research, however, has shown that children can learn new actions and skills from nonhuman agents. Moriguchi et al. (2011) built upon previous research and examined whether children were able to learn words from robots.

To our knowledge, only one previous study reported that children may learn words from a robot. Kanda et al. (2004) examined in a field trial (observation study) whether Japanese school-aged children can form relationships with robots and found that interaction with robots would facilitate children's second-language (English) acquisition. In this study, children interacted with English-speaking robots for 2 weeks in their classroom. The children were given an English test before and after the interaction with the robots. The results revealed that the children who took more time to interact with the robots scored better in the posttest than those who did not. The study by Moriguchi et al.'s (2011) extended these findings in two ways. First, the previous study conducted a field trial and showed the improvement in the English tests. However, it is still unclear whether the improvement was truly due to the interaction with the robots. An interactive child may have interacted with another interactive child and learned the words from that child. Thus, more controlled experiments were needed to assess whether children learn words from a robot. Therefore, this study conducted a basic word-learning experiment in which children observed a robot labeling novel objects with novel words. Moreover, the previous study did not examine whether children learned words from a robot as well as they do from a human.

In this study, preschool children, aged 4–5 years, were shown a video in which either a woman (a human condition) or a mechanical robot (a robot condition) labeled novel objects. After viewing the video, children were asked to select the target objects that had been identified on the tape.

There were two phases in this study. At the beginning of the control phase, the child was informed that he or she was going to learn some words from an agent: "Now, we are going to learn some objects' names from a lady (or a robot) on the video. Before that, the lady (or the robot) will introduce herself (itself). Please watch the video carefully. Okay?" The children then watched the video clip twice. In the video, the agent said, "My name is — [name]. I like bananas!" After watching the video, the child was given some control questions: "What is her (or the robot's) name? What is her (its) favorite food?" The control questions were used to ascertain whether the child focused on the words spoken during the video and whether he or she was able to interpret that information. The child was regarded as having passed the control phase when he or she answered the "favorite food" question correctly.

In the test phase, a child was presented a novel object and asked to name it: "Do you know the name of the object?" Initially, none of the children were able to answer this question with the appropriate name. When the child answered "no," he or she was instructed to watch the new video clips: "Now, she (the robot) is going to label the object. Please watch the video and listen to her (its) voice carefully." During the clip, the agent was presented with a novel object. Gazing at the object, the agent labeled it with a novel word: "This is a — [label]." The child watched each video clip twice. After viewing the video, the child was presented with a new object, and watched the next video clip. The child was given three trials consecutively, then was given a short break. The evaluator then presented the child with three objects labeled in the video clip and gave him or her a few test questions (e.g., "Which is a toma?"). Then the child was asked to point to the correct targets in response to the evaluator's questions. Three test questions were given consecutively. The results are shown in Fig. 9.4.

The results revealed that children tested with the human condition were more likely to select the correct objects than those tested with the robot condition. Nevertheless, children in the robot condition performed significantly above chance

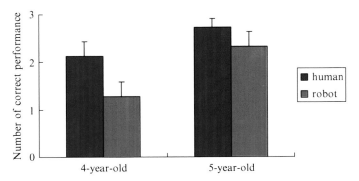

Fig. 9.4 Mean correct answers given in the test phase

levels. This may be the first evidence of its kind to demonstrate that children have the potential to learn words and develop language skills from a robot. The results would contribute to our understanding of how children interact with nonhuman agents.

9.4 The Neural Basis of the Development of Mentalizing

Not all measurements in psychology have such overt behavior as our target. Measures of underlying physiological processes can also be informative, especially in infants and young children for whom overt behaviors are often limited. Recent years have seen some exciting advances in the study of the brain, with techniques that allow researchers to examine not only the brain's anatomy but also its activity while people perform a variety of tasks. These techniques are applicable to social cognition or mentalizing. In this section we review our own experiments and the most recent studies of our colleagues.

Self-recognition is one crucial milestone of the sociability concerns of mentalizing. We attempted to identify the cortical region involved in self-recognition and self-evaluation with self-conscious emotions in adult subjects (Morita et al. 2008). To increase the range of emotions accompanying self-evaluation, we used facial feedback images chosen from a video recording, some of which deviated significantly from normal images. Nineteen participants rated images of their own face (SELF) and those of others (OTHERS) based on how photogenic they appeared. After scanning the images, the participants rated how embarrassed they felt upon viewing each face. As the photogenic scores decreased, the embarrassment ratings dramatically increased for the participant's own face compared with those of others. The SELF versus OTHERS contrast significantly increased the activation of the right prefrontal cortex, the bilateral insular cortex, the anterior cingulate cortex, and the bilateral occipital cortex. In the right prefrontal cortex, activity in the right precentral gyrus reflected the trait of the awareness of the observable aspects of the self, providing strong evidence that the right precentral gyrus is specifically involved in self-face recognition. By contrast, activity in the anterior region, located in the right middle inferior frontal gyrus, was modulated by the extent of embarrassment, suggesting that it is engaged in self-evaluation preceded by self-face recognition based on relevance to a standard self. Although this study was conducted on adult subjects, we need to collect data on brain activity to consider the theoretical and logical aspects of the target issue in adults.

The direction of others' eye gaze is a crucial source of information in social interactions. Eye gaze also provides information about others' communicative intentions and future behavior (Baron-Cohen 1995). Reid and Striano (2005) investigated the functional relevance of gaze cuing in infancy, and presented 4-month-old infants with videos of a face that was directing its eye gaze toward one of two objects. When exposed to both objects again without the face, the infants looked longer at the previously uncued object, indicating that they perceived it as more novel.

On the basis of this study, the authors raised important unaddressed questions: How do infants process the relation between another person and an external object? How do they use the information provided by an adult's eye gaze to guide their attention and process environmental information? Hoehl et al. (2007) employed an event related potentials approach to explore this question. This paradigm allows direct investigation of the neural systems included in information processing even in the absence of overt behavior. Their study assessed how 4-month-old infants process the directedness of adult eye gaze in relation to objects in their field of view, which is the same as the face itself. They presented static photographs of faces with eye gaze averted to the left or right side. One object was presented near the face, either presented on the same side as the direction of the eye gaze or on the other side. Their prediction was as follows: infants would form a stronger memory representation for the cued objects. This would be reflected by an enhanced positive slow wave, which is probably related to stimulus updating or encoding in 4–6-month-olds, during the observation of stimuli depicting eye-gaze-cued objects.

The results suggest that infants differentially process whether an adult's eye gaze is directed at one object or averted from an object. Positive slow wave at frontal sites was enhanced during eye gaze directed toward an object when compared with eyes gazing away from an object. The investigators interpreted this finding as evidence that infants form a stronger memory representation for cued objects than for uncued objects.

9.5 Conclusion

This chapter discusses the development of mentalizing and communication from the perspectives of developmental cybernetics, a recent discipline, and developmental cognitive neuroscience.

Children attributed intention to a robot only when they saw it behaving as a human and displaying such social signals as eye gazing. Children performed the unobserved but intended action only when the robot made Eye Contact with the human. This finding is crucial to the understanding of how children invest intention in nonhuman agents, such as robots. They must detect that the agent has dispatched a social signal and instantiate it as communicative. Such a signal does not need to be complex; a very subtle but strongly impacted one (e.g., eye gazing) is sufficient. Adults might consider social signals complicated, but they are much simpler in some contexts.

New powerful methods and tools have now become available to cognitive neuroscience, which allow questions about mentalizing to be asked more directly than before. One set of tools related to neuroimaging, the generation of "functional" maps of brain activity, is based on physiological changes. Three current techniques are readily applied to development in normal children: event-related potentials, functional magnetic resonance imaging), and near-infrared spectroscopy. These techniques appear especially useful in infants and toddlers for whom overt behaviors are often limited.

It is expected that developmental cybernetics and developmental cognitive neuroscientific approaches will provide a better understanding of the development of mentalizing and communication in the near future.

References

Asada M, Kuniyoshi Y (2006) Robot intelligence, vol 4. Iwanami Shoten, Tokyo
Baldwin DA, Baird JA, Saylor MM, Clark MA (2001) Infants parse dynamic action. Child Dev 72:708–717
Baron-Cohen S (1995) Mindblindness: an essay on autism and theory of mind. MIT Press, Cambridge
Barr R, Hayne H (2000) Age-related changes in imitation: implications for memory development. In: Rovee-Collier C, Lipsitt LP, Hayne H (eds) Progress in infancy research. Lawrence Erlbaum, Hillsdale, pp 21–67
Behne T, Carpenter M, Call J, Tomasello M (2005) Unwilling versus unable: infants' understanding of intentional action. Dev Psychol 41:328–337
Bellagamba F, Tomasello M (1999) Re-enacting intended acts: comparing 12- and 18-month-olds. Infant Behav Dev 22:277–282
Bertenthal BI, Profitt DR, Cutting JE (1984) Infant sensitivity to figural coherence in biomechanical motions. J Exp Child Psychol 37:213–230
Brooks R, Meltzoff AN (2002) The importance of eyes: how infants interpret adult looking behavior. Dev Psychol 38:958–966
Carpenter M, Nagell K, Tomasello M (1998) Social cognition, joint attention, and communicative competence from 9 to 15 months of age. Monogr Soc Res Child Dev 63(4):i–vi, 1–143
Crichton M, Lange-Kuettner C (1999) Animacy and proposition in infancy: tracking, waving and reaching to self-propelled and induced moving objects. Dev Sci 2:318–324
Farroni T, Massaccesi S, Pividori D, Johnson MH (2004) Gaze following in newborns. Infancy 5:39–60
Frith U, Frith CD (2003) Development and neurophysiology of mentalizing. In: Frith C, Walpert D (eds) Neuroscience of social cognition. Oxford University Press, New York
Gergely G, Csibra G (2004) The social construction of the cultural mind: imitative learning as a mechanism of human pedagogy. Interact Stud Soc Behav Commun Biol Artif Syst 6:463–481 (Special Issue: Making minds II)

Gergely G, Nadasdy Z, Csibra G, Biro S (1995) Taking the intentional stance at 12 months of age. Cognition 56:65–193

Gomez JC (2004) Apes, monkeys, children, and the growth of mind. Harvard University Press, Cambridge

Graham SA, Nilsen ES, Nayer SL (2007) Following the intentional eye: the role of gaze cues in early word learning. In: Flom R, Lee K, Muir D (eds) Ontogeny of gaze processing. Lawrence Erlbaum, Mahwah

Hoehl S, Reid V, Mooney J, Striano T (2007) What are you looking at? Infants' neural processing of an adult's object-directed eye gaze. Dev Sci 10:1–7

Itakura S (2006) To what extent do infants and children find a mind in non-human agents? In: Fujita K, Itakura S (eds) Diversity of cognition. Kyoto University Press, Kyoto

Itakura S, Ishida H, Kanda T, Ishiguro H, Lee K (2008) How to build an intentional android: infants' imitation of a robot's goal-directed actions. Infancy 13:519–532

Johnson SC (2003) Detecting agents. Philos Trans R Soc Lond B Biol Sci 358:549–559

Johnson MH (2005) Developmental cognitive neuroscience. Blackwell, Oxford

Johnson MH, Morton J (1991) Biology and cognitive development: the case of face recognition. Blackwell, Oxford

Johnson SC, Slaughter V, Carey S (1998) Whose gaze will infants follow? Features that elicit gaze-following in 12-month-olds. Dev Sci 1:233–238

Johnson SC, Booth A, O'Hearn K (2001) Inferring the goals of nonhuman agent. Cognit Dev 16:637–656

Kanda T, Hirano T, Eaton D, Ishiguro H (2004) Interactive robots as social partners and peer tutors for children: a field trial. Hum Comput Interact 19:61–84

Legerstee MA (1992) Review of the animate/inanimate distinction in infancy. Early Dev Parent 1:59–67

Leslie AM (1987) Pretence and representation: the origins of 'theory of mind'. Psychol Rev 94:412–426

Luo Y, Baillargeon R (2005) Can a self-propelled box have a goal? Psychological reasoning in 5-month-old infants. Psychol Sci 16:601–608

Meltzoff AN (1995) Understanding the intentions of others: re-enactment of intended acts. Dev Psychol 31:838–850

Meltzoff AN (1999) Origins of theory of mind, cognition and communication. J Commun Disord 32:251–269

Meltzoff AN (2005) Imitation and other minds: the "Like Me" hypothesis. In: Hurley S, Cater N (eds) Perspectives on imitation: from neuroscience to social science. MIT Press, Cambridge, pp 55–77

Moriguchi Y, Kanda T, Ishiguro H, Shimada Y, Itakura S (2011) Can young children learn words from a robot? Interact Stud 12:107–118

Morita T, Itakura S, Saito ND, Nakashita S, Harada T, Kochiyama T et al (2008) The role of the right prefrontal cortex in self-evaluation of the face: a functional magnetic resonance imaging study. J Cogn Neurosci 20:342–355

Phillips AT, Wellman HM (2005) Infants' understanding of object-directed action. Cognition 98:137–155

Premack D (1990) The infant's theory of mind. Cognition 36:1–16

Premack D, Woodruff G (1978) Does the chimpanzee have a theory of mind? Behav Brain Sci 1:515–526

Quinn PC, Slater A (2003) Face perception at birth and beyond. In: Pascalis O, Slater A (eds) The development of face processing in infancy and early childhood: current perspectives. Nova Science, Hauppauge, pp 3–11

Reid VM, Striano T (2005) Adult gaze influences infant attention and object processing: implications for cognitive neuroscience. Eur J Neurosci 21:1763–1766

Shimizu YA, Johnson SC (2004) Infants' attribution of a goal to a morphologically unfamiliar agent. Dev Sci 7:425–430

Spelke ES, Phillips A, Woodward A (1995) Infants' knowledge of object motion and human action. In: Sperber D, Premack D, Premack AJ (eds) Causal cognition: a multidisciplinary debate. Symposia of the Fyssen Foundation. Oxford University Press, New York, pp 44–78

Wimmer H, Perner J (1983) Beliefs about beliefs-representation and constraining function of wrong beliefs in young children's understanding of deception. Cognition 13:103–128

Woodward AL (1998) Infants selectively encode the goal object of an actor's reach. Cognition 69:1–34

Chapter 10
Emotion, Personality, and the Frontal Lobe

Satoshi Umeda

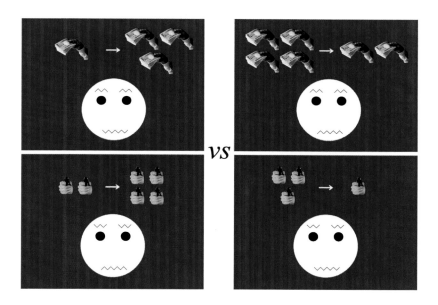

Abstract Previous neuropsychological and functional neuroimaging studies on human emotion and personality have revealed the involvement of the frontal lobe. This chapter presents two original studies, a neuroimaging and a neuropsychological study, focusing on the neural mechanisms for emotional processing and personality.

The first study concentrates on the neural substrates of advanced emotion. Over the past decade, many studies have primarily focused on basic emotions. In social

S. Umeda (✉)
Department of Psychology, Keio University, Mita 2-15-45,
Minato-ku, Tokyo 108-8345, Japan
e-mail: umeda@flet.keio.ac.jp

situations, however, understanding more advanced complex emotions is also important for maintaining successful communications with others. To date, most of the neuroimaging studies on advanced emotions have focused on the neural substrates of recognizing those emotions, such as "theory-of-mind" reasoning. To determine the neural and cognitive mechanisms of advanced emotions, it is essential to focus on the neural substrates of advanced emotion learning, which remain poorly understood. The goal of the first study was to address this question using functional magnetic resonance imaging. The following important findings were obtained in the time-course data of activated brain areas: (1) the medial prefrontal (BA8) and the anterior cingulate cortex (BA32) were strongly activated in negative-based emotion learning, and (2) the right dorsolateral prefrontal cortex (BA9/46) was strongly activated in positive- and negative-based emotion learning. These results suggest that the prefrontal area is important for acquiring the relationship between social situations and complex facial expressions, and that these areas make independent contributions to learning specific emotions.

The second study focuses on the function of the medial prefrontal cortex. Previous functional neuroimaging studies on "theory of mind" have demonstrated that the medial prefrontal cortex is involved when subjects are engaged in various kinds of mentalizing tasks. Although a large number of neuroimaging studies have been published, a somewhat small amount of neuropsychological evidence supports involvement of the medial prefrontal cortex in theory-of-mind reasoning. Findings are presented from a neuropsychological study for two neurological patients with damage to the medial prefrontal cortex. The results indicated that neither patient showed impairment on standard theory-of-mind tests and only mild impairments were seen on advanced theory-of-mind tests. The most striking finding was that both patients showed personality changes after surgical operations, leading to characteristics of autism and showing a lack of social interaction in everyday life. Finally, the possible roles of the medial prefrontal cortex are discussed, with emphasis on the importance of using multiple approaches to understand the mechanisms of theory of mind and medial prefrontal functions.

10.1 Basic and Advanced Emotions

Research on the neural basis of emotion and self has come to be the major topic in social cognitive neuroscience. Many cognitive neuroscientists in this research area have focused in particular on the functional neuroanatomy involved in the cortical midline structures. Numerous different domains in self-processing, including monitoring, evaluation, agency, and theory of mind, are all more or less involved in midline structures, including the medial prefrontal cortex and anterior cingulate cortex (Northoff and Bermpohll 2004).

Over the past several years, a number of neuroimaging studies have tried to clarify the neural mechanisms of "theory of mind," and have demonstrated that the medial prefrontal cortex is activated when subjects are engaged in various kinds of mind-reading or mentalizing tasks (Frith and Frith 1999). Several recent studies in developmental cognitive neuroscience have reported some evidence that people with Asperger syndrome or high-functioning autism show weak activations in the medial prefrontal cortex in comparison with control subjects (Di Martino et al. 2009; Happé et al. 1996; Nieminen-von Wendt et al. 2003). These findings are considered fundamental evidence for deficits in theory-of-mind abilities among individuals with Asperger syndrome or high-functioning autism (Frith 2001).

Concerning types of emotion, no universally accepted typology has been defined, but one widely used division separates basic emotions from more advanced emotions. The emotions of happiness, surprise, sadness, fear, disgust, and anger have been identified as the six basic emotions (Ekman 1992). By contrast, complex advanced emotions such as friendliness, pensiveness, thoughtfulness, and melancholy are all required to modulate social communication with others or to state one's own inner complex mental states (Shaw et al. 2005). In fact, these kinds of advanced emotions are highly associated with mind-reading abilities, because understanding advanced emotions in another is a basis for realizing appropriate theory-of-mind reasoning or empathy-processing for social communications with others. The "basic-advanced" distinction of emotion is also supported by several other findings. For instance, Castelli (2005) reported that people with autism understand the basic emotions but fail to recognize some complex emotions.

10.2 Functional Neuroanatomy of Advanced Emotion

Previous functional neuroimaging studies on human emotion over the past decade have focused primarily on basic emotions, and several neuroimaging studies of advanced emotion have been reported to date. For instance, previous studies suggested that increasing regret in a gambling task enhanced activity in some brain areas including the medial orbitofrontal region (Coricelli et al. 2007). However, the question of how humans learn advanced emotions remains poorly understood at both behavioral and neural levels. The goal of the first study was to address the question using a functional magnetic resonance imaging (MRI) method. Illustrated facial expressions with unknown novel types of emotions were used to examine the learning processes of advanced emotions at the behavioral and neural levels.

In this study, subjects comprised 18 healthy adults (9 men, 9 women; mean age, 22.2 years; mean duration of education, 14.6 years). As materials for this experiment, six figures (two positive-based, two negative-based, and two neutral-based learning conditions) were prepared, consisting of one physical situation with time series and one complex facial expression (Fig. 10.1).

In the learning (MRI scanning) phase, subjects were instructed to lie on the scanner bed and asked to learn the relationships between the physical situations displayed above and the facial expressions displayed below. Each stimulus was displayed for 10 s, followed by the cross-hair presentation. Stimulus-onset asynchrony (SOA) was 50 s. The subjects were also later required to undertake an acquisition test outside

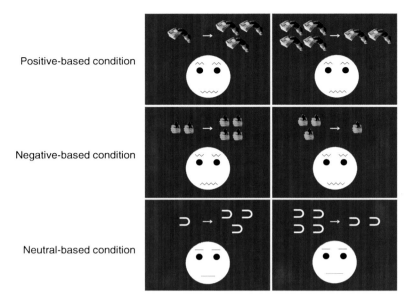

Fig. 10.1 Experimental stimuli used in the positive-, negative- and neutral-based learning conditions

the scanner. In the behavioral test phase after the MRI scanning, subjects were asked to make recognition judgments for figures consisting of one physical situation and one facial expression projected on the computer screen. The recognition test comprised 36 figures, including 12 correct and 24 incorrect combinations.

For acquisition, a 3-T Allegra system (Siemens, Germany) was used to acquire high-resolution T1-weighted anatomic images and single-shot gradient-echo echo planar images (EPIs) with blood oxygen level dependent (BOLD) contrast of 35 axial slices with the following parameters: cubic resolution, 3.5 mm; repetition time, 2 s; echo time, 30 ms; and flip angle, 90°. Image preprocessing and data analyses were performed using Statistical Parametric Mapping software (SPM2; http://www.fil.ion.ucl.ac.uk/~spm/) on a Matlab platform.

For the behavioral results, the mean proportion of correct responses was 0.86 for the positive-based condition, 0.81 for the negative-based condition, and 0.83 for the neutral-based condition. Mean reaction time was 4,394 ms for the positive-based condition, 3,678 ms for the negative-based condition, and 3,991 ms for the neutral-based condition. No significant differences were apparent among these three conditions.

As imaging results, several substantial findings were shown in the time-course data for activated brain areas during the three conditions, as follows: (1) the posterior medial prefrontal cortex (BA8) and anterior cingulate cortex (BA32) were strongly activated in the negative-based learning condition (Fig. 10.2); (2) the right dorsolateral prefrontal cortex (BA9/46) was strongly activated in both positive- and negative-based learning conditions (Fig. 10.3); (3) bilateral parahippocampal areas were strongly activated in the positive-based learning condition (Fig. 10.4).

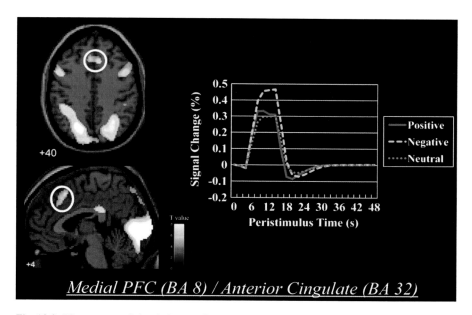

Fig. 10.2 Time course of signal changes for positive-, negative- and neutral-based learning conditions in the posterior medial prefrontal (BA8) and anterior cingulate cortex (BA32)

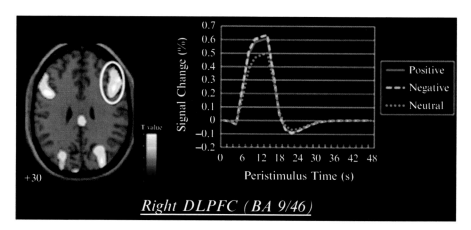

Fig. 10.3 Time course of signal changes for positive-, negative- and neutral-based conditions in the right dorsolateral prefrontal cortex (BA9/46)

The next question is whether any difference in brain activation patterns exists between learning of simple emotion and complex advanced emotion. To answer this question, activation patterns were compared between simple and complex emotion conditions. Simple emotion means stimuli including facial expression with happy eyebrows and happy mouths, or with fearful eyebrows and fearful mouths.

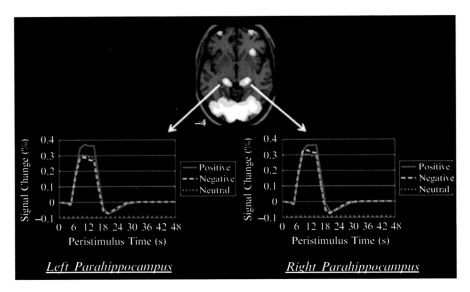

Fig. 10.4 Time course of signal changes for positive-, negative- and neutral-based conditions in *right* and *left* parahippocampal areas

These examples are displayed on the top and middle left panels of Fig. 10.1. Complex emotion means stimuli including facial expression with happy eyebrows and fearful mouths, or with fearful eyebrows and happy mouths. These examples are displayed on the top and middle right panels of Fig. 10.1.

The contrast between brain activations during simple and complex emotion learning indicated that bilateral superior parietal lobules (BA7), bilateral dorsal premotor areas, and the left ventrolateral prefrontal cortex (VLPFC) were greatly activated in complex emotion conditions rather than in simple emotion conditions (Fig. 10.5). Interestingly, the superior parietal lobule and dorsal premotor area are both highly involved in imitation and have been identified as parts of the mirror neuron system (Brass and Heyes 2005).

These overall results suggest that the prefrontal and medial temporal areas are essential for acquiring relationships between situations and facial expressions, and that those areas make independent contributions to learning specific advanced emotions. The findings of this study show that the medial prefrontal cortex (BA8) and anterior cingulate cortex (BA32) are strongly activated in negative-based conditions compared with the other two conditions. This suggests that these areas are critical for acquiring negative-based advanced emotions, which are supposed to be highly involved in empathetic processing (Singer et al. 2004, 2006). As mentioned before, the medial prefrontal cortex has been identified as an area with weak activation among individuals with Asperger syndrome or high-functioning autism. The present result that this area is involved mostly in negative-based conditions could represent further evidence for a lack of theory-of-mind ability in people with Asperger syndrome or high-functioning autism.

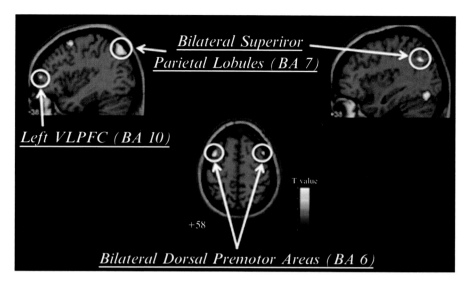

Fig. 10.5 Brain areas showing greater activation during complex emotion learning rather than simple emotion learning

Another important finding was that activations in the right dorsolateral prefrontal cortex were greater in positive- and negative-based conditions than in neutral-based conditions. This finding is consistent with previous results that activation in this area correlates strongly with autonomic activity and activities in the anterior cingulate cortex (Critchley et al. 2001, 2003; Critchley 2004; Fukushima et al. 2010). The present results suggest that bodily responses under the control of the autonomic nervous system accompany learning associations between physical situations and facial expressions with emotional valence.

Moreover, bilateral hippocampal areas showed greater activation in positive-based learning conditions than in the other two conditions. Several previous studies have reported that the hippocampus is greatly activated during the processing of positive emotional stimuli, rather than negative emotional stimuli (Britton et al. 2006; Kuchinke et al. 2005; Prohovnik et al. 2004). Recent morphometric studies using MRI have reported smaller hippocampus volumes in patients with major depression (for a review, see Davidson et al. 2002). The present result showing that the hippocampus is greatly involved in positive-based emotion learning is consistent with the idea that the hippocampus supports positive-based emotional processing.

The contrast between brain activations during simple and complex emotion learning shows that parts of the mirror neuron system are greatly involved during complex emotion learning. Activations in these areas suggest that when people encounter unknown facial expression with a certain situation, parts of the mirror neuron system are used to understand what the facial expression means by actual facial imitations (Leslie et al. 2004). Concerning the greater activation found in the left VLPFC, previous functional imaging data obtained from a false recognition task showed that this area is highly involved when people recognize the presented word as a new word that has not previously been encountered (Umeda et al. 2005). Activation in this area during complex emotion learning thus suggests that understanding of unknown facial expression is based on a saliency detection process for identification as a novel emotion.

The overall findings from the first study provide some evidence that advanced emotion learning is realized by a combination of neural activities for learning, for social and empathetic processing, and for the mirror neuron system.

10.3 Neural Substrates of "Theory of Mind"

A number of neuroimaging studies of "theory of mind" or "mentalizing," the automatic ability to attribute desires and beliefs to other people, have demonstrated that the medial prefrontal cortex or anterior cingulate (paracingulate) cortex is involved in various kinds of tasks requiring mentalizing functions (Brunet et al. 2000; Castelli et al. 2000; Gallagher and Frith 2003; Gallagher et al. 2000; Vogeley et al. 2001). This area has been recognized as one of the "social brain" areas, together with areas like the superior temporal sulcus, amygdala, insula, posterior cingulate (retrosplenial cortex), and fusiform gyrus (Brüne et al. 2003; Wheatley et al. 2007).

Another neuroscientific approach to theory of mind involves neuropsychological investigations to examine the performance of brain-damaged patients. The initial study by Happé et al. (1999) reported that following right-hemisphere damage, patients showed impaired understanding of materials requiring attribution of mental states.

Several other studies examining the effects of brain lesions on theory-of-mind performance have reported that focal frontal lesions impair the ability to infer mental states of others (Bach et al. 2000; Happé et al. 2001; Shamay-Tsoory et al. 2003; Stuss et al. 2001). More careful examinations to understand the roles of different areas within the frontal lobe have indicated that right ventromedial prefrontal lesions impair detection of deception (Stuss et al. 2001) and empathy processing (Shamay-Tsoory et al. 2003), whereas lesions in the internal capsule impair advanced theory-of-mind performance for understanding the thoughts and feelings of fictional characters (Happé et al. 2001). Conversely, other studies have suggested that the ability to understand mental states was found to be intact in patients with orbitofrontal lesions (Bach et al. 2000).

Another comprehensive neuropsychological study of over 30 patients with unilateral (right or left) frontal lobe lesions found that both groups exhibited impaired performance on first- and second-order false belief tests (Rowe et al. 2001). Stone et al. (1998) tested performance on first- and second-order false-belief tests and the faux-pas recognition test in patients with orbitofrontal damage, and found that bilateral orbitofrontal lesions resulted in difficulty only in the ability to recognize a faux pas (see Sect. 10.6 for details of the tasks). Although these studies focused on the effects of damage to the frontal lobe, it remains unclear as to whether selective damage to the medial prefrontal cortex yields any impaired performance on various kinds of theory-of-mind tests.

Some previous studies have actually examined the performance of patients with selective damage to the medial prefrontal cortex. Baird et al. (2006) tested two patients and reported intact intellectual, memory, and language abilities, and visuo-perceptual functions, but showed weak or impaired performance on selective executive function tests. No theory-of-mind performance was tested in this study. Another neuropsychological study addressed the question of theory of mind impairment by testing a patient with a selective lesion in the medial prefrontal cortex (Bird et al. 2004). They carefully examined performance of the patient on various kinds of theory-of-mind tests, but found no significant impairment on tests, and thus stated that extensive medial frontal regions are not necessary for theory-of-mind performance. The findings from both studies have some important implications for our understanding of the effects of damage to the medial prefrontal cortex. Most interesting was the finding that following the damage to this area, patients did not show any severe impairment on test performance, including theory-of-mind performance. However, the data are currently too limited to reach solid conclusions on the effects of medial prefrontal damage.

10.4 Personality Change After Damage to the Frontal Lobe

Another unresolved question is concerned with personality changes in human patients with damage to their frontal lobes. In the classic case of *Phineas Gage*, damage to the orbitofrontal cortex resulted in severe behavioral disturbances in everyday life (Damasio et al. 1994; Harlow 1848). A number of previous studies have reported that patients with damage to the ventromedial prefrontal cortex (VMPFC) show severe sociopathic personality change (Eslinger and Damasio 1985; Saver and Damasio 1991). Following a lesion of the VMPFC, most patients show inappropriate social behavior including emotionally insensitive social interaction, unexpected wandering, alcohol abuse, and confabulation. Despite these sociopathic psychopathological changes, higher-order cognitive functions including general memory performance are largely preserved.

Several hypotheses have already been presented to explain the function of VMPFC. The somatic marker hypothesis proposes that the damage to the VMPFC precludes the ability to use somatic (emotional) signals that are necessary for guiding decisions in a positive direction (Bechara et al. 1994, 1996, 1997, 1999; Damasio 1996). This interpretation, based on findings obtained from the gambling task, leads to the conclusion that the VMPFC mediates decision making. The VMPFC appears to play an important role in social reasoning (Adolphs et al. 1996). The researchers using the Wason selection task found that VMPFC patients show impairment in the reasoning task including familiar cover stories with social laws, whereas they showed no abnormality when reasoning about more abstract material without a cover story. This pattern of results differed from that of normal subjects, i.e., the VMPFC patients had a deficiency in the thematic effect. Adolphs (1999) stated that the VMPFC guides social reasoning and decision-making by the elicitation of emotional states that serve to bias cognition. Furthermore, a recent study of functional MRI using a simple card-playing task indicated that activation of the VMPFC increases as probabilistic contingencies become more complex (Elliott et al. 1999). Elliott and colleagues concluded that the VMPFC mediates guessing.

On the other hand, Schacter et al. (1996) reported that one patient (B.G.) exhibited a striking pattern of false recognition after an infarction of the right frontal lobe including the inferior prefrontal cortex. One possible interpretation for this result is that B.G.'s recognition decision was based on a fuzzy general event description, or a "false gist" which was generated from the presented words.

These accounts for the function of the VMPFC appear to be largely compatible, because a certain type of emotion-based memory processing under complex situations is more or less required in decision making, social reasoning, and guessing. It is notable that emotions pertaining to the VMPFC are assumed to be negative feelings (e.g., fear, anger), because the VMPFC has a direct neural connection to the amygdala (Tranel 2000). Past studies that focus on the function of the amygdala and its related circuits indicate that the amygdala will provide a quick and automatic bias with respect to those aspects of the response that pertain to evaluating the potentially threatening nature of a situation (Adolphs 1999).

However, the question of whether patients with lesions involving the medial prefrontal cortex, located in the vicinity of the VMPFC, shows any personality change remains unanswered. A recent meta-analysis of 39 functional imaging studies for autism-spectrum disorders indicated that the medial prefrontal cortex was less activated during social task performance (e.g., theory-of-mind tests) in autism-spectrum disorders compared with neurotypical controls (Di Martino et al. 2009). Patients with damage to the medial prefrontal cortex may show personality changes, leading to characteristics of autism.

Although previous neuropsychological studies of patients with damage to the medial prefrontal cortex have shed light on the functions of this area, the number of studies remains limited, and actual effects on social functions following damage to this area remain poorly understood (Gallagher and Frith 2003). The second study of this chapter presents two cases of patients with damage to the medial prefrontal cortex, and reports on their performance on various kinds of theory-of-mind tests, whether personality changes were evident, and behavioral disturbances in daily activities.

10.5 Neuropsychological Investigations

The two patients had medial prefrontal damage (Umeda et al. 2010). The first case, T.O., was a 31-year-old man. A full-time employee of a big electronic company in Japan, he had undergone neurosurgery for brain tumor. Magnetic resonance imaging revealed that damage extended through the left-dominant medial prefrontal and anterior cingulate cortices, reaching the left supplementary motor area. The right hand and leg were moderately paralyzed for a few months after surgery, but these symptoms later resolved.

The most striking aspect of T.O. was a reported change in personality. According to his self-report, he noticed that his sense of reality was attenuated after surgery, leading him to feel detached from the world despite being sure of his location, and this feeling often occurred in a manner similar to a panic attack. These symptoms partly resembled the characteristics of depersonalization. He also mentioned that surgery had made him feel depressed, anxious, and withdrawn from everything.

The second patient, H.C., was a 56-year-old man. An employer of a small private company in Japan, he had undergone surgery following rupture of a right pericallosal artery aneurysm. Magnetic resonance imaging revealed that the area of damage included the right-dominant medial prefrontal and anterior cingulate cortices, extending slightly into the right supplementary motor area.

According to self-reports, he noticed that his memory had deteriorated after surgery, with a feeling that most daily episodes could not be clearly remembered. He reported difficulty doing two things simultaneously, and became aware that everything needed a strong effort to be done. He also mentioned that his personality had changed after surgery, leading him to notice that feelings of sadness and anger had been dimmed and that he had become much more depressive, anxious, and withdrawn compared to his previous personality.

Three months after surgery, six of the more frequently used neuropsychological assessments were conducted to examine higher-order cognitive functions in these patients. (1) Wechsler Adult Intelligence Scale-Revised (WAIS-R) for general intelligence, (2) Wechsler Memory Scale-Revised (WMS-R) for memory and attention, (3) Rey Auditory–Verbal Learning Test (RAVLT) for verbal recall ability, (4) Rey–Osterrieth Complex Figure Test (ROCFT) for visuoconstructive skills and visual memory, (5) Wisconsin Card Sorting Test (WCST) for abstract reasoning and ability to appropriately shift cognitive strategies, and (6) Stroop Test for selective attention and inhibition.

On the intelligence test, T.O., with left-sided damage, showed dissociation between verbal and performance intelligence quotient (IQ), with an inferior score on performance IQ. In the IQ test, he showed difficulty in performing subtests of block design, object assembly, and digit symbols. In terms of memory performance, T.O. showed lower scores on some measures for identifying delayed recall performance (e.g., delayed recall on WMS-R, RAVLT, and ROCFT) compared with his relatively higher scores on other measures. The results of mild amnesia were consistent with his self-report regarding daily activities. For instance, he reported often becoming confused in remembering whether he had taken his medication. T.O.'s performances on tests for executive function were all within the normal range. Performance on the standardized aphasic test did not show any difficulties in language activities.

The second patient, H.C. with right-sided damage, showed an intellectual performance within the normal range. In terms of memory performance, he exhibited normal scores on WMS-R, although scores on RAVLT and ROCFT were somewhat lower. In fact, in terms of daily activities he showed difficulty with temporal-order judgments for everyday episodes within a time range of a few days. H.C.'s performance on tests for executive function was also within the normal range. Performance on the standardized aphasic test did not show any difficulties in language activities.

10.6 Theory-of-Mind Performance After Brain Injury

To clarify theory-of-mind performance, patients T.O. and H.C. were tested using four types of story comprehension task: (1) first-order false-belief test (Baron-Cohen et al. 1985; Frith and Frith 1999), (2) second-order false-belief test (Baron-Cohen 1989), (3) strange stories test (Happé 1994), and (4) faux-pas recognition test (Baron-Cohen et al. 1999).

The first-order false-belief test is one of the most famous tests for theory-of-mind reasoning. This test assesses the ability to recognize that others can have false beliefs about the world that can differ from reality, and that people's behaviors can be predicted by the representation of others' mental states. The more complex second-order false-belief test requires participants to understand a second person's concerns about the world, based on social interactions of minds in which people are concerned about each other's mental states.

The last two tests were used to examine the more advanced theory-of-mind reasoning ability in the patients. The strange stories test assesses the ability to infer mental states in a story context for social understanding. A previous study reported that subjects with autism-spectrum disorders show impaired provision of context-appropriate mental state explanations for strange stories, in comparison with normal control subjects (Happé 1994).

As well as the strange stories test, the faux-pas recognition test was used to assess the ability to recognize inappropriate statements in a story context (Baron-Cohen et al. 1999). Subjects were presented with each story and asked whether a faux pas was contained. If a faux pas was detected, they were then asked for an explanation of it. Each subject was also requested to answer two additional questions to test story comprehension in each story, to see whether a comprehensive understanding of each story was achieved. Baron-Cohen et al. (1999) reported that subjects with autism-spectrum disorders show impaired detection of faux pas on the faux-pas recognition test compared with normal control subjects, despite intact story comprehension.

In addition, T.O. and H.C. were required to complete all 50 items in the Autism-Spectrum Quotient (AQ) questionnaire (Baron-Cohen et al. 2001). This questionnaire was developed as a self-administered method of screening for adults with normal intelligence and traits associated with autism-spectrum disorders. Score ranges from 0 to 50 in the questionnaire. Adults with Asperger syndrome or high-functioning autism show a mean score of 35.8, significantly higher than controls with a mean score of 16.4 (Baron-Cohen et al. 2001). Another recent study has shown that the threshold score for suspected Asperger syndrome or high-functioning autism is 26.0 (Woodbury-Smith et al. 2005).

Both patients passed the first- and second-order false-belief tests, providing expected answers suggesting a proper understanding of each story. In terms of advanced theory-of-mind tests, both patients showed good performance on the strange stories test. Overall percentage of providing appropriate explanations for given stories was 85.7% for T.O. and 100% for H.C. on the strange stories test. On the faux-pas recognition test, the percentage of detecting faux pas and having appropriate explanations was 60.0% for T.O. and 100% for H.C. for the provided faux-pas stories. By contrast, the percentage of detecting "no" faux pas was 70.0% for T.O. and 57.1% for H.C. for the provided control stories. Both patients thus reported faux pas even in control stories without any faux pas. In addition, both patients showed higher scores on the two questions for story comprehension (92.5% for T.O. and 89.3% for H.C.).

In terms of the AQ, scores were 31 for T.O. and 29 for H.C., above the threshold score for Asperger syndrome or high-functioning autism of 26 as defined by Woodbury-Smith et al. (2005). Interestingly, both patients spontaneously reported just after completing the questionnaire that they were sure that some of the personality traits focused upon in the questionnaire identified the actual personality changes they felt. They were then asked to complete the AQ again for what they supposed their original personality was before surgery. AQ scores for purported presurgical state of the two patients were 13 for T.O. and 23 for H.C., much lower than the initial scores and below the threshold score for Asperger syndrome or high-functioning autism. Both patients thus appear to have developed some autistic personality traits following surgical operations.

A general finding was that the two patients had developed some characteristics of autism after surgical operations. To specify these characteristics in greater detail, these items were compared with items identified in a two-factor structure model (Hoekstra et al. 2008). Two factors were identified among all 50 items in the AQ, namely "social interaction" and "attention to detail". For T.O., all 18 items fell into the "social interaction" factor while for H.C. six of seven items fell into the "social interaction" factor. The only item falling into the "attention to detail" factor for H.C. was the item that showed he had become fascinated by dates.

In another opportunity separate from this study, 11 patients with damage to other parts of the brain (orbitofrontal lesion, basal forebrain lesion, dorsolateral prefrontal lesion, medial temporal lesion, amygdala lesion, and traumatic brain injury) were asked to complete the AQ, to compare scores and possible personality changes detected by the AQ. Mean score for the 11 patients was 17.0 (range, 9–25), and median score was 17.0. All patients declared that personality traits identified on the AQ were unchanged after surgical operations or closed-head injuries.

10.7 Acquired Autism Trait After Medial Prefrontal Damage

The two patients displayed damage basically limited to the medial prefrontal cortex and showed mild difficulties in memory performance in daily activities, but no serious problems in language activities and executive functions. These patterns of results are basically consistent with previous case studies regarding damage to the same area of the brain (Baird et al. 2006; Bird et al. 2004). Concerning theory-of-mind tests, performance in the first- and second-order false-belief test was perfect in both cases. Performance in the advanced theory-of-mind tests by T.O. was slightly impaired regarding the provision of appropriate explanations in the strange stories test, and was considerably impaired in the identification of inappropriate verbal expressions for given contexts in the faux-pas detection test. By contrast, performance in advanced theory-of-mind tests by H.C. was not impaired at all in either test. An interesting finding in both cases was that a faux pas was often reported even in control stories without any obvious faux pas.

The most notable finding was that both patients showed some difficulties on the faux-pas recognition test. The percentage of detecting faux pas and providing appropriate explanations was 100% for H.C., compared with 60.0% for T.O. Lower performance by T.O. may have been caused by deficits in delayed recall performance as found in WMS-R and ROCFT. Although T.O. showed higher scores on the two questions for story comprehension (92.5%), he experienced difficulty recalling the exact story contents. In fact, T.O. reported the presence of faux pas for all ten faux-pas stories, but could not recall what the exact contents were in each story. Taking these facts into account, his basic performance for detecting faux pas may not have been greatly reduced.

By contrast, both subjects sometimes incorrectly reported faux pas even in stories containing no faux pas, although they could correctly recognize faux pas in stories containing faux pas. Various reasons could explain this pattern of results. First, this pattern could result from perseveration of response in both cases. The response for detecting faux pas could be a prevailing response, as half of the questions in the faux-pas test did contain faux pas. However, no strong evidence of perseveration was found in either case, since very few total perseveration errors were found on WCST. Second, the pattern could result from general difficulty in understanding global contexts in complex situations. If this were the case, the subjects would show some problems in detecting faux pas in stories containing faux pas. However, the results were in direct opposition to this prediction. A final possibility is overcompensation. In a psychiatric sense, this is often defined as an attempt to overcome an actual defect or unwanted trait by exaggerating in the opposite direction. Self-reports from the two patients indicated that personality change extended to abnormal feelings in some emotional dimensions. Unfortunately, no questionnaires examining anxiety traits were conducted, even though both subjects reported anxiety after surgery. These changes may have resulted in the subjects being more sensitive to verbal expressions in comparison to before surgery. The explanation of overcompensation is considered the most plausible for understanding the overdetection of faux pas.

The most interesting finding was that the two patients showed personality changes after surgery, resulting in some characteristics of autism. These tendencies were mainly clarified by findings from the AQ questionnaire. According to the self-reports, both patients showed a lack of theory-of-mind ability in everyday life, reduced spontaneous seeking to communicate with others after surgery, and obsessive focus on a single subject. To elucidate greater detail of those characteristics, these items were compared with the items identified in a two-factor structure model (Hoekstra et al. 2008). As a result, 25 items among the total 26 items for the development of autistic personality traits after surgical operations in both patients fell into the "social interaction" factor. This basically identified acquired functional deficits following damage to the medial prefrontal cortex as a lack of social interaction. Surprisingly, H.C. even reported becoming fascinated by dates, which is considered a strong characteristic of autism. Results for the number of patients with damage to other areas besides the medial prefrontal cortex revealed that these personality changes resulted from damage to the medial prefrontal cortex alone.

However, limitations exist to the interpretation of the present results. Patients were asked to fill out the same questionnaire (AQ) twice, and were requested on the second trial to answer from the perspective of their previous personality before surgery. This obviously represents a "retrospective report" in the postoperative period, and the data are clearly of questionable validity. However, the second AQ trial based on self-reports revealed that some personality traits identified by the questionnaire matched well with actual personality changes reported after surgery. This suggests that results of the second trials were substantially valid. The results were consistent with previous imaging studies for Asperger syndrome, showing that the medial prefrontal cortex is highly involved in understanding theory-of-mind stories compared with understanding control stories in normal control and Asperger syndrome groups, although the level of peak activation was lower in the Asperger group (Happé et al. 1996). Another interpretation of the AQ rise in both patients is the effect of increased depression and/or anxiety. T.O. and H.C. both mentioned feeling depressive and anxious in everyday life. Depression and/or anxiety alone may increase the AQ score. However, most of the control cases with damage to other areas besides the medial prefrontal cortex reported feeling more or less depressed and anxious, but did not show any increase in AQ score after the damage. This evidence suggests that depression or anxiety alone may not greatly affect AQ score.

Previous studies have reported functional abnormality in the medial prefrontal cortex in autism-spectrum disorders during social task performance, such as theory-of-mind reasoning (Di Martino et al. 2009). Based on the model by Hoekstra et al. (2008), autism personality traits detected in AQ were divided into "social interaction" and "attention to detail" factors. Considering the result that nearly all items for the development of autistic personality traits in both patients fell into the "social interaction" factor, the medial prefrontal area does not seem to be involved in the personality trait for "attention to detail." Functional abnormality in the medial prefrontal cortex in autism-spectrum disorders is likely to be associated with a lack of social interaction.

The present study remains preliminary, but some essential implications help in understanding the possible roles of the medial prefrontal cortex. More research is evidently required to confirm the hypotheses discussed in this study. Besides theory-of-mind functioning, several recent neuroimaging studies have shown that the medial prefrontal cortex is involved in moral judgment (Greene et al. 2001), self-referential processing (Kelley et al. 2002; Schaefer et al. 2006), memory for self (Macrae et al. 2004), and detecting the communicative intentions of others (Kampe et al. 2003). If our discussion is expanded by extending regions of interest from the medial prefrontal cortex to the adjacent anterior cingulate cortex, arguments could be made from the perspectives of cognitive control (MacDonald et al. 2000), error detection, or online monitoring (Carter et al. 2000).

Finally, from a comparative neurocognitive perspective, the medial prefrontal cortex is sure to be an essential area in reaching a full understanding of the development or evolution of social communications (Rushworth et al. 2004). A morphological study indicated large spindle-shaped cells in layer Vb of the anterior cingulate cortex in pongids and hominids, but not in any other primate species or mammalian taxa (Nimchinsky et al. 1999). Although the ways in which spindle cells contribute to social functions of the anterior cingulate cortex remain unclear, this observation is obviously of great interest in attempts to clarify the mechanisms of possible evolutionary changes in adapting to social worlds.

References

Adolphs R (1999) Social cognition and the human brain. Trends Cogn Sci 3:469–479

Adolphs R, Tranel D, Bechara A, Damasio H, Damasio AR (1996) Neuropsychological approaches to reasoning and decision-making. In: Damasio AR, Damasio H, Christen Y (eds) Neurobiology of decision-making. Springer, Berlin, pp 157–179

Bach LJ, Happé F, Fleminger S, Powell J (2000) Theory of mind: independence of executive function and the role of the frontal cortex in acquired brain injury. Cogn Neuropsychiatry 5:175–192

Baird A, Dewar BK, Critchley H, Gilbert SJ, Dolan RJ, Cipolotti L (2006) Cognitive functioning after medial frontal lobe damage including the anterior cingulate cortex: a preliminary investigation. Brain Cogn 60:166–175

Baron-Cohen S (1989) The autistic child's theory of mind: a case of specific developmental delay. J Child Psychol Psychiatry 30:285–297

Baron-Cohen S, Leslie AM, Frith U (1985) Does the autistic child have a "theory of mind"? Cognition 21:37–46

Baron-Cohen S, O'Riordan M, Stone V, Jones R, Plaisted K (1999) Recognition of faux pas by normally developing children and children with Asperger syndrome or high-functioning autism. J Autism Dev Disord 29:407–418

Baron-Cohen S, Wheelwright S, Skinner R, Martin J, Clubley E (2001) The autism-spectrum quotient (AQ): evidence from Asperger syndrome/high-functioning autism, males and females, scientists and mathematicians. J Autism Dev Disord 31:5–17

Bechara A, Damasio AR, Damasio H, Anderson SW (1994) Insensitivity to future consequences following damage to human prefrontal cortex. Cognition 50:7–15

Bechara A, Tranel D, Damasio H, Damasio AR (1996) Failure to respond autonomically to anticipated future outcomes following damage to prefrontal cortex. Cereb Cortex 6:215–225

Bechara A, Damasio H, Tranel D, Damasio AR (1997) Deciding advantageously before knowing the advantageous strategy. Science 275:1293–1295

Bechara A, Damasio H, Damasio AR, Lee GP (1999) Different contributions of the human amygdala and ventromedial prefrontal cortex to decision-making. J Neurosci 19:5473–5481

Bird CM, Castelli F, Malik O, Frith U, Husain M (2004) The impact of extensive medial frontal lobe damage on 'Theory of Mind' and cognition. Brain 127:914–928

Brass M, Heyes C (2005) Imitation: is cognitive neuroscience solving the correspondence problem? Trends Cogn Sci 9:489–495

Britton JC, Phan KL, Taylor SF, Welsh RC, Berridge KC, Liberzon I (2006) Neural correlates of social and nonsocial emotions: an fMRI study. Neuroimage 31:397–409

Brüne M, Ribbert H, Schiefenhövel W (2003) The social brain: evolution and pathology. Wiley, Chichester

Brunet E, Sarfati Y, Hardy-Baylé MC, Decety J (2000) A PET investigation of the attribution of intentions with a nonverbal task. Neuroimage 11:157–166

Carter CS, Macdonald AM, Botvinick M, Ross LL, Stenger VA, Noll D, Cohen JD (2000) Parsing executive processes: strategic vs. evaluative functions of the anterior cingulate cortex. Proc Natl Acad Sci USA 97:1944–1948

Castelli F (2005) Understanding emotions from standardized facial expressions in autism and normal development. Autism 9:428–449

Castelli F, Happé F, Frith U, Frith C (2000) Movement and mind: a functional imaging study of perception and interpretation of complex intentional movement patterns. Neuroimage 12:314–325

Coricelli G, Dolan RJ, Sigiru A (2007) Brain, emotion and decision making: the paradigmatic example of regret. Trends Cogn Sci 11:258–265

Critchley HD (2004) The human cortex responds to an interoceptive challenge. Proc Natl Acad Sci USA 101:6333–6334

Critchley HD, Mathias CJ, Dolan RJ (2001) Neural activity in the human brain relating to uncertainty and arousal during anticipation. Neuron 29:537–545

Critchley HD, Mathias CJ, Josephs O, O'Doherty J, Zanini S, Dewar B-K, Cipolotti L, Shallice T, Dolan RJ (2003) Human cingulate cortex and autonomic control: converging neuroimaging and clinical evidence. Brain 126:2139–2152

Damasio AR (1996) The somatic marker hypothesis and the possible functions of the prefrontal cortex. Philos Trans R Soc Lond B Biol Sci 351:1413–1420

Damasio H, Grabowski T, Frank R, Galaburda AM, Damasio AR (1994) The return of Phineas Gage: clues about the brain from the skull of a famous patient. Science 264:1102–1105

Davidson RJ, Pizzagalli D, Nitschke JB, Putnam K (2002) Depression: perspectives from affective neuroscience. Annu Rev Psychol 53:545–574

Di Martino A, Ross K, Uddin LQ, Sklar AB, Castellanos FX, Milham MP (2009) Functional brain correlates of social and nonsocial processes in autism spectrum disorders: an activation likelihood estimation meta-analysis. Biol Psychiatry 65:63–74

Ekman P (1992) An argument for basic emotions. Cognit Emot 6:169–200

Elliott R, Rees G, Dolan RJ (1999) Ventromedial prefrontal cortex mediates guessing. Neuropsychologia 37:403–411

Eslinger PJ, Damasio AR (1985) Severe disturbance of higher cognition after bilateral frontal lobe ablation: patient EVR. Neurology 35:1731–1741

Frith U (2001) Mind blindness and the brain in autism. Neuron 32:969–979

Frith CD, Frith U (1999) Interacting minds—a biological basis. Science 286:1692–1695

Fukushima H, Terasawa Y, Umeda S (2010) Association between interoception and empathy: evidence from heartbeat-evoked brain potential. Int J Psychophysiol. doi:10.1016/j.ijpsycho.2010.10.015

Gallagher HL, Frith CD (2003) Functional imaging of 'theory of mind'. Trends Cogn Sci 7:77–83

Gallagher HL, Happé F, Brunswick N, Fletcher PC, Frith U, Frith CD (2000) Reading the mind in cartoons and stories: an fMRI study of 'theory of mind' in verbal and nonverbal tasks. Neuropsychologia 38:11–21

Greene JD, Sommerville RB, Nystrom LE, Darley JM, Cohen JD (2001) An fMRI investigation of emotional engagement in moral judgment. Science 293:2105–2108

Happé FG (1994) An advanced test of theory of mind: understanding of story characters' thoughts and feelings by able autistic, mentally handicapped, and normal children and adults. J Autism Dev Disord 24:129–154

Happé F, Ehlers S, Fletcher P, Frith U, Johansson M, Gillberg C, Dolan R, Frackowiak R, Frith C (1996) 'Theory of mind' in the brain. Evidence from a PET scan study of Asperger syndrome. Neuroreport 8:197–201

Happé F, Brownell H, Winner E (1999) Acquired 'theory of mind' impairments following stroke. Cognition 70:211–240

Happé F, Malhi GS, Checkley S (2001) Acquired mind-blindness following frontal lobe surgery? A single case study of impaired 'theory of mind' in a patient treated with stereotactic anterior capsulotomy. Neuropsychologia 39:83–90

Harlow JM (1848) Passage of an iron rod through the head. Boston Med Surg J 39:389–393

Hoekstra RA, Bartels M, Cath DC, Boomsma DI (2008) Factor structure, reliability and criterion validity of the Autism-Spectrum Quotient (AQ): a study in Dutch population and patient groups. J Autism Dev Disord 38:1555–1566

Kampe KK, Frith CD, Frith U (2003) "Hey John": signals conveying communicative intention toward the self activate brain regions associated with "mentalizing", regardless of modality. J Neurosci 23:5258–5263

Kelley WM, Macrae CN, Wyland CL, Caglar S, Inati S, Heatherton TF (2002) Finding the self? An event-related fMRI study. J Cogn Neurosci 14:785–794

Kuchinke L, Jacobs AM, Grubich C, Võ ML-H, Conrad M, Herrmann M (2005) Incidental effects of emotional valence in single word processing: an fMRI study. Neuroimage 28:1022–1032

Leslie KR, Johnson-Frey SH, Grafton ST (2004) Functional imaging of face and hand imitation: towards a motor theory of empathy. Neuroimage 21:601–607

MacDonald AW III, Cohen JD, Stenger VA, Carter CS (2000) Dissociating the role of the dorsolateral prefrontal and anterior cingulate cortex in cognitive control. Science 288:1835–1838

Macrae CN, Moran JM, Heatherton TF, Banfield JF, Kelley WM (2004) Medial prefrontal activity predicts memory for self. Cereb Cortex 14:647–654

Nieminen-von Wendt T, Metsähonkala L, Kulomäki T, Aalto S, Autti T, Vanhala R, von Wendt L (2003) Change in cerebral blood flow in Asperger syndrome during theory of mind tasks presented by the auditory route. Eur Child Adolesc Psychiatry 12:178–189

Nimchinsky EA, Gilissen E, Allman JM, Perl DP, Erwin JM, Hof PR (1999) A neuronal morphologic type unique to humans and great apes. Proc Natl Acad Sci USA 96:5268–5273

Northoff G, Bermpohll F (2004) Cortical midline structures and the self. Trends Cogn Sci 8:102–107

Prohovnik I, Skudlarski P, Fulbright RK, Gore JC, Wexler BE (2004) Functional MRI changes before and after onset of reported emotion. Psychiatr Res Neuroimaging 132:239–250

Rowe AD, Bullock PR, Polkey CE, Morris RG (2001) "Theory of mind" impairments and their relationship to executive functioning following frontal lobe excisions. Brain 124:600–616

Rushworth MF, Walton ME, Kennerley SW, Bannerman DM (2004) Action sets and decisions in the medial frontal cortex. Trends Cogn Sci 8:410–417

Saver JL, Damasio AR (1991) Preserved access and processing of social knowledge in a patient with acquired sociopathy due to ventromedial frontal damage. Neuropsychologia 29:1241–1249

Schacter DL, Curran T, Galluccio L, Milberg WP, Bates JF (1996) False recognition and the right frontal lobe: a case study. Neuropsychologia 34:793–808

Schaefer M, Berens H, Heinze HJ, Rotte M (2006) Neural correlates of culturally familiar brands of car manufacturers. Neuroimage 31:861–865

Shamay-Tsoory SG, Tomer R, Berger BD, Aharon-Peretz J (2003) Characterization of empathy deficits following prefrontal brain damage: the role of the right ventromedial prefrontal cortex. J Cogn Neurosci 15:324–337

Shaw P, Bramham J, Lawrence EJ, Morris R, Baron-Cohen S, David AS (2005) Differential effects of lesions of the amygdala and prefrontal cortex on recognizing facial expressions of complex emotions. J Cogn Neurosci 17:1410–1419

Singer T, Seymour B, O'Doherty J, Kaube H, Dolan RJ, Frith CD (2004) Empathy for pain involves the affective but not sensory components of pain. Science 303:1157–1162

Singer T, Seymour B, O'Doherty JP, Stephan KE, Dolan RJ, Frith CD (2006) Empathic neural responses are modulated by the perceived fairness of others. Nature 439:466–469

Stone VE, Baron-Cohen S, Knight RT (1998) Frontal lobe contributions to theory of mind. J Cogn Neurosci 10:640–656

Stuss DT, Gallup GG Jr, Alexander MP (2001) The frontal lobes are necessary for 'theory of mind'. Brain 124:279–286

Tranel D (2000) Electrodermal activity in cognitive neuroscience: neuroanatomical and neuropsychological correlates. In: Lane RD, Nadel L (eds) Cognitive neuroscience of emotion. Oxford University Press, New York, pp 192–224

Umeda S, Akine Y, Kato M, Muramatsu T, Mimura M, Kandatsu S, Tanada S, Obata T, Ikehira H, Suhara T (2005) Functional network in the prefrontal cortex during episodic memory retrieval. Neuroimage 26:932–940

Umeda S, Mimura M, Kato M (2010) Acquired personality traits of autism following damage to the medial prefrontal cortex. Soc Neurosci 5:19–29

Vogeley K, Bussfeld P, Newen A, Herrmann S, Happé F, Falkai P, Maier W, Shah NJ, Fink GR, Zilles K (2001) Mind reading: neural mechanisms of theory of mind and self-perspective. Neuroimage 14:170–181

Wheatley T, Milleville SC, Martin A (2007) Understanding animate agents: distinct roles for the social network and mirror system. Psychol Sci 18:469–474

Woodbury-Smith MR, Robinson J, Wheelwright S, Baron-Cohen S (2005) Screening adults for Asperger syndrome using the AQ: a preliminary study of its diagnostic validity in clinical practice. J Autism Dev Disord 35:331–335

Part III
Emotion, Consciousness and Memory

Chapter 11
Origin and Evolution of Consciousness and Emotion: Has Consciousness Emerged from Episodic Memory?

Takashi Maeno

Abstract This chapter addresses the Origin and Evolution of Consciousness. How evolution has added novel and wasteful functions to species is briefly discussed, and the control models in the brain, which are required for the control of motion

---

T. Maeno (✉)
Graduate School of System Design and Management, Keio University,
Hiyoshi, Kohoku-ku, Yokohama 223-8526, Japan
e-mail: maeno@sdm.keio.ac.jp

and behavior, are described. The assumption is introduced that the functions enabling episodic memory should emerge from slight modifications to the existing control systems of an organism. The system functions that create the functions enabling episodic memory and the functions of consciousness, which are roughly the same, are discussed. Finally, the origin of the phenomenal, rather than functional, aspects of consciousness is addressed.

11.1 Introduction

Evolution is fascinating. It has developed highly sophisticated biological structures over time. At first glance, the mechanism of evolution appears similar to that of a random search. However, since the probability that material objects will accidentally line up to match the cell arrangement of humans is astronomically low, it is very unlikely that mankind has emerged as a result of a random search. Since evolution depends to some extent on mutation and random mating, at first glance it appears to be an extremely time-consuming optimization operation like a random search. This makes it difficult to intuitively accept the fact that highly complicated organisms were produced as a result of evolution. However, sophisticated and complicated humans have in fact evolved from simple material objects. This was possible because evolution, different from a random search, is an appropriate means for the generation of highly sophisticated biological structures. It is important to note here that evolution is a means of creating major changes in the structure and function of individuals by slightly modifying the genetic structure of life.

It is natural to think that not only our bodies, but also our minds, are products of evolution. The same holds true for "consciousness," the greatest mystery of the mind. Then, for what purpose and why has the system of consciousness been generated? From which animal in the evolutionary tree did consciousness begin to emerge? This chapter discusses my assumptions (Maeno 2004) on these questions.

Section 11.2 briefly discusses how evolution has added novel and wasteful functions to species. Section 11.3 discusses the control models in the brain, which are required for the control of motion and behavior. Section 11.4 introduces the assumption that the functions enabling episodic memory should emerge from slight modifications to the existing control systems of an organism. Section 11.5 discusses the system functions that create the functions enabling episodic memory and the functions of consciousness, which are roughly the same. Section 11.6 addresses not the functional aspect of consciousness, but the origin of the phenomenal aspect of consciousness and Sect. 11.7 concludes the chapter.

11.2 What Is Evolution?

First of all, what is evolution? It is hardly necessary to explain, but in order to establish common ground I will touch on it briefly.

One point that has been widely agreed upon regarding evolution is that it is a quasi-optimization means that enables, through the crossing over of genes, mutation, and natural selection, only those species that have successfully adapted themselves to the changing environment to survive. Different from an optimization process, it does not seek an optimal system structure best suited to a given environment, but selects for survival those species that have adapted in their own ways to the given environment. For example, human beings are not the most optimal species in this system, because even cockroaches or worms thrive in their own way. Neither of these species has the optimal system structure, but both have at least adapted to the given environment.

Nor is evolution a system that generates unusual "components." That is, it does give rise to new functions, but individual components forming these functions are substitutes for or an extension of the components of former organisms. In short, evolution is not a brilliant, creative optimization process like designing a living organism from scratch on a blank sheet of paper; rather, it is like rush construction work.

This difference becomes clear when the bones of mammals are compared. While individual bones differ in shape among rats, birds, and humans, the topology of bones, that is to say, the arrangement in which bones are connected to each other, is quite similar. Bird wings were not designed from scratch for flight, but were created by modifying the design of the forelegs of their nonflying ancestors. The hands of humans are not something that was designed from scratch for the skillful use of tools. Hands were created by modifying the design of the forefeet of ancestral animals that did not use their forefeet to hold objects or use tools.

As shown in Fig. 11.1, in evolution, living organisms are designed in an ad hoc, forcible manner, similar to forcibly changing the shape of a ship with a hammer into an automobile or airplane. This applies not only to the body but also to the cerebral nervous system. The cerebral nervous system should be considered as being developed via design changes that enabled the processing of new information by adding multiple connections to the neural network of lower forms of life.

As Fig. 11.1 shows, despite the fact that evolution is rush construction work involving hammering and adding multiple connections, it sometimes produces new functions: wings that developed from hands and legs enabled flight and expanded the range of activities; standing on two legs freed the hands of humans and enabled the use of tools, and the neocortical development of the brain enabled more complex thinking. Evolution is a nonlinear, impactful means that, despite slight change in genes, produces a catastrophic change in bodily functions. Although these new functions can be said to have emerged through the repetition of more gradual and minor evolutionary processes, this chapter focuses on major changes that clearly suggest the birth of new functions.

The new value produced through evolution always accompanies some waste or useless functions. While the process of optimization adds only those functions necessary and sufficient for the emergence of new functions, the process of evolution is more redundant, that is to say, it contains waste. However, it is this waste that becomes important for the adaptation or the next evolution of individuals. In other

Fig. 11.1 Difference between design of artifacts and evolutionary design of organisms. (**a**) Artifacts: structure and functions both change in a discontinuous manner. (**b**) Organisms: genetic change is small/ functional change is great

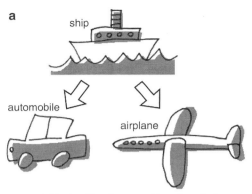

Artifacts: Structures and functions both change in a discontinuous manner

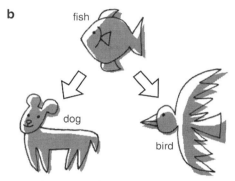

Organisms: Genetic change is small /functional change is great

words, this waste can be considered leeway. Redundancy leads to robustness. Since environmental changes are the driving force of evolution, evolution adopts not an optimization process that aims to satisfy a static objective function, but a quasi-optimization process that enables adaptation to changes in objective functions or the conditions of constraint.

The waste or leeway that results from evolution can often be mementos of former organisms. For example, the appendix and the coccyx are functions of former animals and despite their uselessness for humans, it is believed that they simply happened to remain without disappearing.

It should be noted that the waste and leeway produced via evolution must either accompany new functions acquired through evolution or from the result of slight modifications to the shapes or structures of former living organisms. When new waste or leeway emerges, it should be remembered that brand new waste or leeway never emerges independently of effective functions added by a brand new structure.

Such a view can be considered one type of teleological functionalism (or extension to the level that contains waste and leeway), which holds that life basically acquired only those functions that adapted to the selective pressure of evolution. Henceforth

11 Origin and Evolution of Consciousness and Emotion... 249

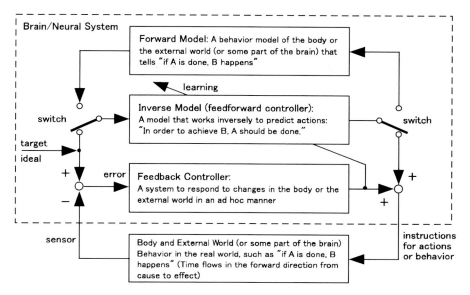

Fig. 11.2 Schematic view of neural control system

in this chapter, teleological functionalism is treated as including the above view of waste and leeway.

Allow me to state my major point here before further discussion: the human function called "consciousness" has emerged from slight modifications to the functions of former living organisms. It is inconceivable that this function was brought by a sudden drastic change in structure when a species evolved from a living organism that did not have similar functions.

In the next section I explain the inevitable path to the generation of the function called consciousness.

11.3 Evolutionary Journey of Control Systems Among Living Organisms

How did the control systems of living organisms evolve before the emergence of "consciousness?" Explained very briefly from my familiar, engineering point of view, it evolved from feedback control to feedforward control.

Fig. 11.2 shows the flow of control by living organisms, that is to say, how the following four types of control or the control system is learned:

1. Feedback control that responds to bodily or environmental changes in an ad hoc manner
2. Feedback error learning by an inverse model (feedforward controller) that predicts causes from effects, such as "In order to achieve A, B should be done"

3. Inverse model-based feedforward control (feedforward controller)
4. Images or thoughts (internal activity) obtained by connecting in a circular manner a forward model of bodily or external (or part of the brain) behavior ("If A is done, then B happens") with the aforementioned inverse model

First of all, I will explain the four types of control using an example of motion control. When you extend your arm to reach for something, how is the motion of the arm controlled?

At first, the feedback control mentioned in #1 must be at work. You reach for something and if you cannot get it, then you try to stretch your arm further. If you overreach, you take your arm back a little. Information about not reaching or overreaching is obtained via sensory information, such as visual sensory information. This is feedback information. Feedback control based on such feedback information is done in an ad hoc manner.

Reflexes are a good example of feedback control. The action or behavior of insects is believed to be basically reflex-based feedback control. When it is bright in front of one eye of a moth and dark in front of the other, the moth displays the reflex action of increasing the flapping of the wing on the dark side, leading to flying to the bright side. In other words, moths are destined by a simple neural circuit to fly into an insect repellent light: the behavior is not the result of high-level decision making.

By contrast, when, for example, the distance and the intensity of force applied to the arm to reach something is calculated in the brain and a force is applied to the arm based on the results of the calculation, this is called feedforward control based on an internal model. The feedforward controller is located in the cerebellum and as mentioned in #2, it is known to learn through feedback control errors (Kawato 1996). Since the feedforward controller solves inverse problems, such as "moving the arm that far requires applying X force to the arm muscles," which is the opposite of an ordinary cause–effect flow, such as "if some force is applied to the arm muscles, the arm will move," it is also called an inverse model. Whereas feedback control may overrun, over-return, or vibrate, and require some time to convey information to the brain, feedforward control (#3) based on an internal model (inverse model) has the advantage of accurate high-speed control, since the arm, in this case, can reach the target smoothly based on the prediction. However, since changes in the position of the target or the arm are not being monitored continuously via sensory information, such as visual and somatic input, feedforward control has the disadvantage of not being able to handle the situation once the target moves or the arm becomes immobile for unknown reasons.

It is easily conceivable that structural evolution can take place from living organisms with a feedback control function to those with a feedforward controller (inverse model) capable of feedback error learning. This is because a feedback controller is a neural network connecting sensory neurons with cortical neurons and motor neurons and the feedforward network, and the neural network of its learning paths can also be constructed easily by adding connections to the neural network. In other words, just as bodies have been modified with a "hammer," a feedforward model based on an inverse model can be created by making the connections of the neural network a little more complicated.

I mentioned that the control system of insects is based mainly on feedback. It is widely known, however, that insects are capable of some learning. Insects can somehow remember such information as the location of food or the nest. This is a type of inverse model. Insects, which have learned an inverse model in that if they go to a certain location A, they can get some food and thus use the inverse model to create an action plan that works in the opposite direction from the real world (a fact), such as "in order to get some food, I should go to location A." Thus, it seems that simple organisms like insects also perform not only feedback control, but also inverse model-based feedforward control.

It is clear that living organisms capable of feedforward control using the inverse model have an advantage over those capable of only ad hoc feedback control in terms of environmental adaptation (evolution takes place because it is advantageous). As mentioned above, it is because the inverse model is a type of memory used to create an action plan, "in order to do A, I should do B" and because it stores past environmental information in the form of feedback error learning for use in the next action or behavior.

Then, what will the living organisms that have evolved to the next level be able to do? It is the acquisition of a forward model and the emulation (#4) of action and behavior based on a forward-inverse model. The forward model, literally as opposed to the inverse model, reproduces a phenomenon in the brain in the same direction as the real-world cause and effect flow. I can explain this by using the aforementioned reaching action as an example. The forward model is a model that reproduces in the brain the actual action of the arm that applying some force from the arm muscles will stretch the arm toward the target. It is believed that mirror neurons are somehow associated with this model. The presence of such a neural network in the brain enables the quasi-optimization of the action within the brain. While the inverse model in the brain was originally connected to the real world where time flows in the forward direction, the inverse and forward models can be connected to each other in a circular manner by changing the switch settings at both ends as shown in Fig. 11.2. Such connections will enable the following simulations in the brain: instead of actually attempting the estimate of the inverse model, "in order to achieve A, B should be done," it can be tried in the brain using the forward model and if the estimate does not work, the error can be recalculated and corrected in the inverse model and the result can be tried again in the forward model. Kawato calls this model a forward inverse relaxation-type neural circuit model (Kawato 1996).

As the inverse model and the feedback controller, the forward model can also be formed by building neural networks. Consequently, it is easy to assume that the forward model in the brain can be designed using a technique of multiple connections called evolution.

This means that the living organisms that have acquired the forward model can perform not only ad hoc feedback control, but also feedforward control capable of making a smarter calculation, "in order to achieve A, B should be done," and even a forward inverse relaxation-type calculation like "I will think about what should be done to achieve A in my head," which is a more advanced calculation than that of the former two controls. Thus, these organisms are capable not only of implementing the inverse model, that is to say, performing actions or behavior in accordance with past memory, but also of running memory-based simulations in the brain in search of better actions or behavior.

It seems that in the brain, a number of forward-inverse models are connected in a hierarchical series–parallel structure, where a large number of neurons repeat an on/off firing action like restless waves in response to external and internal information. Although Fig. 11.2 shows only one of each forward model, inverse model, and feedback controller, in reality a large number of these models and controllers are connected to each other in a series–parallel structure.

It should be noted here that "consciousness" has not yet been discussed. It may be argued that some terms associated with consciousness, such as memory and thinking, have been used. The terms I used are not related to consciousness. In "memory", there is implicit knowledge or nondeclarative memory, which is the memory that the body has stored. There are many cases where memory is utilized in ways the body has learned without the intervention of consciousness, such as automatically reaching out for an object. When I discussed the forward inverse relaxation model, I used the expression, "I will think about what should be done to achieve A in my head." Here, I used the term "think" as a metaphor. While thinking may be a conscious behavior when one is aware of it, "think" in this expression is used to describe the circulation of information in the brain that occurs when forward-inverse computation is repeated among neural network modules working unconsciously and that looks as if one is thinking.

In short, ad hoc feedback control, feedforward control using the memory of "if A is done, then B happens," and forward inverse relaxation calculations that figure out "What should be done to achieve A?" all do not require conscious intervention. That is, for the living organisms that live mainly by feedback control, such as those that eat if there is food, sleep if they get sleepy, and fight or escape from other organisms if attacked, consciousness is not necessary. Nor do those living organisms need consciousness, which can store such information as the locations of food and the nest and the appearances of enemies in inverse memory. Moreover, those living organisms that can solve simple inverse problems (which location has food, which way is shorter to the nest, and whether they should fight against organisms with certain appearances or escape from them) by forward inverse relaxation computations do not need consciousness.

At first sight, it seems advantageous to have awareness of one's own qualia of eating food, qualia of making decisions using the results of an inverse model, and qualia of weighing two different paths. On second thoughts, however, there seems to be no evolutionary advantage in having consciousness among these species. As long as the neural network in the brain does the same computations and provides the same results as consciousness, they do not have to consciously feel the presence of self as qualia when performing these computations. When standing, we unconsciously strike a balance and when walking, we unconsciously move our legs alternately. Just imagine living organisms whose entire actions are performed unconsciously like these actions. As already mentioned, since evolution is a quasi-optimization process that adds only those functions that make the emerging species more advantageous in terms of environmental adaptation than the former species, consciousness does not emerge unless its addition has some advantage for the emerging species.

11.4 Acquisition of Episodic Memory Through Evolutionary Processes

Episodes can be categorized into two types: personal episodes, such as "I came back from a trip yesterday" and "I had a fried egg this morning" and other episodes, such as "Prince Shotoku issued the Taika Reform Edicts," "The LDP won a great victory," and "A grim incident was reported on the news." In this chapter, episodic memory refers to the former type of memory.

What are the differences between living organisms capable and incapable of episodic memory?

Both organisms are similar in that they are capable of semantic memory. Let us suppose that there is some good food at a certain place. Even those living organisms incapable of episodic memory are capable of performing a memory task that leads to an inverse action, "in order to find food, I should go to that place," based on past experiences of obtaining food at the place. When hungry, the organism can of course take the action of going to that place and eating based on memory. The living organisms capable of episodic memory can remember experiences of when they went to a certain place, they found good food. When hungry, they can go back to the place based on that experience. Moreover, based on these experiences, they can also perform brain processing to create semantic memory. Thus, for memorizing the places where good food can be found, there is not much difference between the two types of living organisms.

What happens if they are in an unstructured environment, that is to say, in a changing environment where there are many food places, but the amount of food at each place changes often or food at some places spoils more quickly than other places? Living organisms incapable of episodic memory can also alter their behavior according to the occasion if they have developed sophisticated inverse models based on experience. They can select their behavior: if there is little food at one place, they go to another place or if food at one place spoils easily, they choose not to eat there, but to go to another place. However, since they are not capable of episodic memory, they cannot handle detailed information with a complicated schedule, such as the amount of food they had yesterday at which feeding place or the food at one place was fresh the day before yesterday but was rotting today. On the other hand, living organisms capable of episodic memory can select the behavior of going to place C based on episodic memories that all of the food they ate at place A yesterday was good and that they went to place B this morning, but almost all of the food there was rotten. Thus, it seems that episodic memory provides more sophisticated action alternatives.

The difference between these two types of living organisms becomes more obvious when communication is considered.

Living organisms capable only of semantic memory can memorize such meanings as who constitutes relatives, enemies, or friends. Therefore, they can take simple actions such as rejoicing when a parent approaches, escaping when an enemy approaches, and eating when a parent brings food to their mouths.

However, they are incapable of performing an episodic "what, where, when, and with whom" memory task. In other words, they cannot memorize their experiences, such as eating something good with their parents somewhere, accomplishing something with friends, or fighting against an enemy, as episodes. Since they are capable of performing only simple memory tasks, such as "Person A is XXX" and "Object B is YYY," all they can memorize are simple meanings, such as "Parents are gentle," "It is fun to play with friends," or "I don't like him/her." They have no reminiscences.

In short, they are in a condition similar to dementia. They cannot recall what they had for breakfast and with whom today, what kind of conversation, they had with friends and what information they got from the conversation or what was written in the first half of the book they are reading. Even worse, they cannot remember whether or not they had breakfast this morning, met their friends, or read the first half of the book. It is said that a person suffering from episodic memory loss for 20 years always behaves as if he/she is his/her old self at the time when he/she lost episodic memory. This is the same situation as living organisms capable only of semantic memory: they cannot remember any past episodes.

It is readily understood that living organisms capable of episodic memory have a greater advantage over those incapable of it in adapting to the environment. Being capable of episodic memory means to be able to describe the chains of cause and effect that develop over time with complicated situational changes comprising "what, when, where, and with whom" memory tasks and the subsequent "what, when, where, and with whom" memory tasks in the brain in the form of forward models. Compared with semantic memory capable of memorizing only meanings such as "A is B," the content of episodic memory, which can be understood and expressed, is more intricate and sophisticated.

The life of living organisms incapable of episodic memory is an easygoing one, where they eat when hungry and sleep when sleepy, although they cannot recall episodes like what time they got up this morning or whether or not they had breakfast this morning. On the other hand, living organisms capable of episodic memory can remember what time and where they woke up this morning, what they ate, and the content of the newspaper they read while having breakfast. It is clear that they can lead a more intellectual life than the easygoing organisms.

Moreover, living organisms capable of episodic memory can run various cause–effect simulations by conducting forward inverse relaxation computations. In other words, they can carry out an act of thinking that examines the feasibility of a variety of actions based on the results of different experiences. Thus, it seems that living organisms capable of episodic memory have acquired it because it is evolutionarily advantageous.

11.5 Relationship Between Consciousness and Episodic Memory

What functions are required for living organisms to become capable of episodic memory?

I have mentioned that in the brains of living organisms incapable of episodic memory, the neural networks governing feedback controllers, forward models, and inverse models seem to be connected in a hierarchical series–parallel structure, where a large number of neurons repeat an on/off firing action like restless waves in response to the flow of external and internal information.

In my book, *Why has the brain created the mind?—Hypothesis of passive consciousness to solve the mystery of the self* (Maeno 2004), I compared modules in brain neural networks to busy dwarves performing their roles. While it may not be a good comparison, "dwarves" are used as a metaphor for modules in the brain that achieve simple functions.

The dwarves responsible for feedback controllers, forward models, and inverse models process information in an autonomous, decentralized manner, for which consciousness was not required. This is because human behavior is determined not by a dictator called "consciousness" in a top-down manner, but by dwarves of "unconsciousness" in a democratic, self-organized, bottom-up manner. For living organisms that remember only such meanings as "A is B," a system to observe what one is doing now, such as consciousness, is nothing but useless.

By contrast, episodic memory is the memory of what one did. In order to remember what one did, a function must exist in the brain that sends the preceding information about what one is doing now into episodic memory.

For convenience, let us call the system in the brain that sends the information about what one is doing now to episodic memory "module A." This is a system that monitors what one's body is doing—walking, talking or eating—as a result of work done by dwarves (neural networks in the brain) in an autonomous, decentralized manner, and memorizes it as an episodic memory of "I did such and such."

Such a system can be seen as an extension of the system shown in Fig. 11.2. In other words, it can be easily created by evolution, which is a quasi-optimization computation that performs rush work with a hammer or by adding multiple connections. That is, episodic memory, a model that memorizes cause-and-effect relationships that develop over time, can be generated relatively easily by slightly modifying forward models that memorize simple cause-and-effect relationships. It can be inferred, then, that module A, a system to sum up what one is doing as a result of what subsystems did and send it to episodic memory, would be a system related to short-term memory. It is because short-term memory is a system for memorizing events that happened in the recent past. What one is doing now is not an event independent of the past, but a phenomenon with a time interval that is long enough to build context before and after it. What short-term memory does is to memorize the neighboring events that happened in the recent past.

In the above, I have discussed that module A is required to perform an episodic memory task.

Next, I would like to discuss "consciousness." Although I have already used the term several times, let me confirm the definition of consciousness again. It is said that there are two types of consciousness: awareness from focusing attention on things or events and self-awareness, which is a sense of self. The difference between the two is the target of attention, things/events or one's self. They are the same in that they focus attention on what one is doing now.

Since Chalmers (1996), philosophers of the mind have tended to approach consciousness from two different aspects: the functional aspect and the phenomenal aspect. The functional aspect focuses on the functions of consciousness, such as when we say that the system of consciousness is a module that achieves the function of "focusing attention on what I am doing now." On the other hand, the phenomenal aspect addresses the vivid conscious phenomenon surrounding "I am focusing my attention on what I am doing now." It can be described as qualia, the texture of consciousness. Many philosophers of the mind agree that no matter how clearly the functional aspect of consciousness is elucidated, the phenomenal aspect may still remain a mystery. In other words, even if we can describe the functions of consciousness using a diagram like Fig. 11.2, we still cannot describe the qualia of feeling the freshness of an apple in front of us or that of feeling we are alive at this moment on a piece of paper.

Here, I would like to focus my discussion on the functional aspect of consciousness. I will discuss the phenomenal aspect later. The functional aspect of consciousness refers to a function of "focusing attention on what I am doing now." To focus attention on what one is doing now means to turn conscious attention to the most important, noteworthy point of what one is now experiencing as a result of unconscious, autonomous-decentralized information processing by dwarves.

Since module A serves to send information regarding what one did to episodic memory, it is likely that the functional aspect of consciousness is part of module A. The role of module A includes turning attention to and experiencing part of the behavior of autonomous decentralized dwarves and sending it to episodic memory. Of this role, consciousness works at turning attention and experiencing.

That is, module A is indispensable for performing episodic memory tasks and consciousness constitutes its first half. In other words, the function of consciousness is required to perform episodic memory tasks.

However, further examination is required to see if the function of consciousness is a sufficient condition for the functioning of episodic memory. If we can deny the possibility that the function of consciousness is useful for something other than episodic memory, we can safely state that the function of consciousness is a necessary and sufficient condition for episodic memory, that is to say, the function of consciousness emerged through evolutionary processes to enable episodic memory.

What advantages does consciousness provide for the environmental adaptation of living organisms? Although I have been considering this point for many years, I have not found any advantages of consciousness other than enabling episodic memory. It seems to me that consciousness, which is a system to focus on and experience what one is doing now, has no other advantages than making preparations for episodic memory to memorize the experience.

There is a group of psychologists who claim that the function called consciousness emerged to become aware of using language. If there had been no consciousness, at least languages as sophisticated as those used by humans might not have existed. If this is true, it can be said that consciousness is a necessary condition for language. However, since there are living organisms that are capable of episodic memory but do not have language, consciousness cannot be a sufficient condition for language. Moreover, since we humans can become aware of those experiences that cannot be verbalized, we intuitively know that consciousness did not emerge for language.

There are many who think that there may be living organisms that have consciousness, but are incapable of episodic memory. These people claim that while consciousness is said to be a necessary condition for episodic memory, if living organisms incapable of episodic memory, such as insects and more primitive amoebae, had consciousness, consciousness would no longer be a necessary condition for episodic memory. Since we cannot ask insects if they have consciousness or not, it is difficult to prove that they do. However, what I have emphasized repeatedly in Sects. 11.2 and 11.3 denies the possibility: evolution is a process of adding functions that enhance environmental adaptation and in such evolutionary processes, functions that do not contribute to environmental adaptation do not emerge. Waste and leeway, of course, are sometimes added, but these additions occur only in association with the addition of some other effectual functions. For example, it is very unlikely that consciousness, which does not seem to be required for the environmental adaptation of insects, was added when insects emerged. On the other hand, since living organisms capable of episodic memory have an advantage over organisms incapable of it in terms of evolution and the function of consciousness is required to perform episodic memory tasks, adding consciousness to enable episodic memory is evolutionarily plausible.

Thus, I believe that only those living organisms that are capable of episodic memory have consciousness. It has been found that birds and some mammals are capable of episodic memory. It is, therefore, very likely that these are the species that have consciousness. On the other hand, it is questionable whether birds have consciousness.

Let us persue the following experiment on the episodic memory of birds. Birds hid fresh food in site A and site B. While food in site A tastes better, it spoils more easily. One bird kept eating food hidden in site A for some days, but moved to site B when the food at site A started rotting. It was claimed that this behavior showed that birds are capable of memorizing time-series information, such as "since food at A was hidden X days ago, it would be rotten by now," as an episode. However, this could be a case where an episodic memory task was performed without using consciousness or module A. Birds do not have to consciously experience the behavior of "I am hiding some food now," but still can use their episodic memory and perform in the aforementioned manner if they have acquired the meaning that it takes 3 days for food to become rotten and have an unconscious timer. Therefore, while the function of consciousness is required to perform sophisticated episodic memory tasks such as those of humans, there can be some other ways for simpler episodes.

Therefore, strictly speaking, I believe that the function of consciousness is a necessary and sufficient condition for performing episodic memory tasks that are too sophisticated for other methods (that do not use consciousness) to handle.

11.6 Is the Phenomenon of Consciousness Evolutionarily Necessary?

The above discussion will not satisfy the interest of philosophers of the mind, because it was about the functional aspect of consciousness. What philosophers of the mind truly wish to know is whether or not the phenomenal aspect of consciousness is an evolutionary necessity.

There seem to be two possible situations where the phenomenal aspect of consciousness takes on evolutionary significance. One is when the functional and phenomenal aspects of consciousness are inseparable. That is, situations where organisms have functional consciousness but do not have phenomenal consciousness are impossible. This is the view the identity theory holds. Another is when the functional and phenomenal aspects of consciousness evolve separately. In this case, it must be explained that there is a situation where having phenomenal consciousness provides the advantage of improving environmental adaptation.

Although I think the former is intuitively correct, controversy over the identity theory has not been settled. Rather, the odds seem to be against it. Unfortunately, I do not have clear explanations for it either, but since I disagree with Chalmers and the now popular post-Chalmers zombie argument, I will start my discussion from this point.

Chalmers (1996) argues that it is the phenomenal aspect of consciousness that is problematic and incurs a contradiction in the identity theory. Chalmers cites a story of a zombie as an example to point out the contradiction in the identity theory. Originally, a zombie in philosophy meant a being that looks like, talks like, and behaves like a human being, but does not have phenomenal consciousness. Can you imagine such a being? I can.

It is very likely that future robots will resemble such zombies. In a future society with highly developed information processing and neural network technologies, robots that have a mind like that of Astro Boy may come into existence. In such a society, the aforementioned system with a function to focus attention on some part of the results of its own unconscious information processing, experience it, and memorize it as an episode, may be easily created. However, if the mechanism to create the phenomenal aspect of consciousness has not yet been elucidated by then, the robot would have the same function of consciousness as humans, but lack the qualia of phenomenal consciousness that humans have.

If you asked such a robot the question, "Is the apple good?" it would be able to answer that the apple you cut for it is the sweetest ever and tastes great. It could reflect on last year's sweet memory with you and say that the sunset at that time was very romantic. However, even if the robot says that the sweet apple is pleasant to the palate or that it feels wistful every time it recalls the sunset, it only means that a program designed to activate functions to say such things (or lies) is at work (or neural networks fire and generate meaningful patterns). The qualia of sweetness, happiness, or wistfulness did not exist in its brain or sensory organs 1 year ago nor would they exist at present.

On the other hand, Chalmers takes this zombie story one step further. A zombie, he says, is a being that not only looks, talks, and behaves like humans and has the same facial expressions, but also has the identical firing distribution in the neural networks of the brain. The only difference is that it does not have the phenomenal aspect of consciousness that humans do. Chalmers further claims that he can imagine such a zombie and that all humans must also be able to imagine it as he. Can you?

There seem to be many philosophers who say they can, but I just can't.

To be able to imagine a zombie whose firing distribution in the neural networks of the brain is identical to that of humans and yet whose consciousness does not have a phenomenal aspect is equivalent to saying that the phenomenal aspect of consciousness is not a result of the workings of neural networks. This contradicts teleological functionalism, which holds that the phenomenal aspect of consciousness must have emerged as a result of the structural evolution of neural networks in the brain. I just cannot imagine such a zombie.

Is it possible then for the former (Chalmers) to refute the latter (teleological functionalists) or vice versa?

It seems impossible. While the former represents mind–body dualism, the latter represents the identity theory. The dispute as to which is true is an issue that metaphysics should address. Moreover, the logical structures of the two theories make it impossible for them to refute each other. Logically, therefore, such a dispute will only lead to a catch-22 situation.

Since Chalmers grounded himself in mind–body dualism at the moment he said he could imagine such a zombie as he had described, it is entirely natural for him to think that the identity theory is a contradictory theory and deduce mind–body dualism from such a zombie view. Since the identity theory and mind–body dualism differ in their premises, he cannot refute the identity theory even if he can point out its shortcomings.

Chalmers dismisses the identity theory as being unable to explain qualia by neural networks. However, identity theorists ground themselves in the identity theory not because it can already explain every phenomenon in the world: the identity theory is their premise or framework from which to examine various phenomena. Therefore, even if Chalmers can point out problem areas of the identity theory, he cannot distinguish cases where the theory itself is false from those that just indicate unknown areas that the theory has not yet elucidated.

The opposite holds true: while identity theorists can counter-argue against Chalmers' dualism based on the teleological functionalism of the identity theory, they cannot completely refute his dualism. Therefore, I limit my discussion here just to mentioning ways to counter-argue. Needless to say, while those grounded in teleological functionalism will agree with the following argument, Chalmers and those who agree with his view will not.

Although both functional and phenomenal aspects of consciousness were acquired through evolutionary processes, let us examine cases where neural networks are not involved, as Chalmers claims, in the phenomenal aspect of consciousness. This means that one day, probably when the human species emerged, a new mechanism called phenomenal consciousness, which cannot be explained by material behavior like the firing of neural networks, has suddenly and evolutionarily emerged.

As already mentioned, evolution is a process of quasi-optimization via rush construction work, where shapes are forcibly changed with a hammer or work is completed by forcibly adding multiple connections. I wonder how a new mechanism of phenomenal consciousness, which cannot be explained by material behavior, could possibly emerge through such an ad hoc, unrefined method. It is an utter mystery to me. Moreover, what advantages would the mysterious phenomenal consciousness have in the environmental adaptation of living organisms? It is also an utter mystery. As Chalmers said, it is a mystery with no clues. In imitation of Chalmers' zombie argument, more than a few philosophers and neuroscientists now go as far as to say when faced with a difficult problem that this is a mystery that cannot be solved in the near future.

On the other hand, viewed from the standpoint of teleological functionalism, it is very unlikely that the phenomenal consciousness independent of material behavior emerges through evolutionary processes.

Based on identity theory, I think that the phenomenal aspect of consciousness is formed by neural networks in the brain. Whether to ground oneself in the identity theory or mind–body dualism is a matter of one's beliefs, Similar to questions like "Do you believe in God?" and "Which do you think is more essential, liberalism or communism?" Unfortunately, there is no other way to bring a metaphysical dispute to an end but for each individual to select an intuitively plausible position induced from his/her rich past experiences accumulated over time. It is self-evident that no correct answer exists in the world after structuralism that does not have an absolute value system. In such a context, my beliefs tell me that it is more plausible to think within the simple view of the identity theory common among present Japanese scientists than to fall into dualism and think that the yet to be defined qualia of consciousness exists independent of materials. I believe that the qualia of consciousness should be created, not by something unknown and independent of materials, but by neural networks, even though the mechanism whereby the phenomenal aspect of consciousness is generated by neural networks remains unclear.

It may be better to use the term "intuition." That is, I intuitively think that the phenomenal aspect of consciousness is created by neural networks in the brain (although its generation mechanism is not fully understood). In other words, I can imagine how the phenomenal consciousness is created by neural networks. Connectionists would agree with me. Since neural networks are systems that can generate patterns that cannot be represented by letters (signs), such as actions or behaviors of the body or the external environment, as a forward model, I can vaguely but intuitively imagine how phenomenal consciousness, a forward model governing other areas of the brain, is created as a field. (I know this explanation is not good enough. I intend to make it clearer in the future, but meanwhile, those who are not satisfied with the above discussion are recommended to use artificial neural networks, which will help them understand why I say "I can imagine.")

What people on the other side claim is also merely based on their own beliefs or intuition, that is to say, a belief that they can imagine zombies that have the same neural activities as humans, but do not have the qualia of consciousness or that it is impossible for neural networks to generate the phenomenal aspect of consciousness. In fact, Chalmers states flatly that he knows no ways other than intuition to explain why he can imagine zombies.

I will stop the unproductive, metaphysical argument against Chalmers here and shift the subject to the difficulty of conceiving a situation where organisms have developed the functional aspect of consciousness, but not the phenomenal aspect of it. In other words, I will discuss why I believe that the phenomenal consciousness emerged concurrently when the functional consciousness emerged through evolutionary processes. Needless to say, I will discuss it within the framework of identity theory.

I mentioned previously that I could imagine a robot with functional consciousness, but without phenomenal consciousness. When such a robot says it feels wistful, it is not actually feeling wistfulness that twinges the heart. When it says it hurts, it is not actually feeling pain. Then, is it telling a lie by saying it feels the qualia that it actually does not feel? When a future robot says that it feels pain or wistfulness, every response, including facial expressions (if they exist), pulse, and the sensation of warmth on the face, must be identical to those of humans. If someone said, "You may be telling a lie. I don't think it really hurts," it will reply, "Are you kidding? It really hurts! I can't stand it. Give me some medicine" and start thrashing about with pain. The robot does not think at all that it is not painful or that it is telling a lie. Since it has the same function of consciousness as humans, it acts as if it really feels the pain. If asked how painful it was 1 h ago, it may answer that it was worse and so painful that it almost fainted. Despite the fact that it did not feel the pain qualia 1 h ago as it does not now, it insists it felt the pain 1 h ago with a straight face and acts exactly like a person who really means it. Naturally, such a robot passes a Turing Test since it says it hurts with full conviction. It does not even think that it is telling a lie. We can no longer detect the lie from its appearance. The robot is so realistic that it almost makes us question the fact that it does not have pain qualia. It can explain the pain it felt 1 h ago in detail with a straight face. With a grimace, it points to where it hurt most. I think that if one could create such a highly realistic robot, one should actually make it feel pain. Unfortunately, however, it cannot be done at present, since we do not know how to create the qualia of the phenomenal consciousness.

Now, let me replace the robot with an organism similar to a human that has emerged through evolutionary processes. In other words, let us suppose that there happened to be a living organism somewhere on earth or on another planet, which has the functional consciousness that has evolved to enable episodic memory, but does not have the phenomenal consciousness. Is it likely for such an organism to emerge through evolutionary processes?

Here, the discussion is based on the identity theory, that is to say, based on the union of connectionism and teleological functionalism within physicalism. Therefore, it is premised on the following: this world is a field where the phenomenal consciousness can be created by neural networks through evolutionary processes.

Now, is it possible for an organism to emerge through evolutionary processes, which is not actually feeling pain qualia despite the fact that it sincerely explains the pain of 1 h ago or the current pain in detail with a grimace while pointing to the place where it hurts?

In the case of robots, Since we humans do not have the knowledge to create qualia, we could not help but create such weird robots. If phenomenal consciousness could be created by neural networks (although mankind at present does not yet know how), the creation of organisms with true pain qualia would be easier to accept than that of "roundabout" organisms that pretend to be in pain, because these organisms with pain qualia do feel the pain. It is intuitively much easier for those of us with qualia to understand such organisms (they are easier to understand, but do not constitute proof of inevitability).

Let us examine the mechanism of pain qualia among humans. When some big distortion occurs in a fingertip and pain receptors start firing, this information is conveyed to the cerebral cortex. There is a map of the entire body in the sensory area of the cerebral cortex and when a finger hurts, the neurons at the site corresponding to sensations at the fingers fire. While it seems that qualitative information about pain is generated at a corresponding site in the sensory area of the cerebral cortex, the actual pain qualia is felt at the fingertips, despite the fact that there are only pain receptors at the fingertips and no such devices exist that generate the phenomenon of pain qualia. The phantom limb sensations introduced by Ramachandran and Blakeslee (1998) may sound even stranger. A person who has lost his/her right arm vividly feels the pain qualia of fingernails of the tightly closed fist biting into the palm despite the fact that there is no right hand.

How could things like this happen? While pain is supposed to be created by neural firing patterns in the brain, the phenomenon of phantom limbs seems to show, at least to me, that some structure or mechanism must be existent that somehow causes pain, a phenomenal aspect of consciousness, in certain areas of the body. As I wrote in my book (Maeno 2004), it seems to me that we have no choice but to think such phenomena occur because we humans are made that way, that is to say, to have delusions or illusions. Although we do not know yet how such delusions or illusions are created, since the organisms that feel pain qualia at the very site of the body are intuitively easier to understand than the aforementioned "roundabout" organisms, we must be made that way.

The neural firing patterns of pain in the brain and pain qualia in different body areas are inseparable. While Kripke claims that such physicalism does not make sense, I still think they are inseparable. Kripke argues that if they were inseparable, it would mean that when the neural firing patterns of pain are reproduced in the bran of a dead person, he/she will feel the pain, which is not what happens. I think his preconditions are wrong. The neural firing patterns of pain in the brain and pain qualia in body areas are inseparable when the switch of consciousness is on: they are conditionally inseparable and this conditionally inseparable situation is defined by the operation of other neural networks. Since this condition is not met when one is sleeping or dead, the area of the body does not feel pain even if the neural firing patterns of pain are reproduced in the brain.

In summary, while the discussion in this chapter could not provide explanations for evolutionary validity of the presence of pain qualia, I argued that under the premise (belief) that the phenomenal consciousness is created (must be created) by neural networks, organisms that can feel delusional or illusional qualia are more likely than the conventional zpmbie type organism to exist.

Nevertheless, one thing that is clear about the phenomenal consciousness at present is that while the framework of the identity theory on which I base my beliefs has not yet elucidated how it is created by neural networks, Chalmers' dualism cannot even expect to find a clue to this mystery. Which theory one chooses depends on one's beliefs.

11.7 Death Belongs to Evolution

Lastly, I would like to make a comment on my view of life and death. This is not a comment by an engineer, but an emotional comment by one human being with consciousness. The reason why I took an interest in consciousness to begin with is that I personally wanted to know from where consciousness has come and to where it is headed. At present, I think that consciousness came into being out of nothing after my physical body was born, probably when I was around 2 years old, and will disappear with my physical death.

Speaking in an evolutionary context, it is natural to at least think that since the role of the functional consciousness is to prepare for the creation of the episodic memory of individuals, this role of consciousness ends with the physical death of individual organisms. Needless to say, the death of organisms is for evolution to take place. If organisms continued to live and never died, there would be no room for new lives to live. If new lives were not born, the mechanism of evolution that generates new species best suited to a given environment would not be able to function. As Dawkins (2006) stated, therefore, individuals are only vehicles for genes and consciousness is only a disposable part of the vehicle. Thus, it may be possible to logically understand the origin of the functional consciousness from the standpoint of an outsider, but my personal phenomenal consciousness resists accepting such an explanation. I don't want to die. It all sounds empty. Isn't there a more comforting conclusion?

The other day, I had a chance to talk with an old lady about the mind. Her comments were impressive. She said, "I think human beings want to become like the wind. The wind never thinks, it just keeps blowing. Insects also live without thinking or worrying about anything. It is only humans that have worries. I think that the reason why we humans get old, lose our memory and consciousness, and become senile is to become happy like the wind and insects."

Come to think of it, I was happy when I was a kid as I was protected by my parents. Before I knew it, my knowledge, memory, and worldly desires increased. In the twilight years of my life, my knowledge and memories will gradually disappear, my functional and phenomenal consciousness will weaken, and my being will gradually return to nature. The activities of humans as individual members of the human species transit from coming into being out of nothing to making a soft landing back to nothing.

References

Chalmers DJ (1996) The conscious mind: in search of a fundamental theory (philosophy of mind series). Oxford University Press, Oxford
Dawkins R (2006) The selfish gene, 3rd edn. Oxford University Press, Oxford
Kawato M (1996) Computational theory of the brain. Sangyotosho, Tokyo
Maeno T (2004) Why has the brain created the mind? – Passive consciousness hypothesis to solve the mystery of the self. Chikumashobo, Tokyo (in Japanese)
Ramachandran VS, Blakeslee S (1998) Phantoms in the brain: probing the mysteries of the human mind. Harper Perennial, New York

Chapter 12
The Logic of Memory and the Memory of Logic: Relation with Emotion

Philippe Codognet

Photo © Bibliothèque Nationale de France, Paris

Parts of this paper has been published in German as "Transgene Archive", in: *Leidenschaften der Bürokratie. Kultur- und Mediengeschichte im Archiv*, Sven Speiker (Ed.), Kadmos Verlag, Berlin, 2004 and in French as "La pensée aveugle", in: *L'art a-t-il besoin du numérique?*, J-P. Balpe & M. de Barros (Eds.), éditions Hermes, Paris, 2006

P. Codognet (✉)
JFLI-CNRS/UPMC/University of Tokyo, Information Technology Center,
2-11-16 Yayoi, Bunkyo-ku, Tokyo 113-8658, Japan
e-mail: codognet@jfli.itc.u-tokyo.ac.jp

Abstract This chapter considers some historical background in the Western history of ideas related to digital technologies, biotechnologies, and the links between them. A series of artificial systems are described and recalled that have been invented throughout the years to help memorizing, sometimes in very strange ways. Indeed the scholarly tradition of the Art of Memory, also called in Medieval times *ars memorativa*, *ars memoria* or *artificiosae memoriae*, was a strand of classical studies that persisted until the Baroque Era but that in fact extended back to Antiquity, as Cicero himself is considered to be the founding father of this discipline. If the effectiveness of those mnemonic systems can be questioned, it is interesting to note that they were designed in a rather systematic, nearly rational manner, but were based in some way on emotional memory. This awkward and rather forgotten branch of philosophy (in the medieval sense) is also linked to the tradition and the quest for a universal or perfect language that created a vivid topic of study from the medieval times up to the utopian seventeenth century. Binary notation, for instance, was invented long before its use in digital computers but was already considered as a universal language. Also discussed is the logical next step after cybernetics and the digitalization of the world for archiving purposes: life sciences and biotechnologies. this new vision is illustrated by a utopian project of using cockroach DNA as the perfect self-duplicating and self-preserving archive.

12.1 Introduction

In the last decade, major advances have been made in understanding the mechanism of memory in both animals and humans. Research studies clearly show that emotions have a key influence on both the way we memorize facts and the way those memories are recalled. Since the 1950s, well-known experiments in animal psychology such as Pavlovian fear conditioning have shown that memory is conditioned by emotions in animals. The same is certainly true in humans, but recent research identified two different mechanisms: memory of emotions (and thus conditioning) and emotional memory, that is, the fact to better remember something if it is associated with some emotional stimulus (Ledoux 1996, 2000; LaBar and Cabeza 2006). Studies in cognitive neuroscience using functional magnetic resonance imaging (fMRI) showed that the area of the Amygdala in the brain is central in controlling emotion (Cardinal et al. 2002; Dolcos et al. 2006) and could work together with other brain areas involved in learning, such as the hippocampus, for processing memories. Another related research, also based on psychological experiments monitored by fMRI, showed that it seems that the same areas in the brain are responsible for imagery and remembering in the so-called subjective time involved in the mental experience of remembering the personal past or imagining the personal future (Nyberg et al. 2010). Indeed it seems that the activities of imagining a walk in a familiar environment in the past or future, or remembering an actual walk, trigger similar areas located in the left frontal cortex, cerebellum, and thalamus, and these are different from those activated when imagining it in the present.

Of course, the mechanisms of memory are much more complex and are gradually becoming more understood through research in various fields (psychology, neurology, cognitive sciences, and so forth) showing in particular the dynamic character of memory. However, the few remarks above might shed a new light on some old tricks used for mnemonics (or artificial memory) that have been in use since Antiquity, such as associating fearful or strange images to facts or sentences that someone wants to remember. As strange as this prescientific domain could be, the scholarly tradition of the Art of Memory extends back to Antiquity, with Cicero himself as founding father, and persisted until the Baroque Era. For basic references on this topic see Yates (2001) and Rossi (2000), while a slightly different perspective is addressed by Carruthers (1990), which contains interesting translations of some medieval treatises on artificial memory. More translations of original Latin text can be found in Carruthers and Ziolkowski (2002). This discipline was concerned with mnemonics and the ability to memorize anything at will, at a time when paper and other writing support were rare, and, even if the effectiveness of those methods cannot be assessed, this nevertheless appears to be an interesting curiosity in the Western history of ideas. Moreover, this tradition could be linked to more contemporary topics in the digital era. The current success of the computer as a universal information-processing machine, and the hope of DNA for universal information-archiving storage, essentially lie in the fact that there exists a universal language in which many different kinds of information could be encoded and that this language could be fully mechanized. Computers are artifacts aimed at storing and manipulating information encoded in various ways, information being basically anything that can be algorithmically generated. Indeed, this idea of performing, within a given symbol system, operations on symbolic objects in a purely syntactical manner, without reference to their usual meaning or modus operandi (for instance because their meaning is unknown) was proposed by Leibniz in the seventeenth century as an important source for scientific discovery, and he named it the *cogitatio coeca*, that is, "blind thought." Moreover, it is worth noting that Leibniz is also the inventor of the binary notation at the core of modern computers and general coding theory (including DNA).

12.2 Memory in the Flesh

The German mathematician and philosopher Leibniz (1646–1716) is well known both for his fundamental work in mathematics (for instance the discovery of the infinitesimal calculus) and in philosophy (e.g., his *Metaphysics* and *The Monadology*). Leibniz historically belongs to the "classical age" but was influenced by preclassical lines of thought coming from Raymond Lull and Giordano Bruno (Yates 1982). He was, however, also ahead of his epoch and somehow also belongs to our modern times, as he is considered as the founder, on the one hand, of modern symbolic logic (Siatzhy 1967) and, on the other hand, of modern semiotics (Dascal 1978).

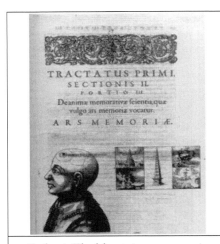
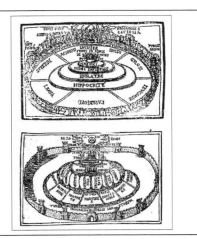

| Robert Fludd, *utriusque cosmi* (…) *historia,* Londres, 1619 | Cosmas Rossellius, *Thesaurus Artificiosae Memoriae,* Venice, 1579 |

Fig. 12.1 Examples of treaties on "The Art of Memory"

Leibniz considered that erudition or "perfect knowledge of the principle of all sciences and the art of applying them" could be divided into three equally important parts: the art of reasoning (logic), the art of inventing (combinatorics), and the art of memory (mnemonics). He even wrote an unpublished manuscript on the *ars memoriae*. The main idea of the *ars memorativa* (i.e., mnemonics or "artificial memory") is to organize one's memory in "places" (*loci*) grouped into an imaginary architecture, for instance the rooms of a house or the buildings of a city. This basic architecture must be well known and familiar, in order to let oneself wander easily within it. Then, to remember particular sequences of things, one will populate these rooms with "images" (*imagines*) that should refer directly or indirectly to what has to be remembered. The main assumption here, roots of which could obviously be traced back to Plato (Couliano 1987), is that (visual) images are easier to remember than words. With its emphasis on the power of images, this tradition naturally leads to the notion of a perfect language based on images instead of words, as images "speak more directly to the soul." These endeavors to organize memory as a pictorial "Theatre of the World" are best exemplified in the systems of Giordano Bruno (*De umbris idearum*, Paris, 1582) (see, e.g., Yates 1964), in the utopia of Tommaso Campanella (*The City of the Sun*, Frankfurt, 1623), in the *Memory Theatre* of Giulio Camillo Delminio (*Idea del Teatro*, Venice, 1550), or in the attempt of Paolo Lomazzo to create a *Temple of Painting* (*Idea del Tempio della pittura*, Milano, 1590) (see also Yates 2001, Chap. 7) (Fig. 12.1).

However, a basic memory architecture to be used as a personal archive is obviously one's own body. Several treatises indeed make reference to this, for instance the image of "the body as a series of memory places" in Filippo Gesualdo's

12 The Logic of Memory and the Memory of Logic: Relation with Emotion

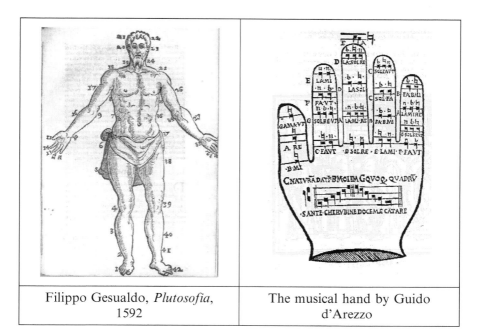

| Filippo Gesualdo, *Plutosofia*, 1592 | The musical hand by Guido d'Arezzo |

Fig. 12.2 The body as memory

Plutosofia (Padova, 1592) shows a body deconstructed in 42 *loci* to be used as a mnemonic model, or "the hand as bodily mnemonics" in Stephan Fridolin's *Schatzbehalter der wahren Reichtümer des Heils* (Nuremberg, 1491), which deconstructs the hand as a "treasure chest" (a traditional image of memory architecture). These illustrations can be found in the exhibition catalogue (Sherman 2000). The hand was also used in medieval times to remember melodies and songs, as music was most easily remembered if notes could be associated with positions on the hand. Guido d'Arezzo, who is credited with developing modern musical notation, also devised a hand-based notation for music, using the different joints of the five digits. This method was called the Guidonian hand (Fig. 12.2). Many descriptions and depictions of this system can be found in manuscripts and books from the twelfth century to the seventeenth century, evidence of its popularity at this time (Weiss 2000).

Not only music but also words could be remembered with the use the hand: a manual alphabet using different positions of the hand and the fingers was designed by Cosimo Rosselli in his *thesaurus artificiosae memoriae* (Venice, 1579), and another one using each phalange and designated areas on the palm was later designed by George Dalgarno (Sherman 2000), better known for his *ars signorum* (1661), one of the first attempts to design a universal language in the seventeenth century.

However, using a fixed memory system such as the body or some hand-based notation is not flexible enough for the ambitious purposes of the Art of Memory. The medieval and Renaissance treatises on the *Ars Memorandi* thus allowed each

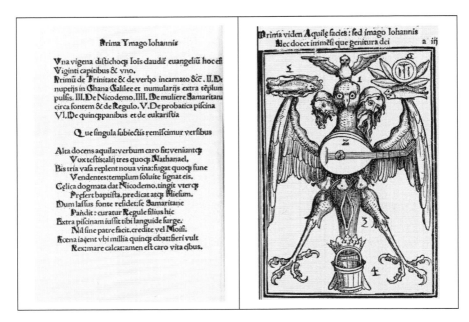

Fig. 12.3 Peter von Rosenheim, *Ars Memorandi*, Pforzheim, 1502

person devise his or her own customized system by using carefully designed images as memory aids. In order to be better remembered, images should be striking and incongruous and the use of animals, especially fantastic ones, is of prime choice for this purpose. Thomas Bradwardine, in his *De memoria artificiali* (circa 1335), depicts for instance how to compose a colorful image in order to remember the 12 zodiacal signs: an incredible bloody melee of a ram fighting a bull, a woman giving birth to two twins playing with a horrible crab, and so forth. An English translation of this text can be found in Carruthers (1990), Appendix B. The more fantastic, violent, or even erotic will be the image, the better it will be remembered. Another way of designing such striking images is the creation of chimerical beasts, the different parts of which refer to different objects or things to be memorized. Several wonderful examples can be seen in *Ars Memorandi*, a book printed in 1502 in Pforzheim, consisting of a Latin text by Peter von Rosenheim (*Roseum memoriale*, an aid for the study of the bible written in 1420–1430) and amazing woodcuts representing memory images (Fig. 12.3). The Houghton Library of Harvard University published a modern reprint of this book in 1981. For instance, the "first image of John" depicts a three-headed bird (to recall the Trinity) playing a mandolin (the Wedding at Cana) with various other smaller attributes referring precisely to the gospels of John, as each image contains several labels also appearing in the text on the opposite page.

In a similar manner, the use of Bestiaries as basic sources for memory aids seems to have been frequent in medieval times (Carruthers 1990, Chap. 4), as those texts were widely disseminated, even if most of the time they were in purely textual forms without images. They nevertheless contain the description of habits and physical features of many animals and monsters ready to be used to compose imaginary images for artificial memory purposes.

Thus images of animals have been used an aid for memorizing knowledge since the *Ars Memorandi* but could we, in our modern times where every idea calls for its reification, try to use real animals to help us for archiving purposes? Could we design some specially engineered beasts and use them as some sort of living archive?

12.3 Memory in the Genes

Among the various celebrations for the new millennium, the New York Times and the Museum of Natural History in New York proposed in 1999 the construction of a metaphorical Time Capsule (*the New York Times Capsule*) aimed at archiving a snapshot of the state of the world at the end of the second millennium. Such storage should be kept undisclosed until the year 3000, the end of the third millennium, when our descendants would open this preserved and mummified cross section of life. The proposal of the Spanish architect Santiago Calatrava won the favors of the jury with a stainless-steel capsule based on curvilinear abstract forms that contains objects from everyday life: mobile phone, dog food, condoms, photographs and noises of New York City, and some key editions of the *New York Times* (who was obviously sponsoring the project).

But let us rather detail a competing proposal for the New York Times Capsule project (NY Times 1999), which is conceptually more interesting but also more frightening: the proposal by Jaron Lanier (Lanier 1999) of a bioengineered living archive, which was in fact rejected. Jaron Lanier is a mythical hero of Silicon Valley, a pioneer in the fields of Virtual Reality and Tele-Presence, but his design was more concerned with biotechnology than advanced computer science. It consisted in digitizing all issues of the New York Times from 1999 onwards and inserting them by genetic engineering into the DNA of New York cockroaches, coding thus the binary computer encoding with the ternary DNA code using the four bases: A, T, G, C (adenine, thymine, guanine, and cytosine). The interesting point here is perhaps not so much the system in itself but the fact that it can be seen in the tradition of the quest for a universal language, and that it perfectly exemplifies the links between computer science and biotechnology that will be developed later.

To quote Jaron Lanier, who worked on this project together with a biologist from Columbia University in New York: "Once an archive is selected, it will be written into a computer file and coded into DNA base pairs. The sequences will then be synthesized by conventional protocols. Then the archival DNA will be ligated into cockroach intron DNA via injection into eggs." There are large portions of the DNA, called introns, that seem to serve no known purpose, but they are nevertheless copied from generation to generation. This could be a compact form of storage indeed, self-replicating and self-repairing. Why not use this DNA substrate as a giant Babel Library, in perpetual duplication and regeneration? Surprisingly, but perhaps not so much in fact, this project is technically feasible with today's biotechnologies, and can even be budgeted: bio-tech companies charge about half a dollar per base pair for creating DNA sequences. One letter can be coded in ASCII with eight bits within a computer or four base pairs within DNA, and thus will cost $2. A page will be about $2,000. How much knowledge could we feed the bug with? As Jaron Lanier said: "A single cockroach's introns will easily be able to contain the articles, letters, and other primary texts of one full year's editions of the *Times Magazine*. ...The cockroach easily has over a billion base pairs in its introns, which will have a capacity to represent over 250 million letters. That is far in excess of what is needed for the archive, even with the requirement of redundancy." Recall that genes only use 3% of the whole DNA, leaving 97% for supposedly garbage introns. Such archiving might be seen as unstable because it will be written over and over in the DNA of each generation. There is therefore a potential risk of mutation that will make it unreadable, but it is quite low indeed because "cockroach's genes are extremely stable and have not changed substantially for millions of years," and also because it is easy to have multiple copies that will be compared to give back the original text. Medieval Libraries did not work differently in some sense, with huge Scripture rooms and copyists who were constantly rewriting ancient text to have them better preserved. Cockroaches are very resistant creatures; they have survived all natural cataclysms and could also most probably survive potential human cataclysms like a nuclear war. These transgenic archival cockroaches, multiplied within a few decades or centuries in millions of (nearly) perfect copies, would be the quasi-indestructible receptacles of the memory of humanity. "Within approximately 14 years, the archival roaches will inexorably become so endemic as to become an ubiquitous and permanent feature of the Manhattan Island," said Lanier.

Thus, can we see cockroaches as basic blocks of a virtually infinite, distributed and self-replicated archive? What a librarian dream! A sphere whose center is everywhere and circumference nowhere, to paraphrase Nicolas of Cues, who was considering in his book *De docta ignorantia* in 1440 an infinite universe, expanding thus tremendously the closed Aristotelian universe. The learned ignorance of those archival cockroaches, more knowledgable than the most erudite scholar but unable to decode and access this immense knowledge, brings up a bitter paradox. But could these memory machines, unaware of the information data they contain, marked with the seal of genetic manipulations, with their DNA cut and pasted by wizard doctors up in the NASDAQ, be the future of our memory? Bugs of knowledge, fed with the dustbin of our commodities and the garbage of our spectacle, mass-storage in search of an improbable interface, they indeed reify in themselves, in their own flesh and blood, the state of our society at the turn of the last millennium.

12.4 Binary Notation

But another question is worth asking. Do we have now, with the genetic code, the ultimate universal language? This tradition of a universal or perfect language, so vivid in the utopian seventeenth century, indeed goes back to medieval times (Eco 1995; Pellerey 1992; Knowlson 1975). The current success of the computer as a universal information-processing machine, and Lanier's hope of the DNA for universal information-archiving storage, essentially lies in the fact that there exists a universal language in which many different kinds of information could be encoded and that this language could be fully mechanized. Computers are artifacts aimed at storing and manipulating information encoded in various ways, information being basically anything that could be algorithmically generated. Information theory, as proposed by Claude Shannon more than half a century ago (Shannon 1948; Shannon and Weaver 1949), can be thought of as some sort of simplified or idealized semiotics: a cyphering/decyphering algorithm represents the interpretation process used to decode some signifier (encoded information) into some computable signified (meaningful information) to be fed to a subsequent processing step. But the mechanization of this process by the computer is much eased by the use of a universal language in which all information is encoded: the binary notation, due to G.W. Leibniz. He published his "discovery" in 1703 (Leibniz 1703), and this started a growing interest in nondecimal numerical systems, but Leibniz's invention can be traced back to 1697, in a letter to the Duke of Brunswick detailing the design of a medallion on which the binary notation was praised as *imago creationis*, that is, similar the creation of the world by God because all numbers can be generated from Nothing by the One (*unus ex nihilo omnia*), cf. the front page of this chapter. He delayed its publication until finding an interesting application, and the one he chose was the explanation of the Fu-Hi figures, the hexagrams of the *I-Ching* or *Book of changes* from ancient China, communicated to him in 1700 by Father Bouvet, a Jesuit missionary in China. Two and a half centuries later, binary notation found another application with a much broader impact: digital computers. Although the first electronic computer, the Electronic Numerical Integrator and Calculator (ENIAC) created in 1946, made use of a hybrid of decimal and binary notations, subsequent machines from the early 1950s used binary notation only. This would indeed concretize the well-known dream of Leibniz of a universal language that would be both a *lingua characteristica*, allowing the "perfect" description of knowledge by exhibiting the "real characters" of concepts and things, and a *calculus ratiocinator*, making it possible for the mechanization of reasoning.

12.5 From Computing to Biology

But at the same time, another revolution was on its way: 1953 was the date of the discovery of the double-helix structure of the DNA by James Watson and Francis Crick. As shown in Kay (2000), genetics is the daughter of Computer Science and Cybernetics. The scientific concepts put forward in the late 1940s with the development of information theory by Claude Shannon and of cybernetics by Norbert Wiener, at the core of the development of computers, were fundamental in changing the concepts of molecular biology and making it enter the age of information. DNA was thus seen as a "word," a string of symbols taken from an "alphabet" consisting of the four bases A, T, G, C; and the problem of understanding how the DNA can produce the 20 amino acids at the basis of life was considered as a deciphering problem: breaking the genetic code. Symposiums on "Information Theory in Biology" were held, information-theoretic tools were used for a few years by many researchers, but the code resisted until 1961 when Marshall Nirenberg (later Nobel laureate, as Crick and Watson) showed that the base sequence TTT was coding phenylalanine (one of the 20 amino acids). It would take several more years until the genetic code will reveal all its secrets and thus open the way for the development of biotechnologies, as we know them now.

It is interesting to detail this coding of the amino acids in the DNA by three-letter words calls codons, as this provides a good example of the use of some information-theory concepts introduced by Shannon and helps us to assess its impact on molecular biology (Adami 1998). Let us consider the set of all amino acids as a 20-letter alphabet, to be coded by words (i.e., sequences of letters) from a 4-letter alphabet (A, T, G, C). To code 20 different elements, using 2-letter words is not enough (as they can code up to only 16 different elements) and 3-letter words have to be used. But these triplets can code up to 64 different elements, which means that this code will be redundant and thus several words will denote the same element of the encoded initial alphabet. Shannon introduced the notion of entropy to measure the degree of uncertainty of some information, computed from the probability distribution of the values the information can take. If we consider that our 20 amino acids occur with the same probability, the entropy of our "information source" (the set of 20 elements representing the amino acids) is given by $H = \log_2(20) = 4.322$. However, all amino acids are not equally distributed in Nature, and their respective abundance is not 5% but varies from 1% to 10%. Computing again our entropy H with this probability distribution would lead to a value of 4.21, not far from the approximated one. Now, Shannon's noiseless coding theorem states that the length of an optimal code for this set using a 4-letter alphabet (A, T, G, C) is between $H/\log_2(4)$ and $H/\log_2(4) + 1$, i.e., between 2.10 and 3.10. Thus with 3-letter codons, we have an optimal code. Another important concept of information theory is than of error than could occur when information is transmitted from a source to a destination through a (possibly noisy) channel. Encoding and decoding should help in error recovery whenever possible, and redundant codes are thus important because the same element can appear in multiple ways. For instance, in our DNA code the arginine amino acid (ARG) is coded by six DNA codons: CGA, CGC, CGG, CGT, AGA, AGG. Thus any triplet of the form CGX, where X stands for an unknown value (for instance because of an error) indeed represents Arg. If we look at the details of this DNA, each of the 20 amino acids is coded by one, two, three, four, or six 3-letter codons (for a total of 64 codons), and the most abundant amino acids usually have the most redundancy.

The fact that these researchers were convinced of discovering the universal language of life is obvious in their own words. The French biologist Jacques Monod, Nobel laureate in 1965 and champion of the cybernetic approach in molecular biology (he proposed for instance the models of "Cybernétique Enzymatique" and "gène informateur") said in his inaugural address at the Collège de France in 1965: "The surprise is that the genetic specificity is written not with ideograms as in

Chinese but with an alphabet as in French, or rather as in the Morse code" [«*La surprise, c'est que la spécificité génétique soit écrite, non avec des idéogrammes comme en chinois, mais avec un alphabet comme en français, ou plutôt en Morse*»]. This citation is reproduced in Kay (2000). Indeed several biologists in the 1960s considered the relationship between the Chinese characters of the *I-Ching* (64 hexagrams, each composed of three diagrams, i.e., three couples of binary signs) and the coding of the amino acids with three bases in the genetic code (the four bases being coded by two binary numbers) (Kay 2000). We are back to Leibniz again, who used binary notation for explaining the *I-Ching* in his seminal paper.

But reading the DNA and deciphering it was not enough. David Jackson, a champion of biotechnology said: "To be fluent in a language, one needs to be able to *read*, to *write*, to *copy*, and to *edit* in that language. The functional equivalents of each of those aspects of fluency have now been embodied in technologies to deal with the language of DNA" (Jackson 1995).

Thus he was not only looking for a universal language, he also wanted a universal word processor!

12.6 Conclusion

The computer may seem today, especially with the World Wide Web, the perfect archiving machine, and the hegemony of digital coding (binary notation) in digitalizing the world (texts, books, images, music, video, and so forth) does not seem to reach any limit. Data-centers are created everyday with *Peta-* or *Exa-*bytes of (more or less useful) information, but more are always needed as the size of digital information created every day is growing in an exponential manner. This gigantic and distributed Archive can be seen as the reification of old ideas, now buried in some obscure dead-end of the Western history of ideas discussed in this chapter such as the Art of Memory, Memory Theatres, and the quest for a universal and/or perfect language. But just after the invention of the digital computer, the advances in molecular biology in the second half of the twentieth century, and the discovery of the DNA code and its manipulation, made life sciences and biotechnologies the logical next step after cybernetics and the digitalization of the world for archiving purposes. The computer and its data storage we possess today might soon be replaced by more compact substrates for storage and even for computation, as wetware will replace hardware. Live organic storage might quickly become more handy and useful than digital archives and the plastic support of CDs or DVDs, or "memory sticks." This novel and somewhat frightening vision is illustrated by a utopian and (thankfully) yet unrealized project that was proposed some years ago by Virtual Reality pioneer Jaron Lanier in order to create a new type of archive for *the New York Times* newspaper.

Computer, Memory, Images, Emotion, Logic, Binary Notation, Universal Language, DNA, Living Organism, Animals: these are the keywords of our study, and the links between them are sometimes deeper than we think.

References

Adami C (1998) Introduction to artificial life. Springer, Berlin
Cardinal RN, Parkinson JA, Hall J, Everitt BJ (2002) Emotion and motivation: the role of the amygdala, ventral striatum, and prefrontal cortex. Neurosci Biobehav Rev 26:321–352
Carruthers M (1990) The book of memory. Cambridge University Press, Cambridge
Carruthers M, Ziolkowski JM (2002) The medieval craft of memory: an anthology of texts and pictures. University of Pennsylvania Press, Philadelphia
Couliano JP (1987) Eros and magic in the renaissance. University of Chicago Press, Chicago
Dascal M (1978) La sémiologie de Leibniz. Aubier-Montaigne, Paris
Dolcos F, LaBar KS, Cabeza R (2006) The memory-enhancing effect of emotion: functional neuroimaging evidence. In: Uttl B, Ohta N, Siegenthaler A (eds) Memory and emotion: interdisciplinary perspectives. Wiley-Blackwell, Hoboken, pp 107–134
Eco U (1995) The search for the perfect language (making of Europe). Wiley-Blackwell, Hoboken
Jackson DA (1995) DNA: template for an economic revolution. In: Chambers D (ed) DNA: the double helix: perspective and prospective at forty years. New York Academy of Science Press, New York, pp 356–365
Kay LE (2000) Who wrote the book of life? A history of the genetic code. Stanford University Press, Palo Alto
Knowlson J (1975) Universal language schemes in England and France 1600–1800. Toronto University Press, Toronto
LaBar KS, Cabeza R (2006) Cognitive neuroscience of emotional memory. Nat Rev Neurosci 7:54–64
Lanier J (1999) Interview in the New York Times. http://www.nytimes.com/library/magazine/millennium/m6/design-lanier.html. Accessed 18 May 2011
Ledoux J-E (1996) The emotional brain. Simon and Schuster, New York
Ledoux J-E (2000) Emotion circuits in the brain. Annu Rev Neurosci 23:155–184
Leibniz GW (1703) Explication de l'arithmétique binaire. Mémoires de l'Académie Royale des Sciences, Paris
Nyberg L, Kim ASN, Habib R, Levine B, Tulving E (2010) Consciousness of subjective time in the brain. Proc Natl Acad Sci U S A 107:22356–22359
NY Times (1999) New York Times, The Times Capsule. http://www.nytimes.com/library/magazine/millennium/m6/index.html. Accessed 18 May 2011
Pellerey R (1992) le lingue perfette nel secolo dell'utopia. Laterza, Roma
Rossi P (2000) Logic and the art of memory: the quest for a universal language. University of Chicago Press, Chicago (original edition in Italian: Milano, 1960)
Shannon C (1948) A mathematical theory of communication. Bell Syst Tech J 27:379–423, 623–656
Shannon CE, Weaver W (1949) The mathematical theory of communication. University of Illinois Press, Champaign http://cm.bell-labs.com/cm/ms/what/shannonday/paper.html. Accessed 18 May 2011
Sherman CR (2000) Writing on hands, memory and knowledge in early modern Europe. University of Washington Press, Seattle
Siatzhy N (1967) A history of modern logic from Leibniz to Peano. MIT Press, Cambridge
Weiss SF (2000) The singing hand. In: Sherman CR (ed) Writing on hands. Memory and knowledge in early modern Europe. University of Washington Press, Seattle, pp 35–45
Yates F (1964) Giordano Bruno and the hermetic tradition. University of Chicago Press, Chicago
Yates F (1982) Lull and Bruno: collected essays. Routledge and Kegan Paul, London
Yates F (2001) The art of memory. University of Chicago Press, Chicago (first edition: London, 1966)

Chapter 13
Conclusions: Emotions (and Feelings) Everywhere

Stan A. Kuczaj II

Those of you who have read each of the chapters in this book will undoubtedly have noticed that the study of emotions is far from straightforward. Part of the problem is definitional in nature. The terms "affect", "emotion", and "feeling" are defined in different ways by different authors, a persistent problem in the study of emotion throughout history (see the chapter by Kuczaj, Highfill, Makecha and Byerly for a brief consideration of definitional differences and similarities in the history of the study of emotions). Professor Watanabe and I elected not to force an arbitrary set of definitions on the authors of the chapters in this book, but instead allowed the authors to follow the tradition of providing definitions for their use of terms such as "feeling" and "emotion". We hope that this approach has illustrated the difficulty of deciding on set definitions for terms such as "affect", "emotion" and "feeling". The study of emotions is sufficiently complex to weather the historical and contemporary semantic differences, but common usage and definitions across theorists and empirical investigators would facilitate comparisons of theoretical approaches and empirical findings. This will undoubtedly happen, the result being a clearer scientific understanding of the extent to which terms such as affect, emotion, and feeling are synonymous or different components of emotional phenomena.

In addition to the myriad definitions, there exist both numerous approaches and numerous topics in the literature on emotions. One contemporary approach focuses on the relationship between emotional experiences and biochemical changes in the central nervous system, an approach that is well represented in Shinozuka's chapter on possible roles of neural peptides in aggression, Steklis and Lane's chapter on the

---

S.A. Kuczaj II (✉)
Department of Psychology, The University of Southern Mississippi,
118 College Dr. #5025, Hattiesburg, MS 39406-5025, USA
e-mail: s.kuczaj@usm.edu

possible uniqueness of human *awareness* of emotions, and Umeda's chapter on the use of neuroimaging techniques to better understand emotional processing.

Another approach that has a long history in the study of emotions is the comparison of animal emotions and human emotions (Darwin 1872). A number of the chapters in this book carry on this tradition. Pepperberg's chapter on the role of emotion in the cognitive processing of African grey parrots illustrates how careful observation of animals can provide insights into the ways in which emotions and cognition interact. The chapter by Kuczaj and Horback examines the relationship between emotions and play in humans and animals, including the possible roles of play in the ontogeny of emotional processing skills and the ontogeny of flexible problem solving skills. The possibility of dolphin emotions and the related possibility that dolphins are sensitive to the emotional states of other dolphins (and perhaps some other species) are explored in the chapter by Kuczaj, Highfill, Makecha and Byerly. The chapter by Watanabe concerns the issue of animal aesthetics. After noting that some songbirds prefer some forms of music over others, that pigeons can discriminate "good" and "bad" paintings, and that zebra fish prefer some forms of painting over others, Watanabe suggests that at least some animals can discriminate and "enjoy" particular aesthetic stimuli.

In his chapter, Kotrschal emphasizes that emotions are essential components of social organization in humans and animals, noting that effective communication and social cognition depend on the ability to read others' emotional states, communicate one's own emotional state, and satisfy the emotional needs of both oneself and others. If emotions are involved in "virtually any decision made about others", as Kotrschal suggests, then understanding the role of emotions in social interactions is essential for correct interpretations of the social behavior that occurs both within and across species.

Steklis and Lane argue that only humans are *consciously aware* of their own and others' emotions, a capacity that they suggest allows humans to both reflect on their own (and others') emotions and to communicate these reflections with others. In their view, even though humans and other mammals may experience emotions that are based on similar visceromotor and somatomotor foundations, only humans can consider what these emotions might mean and ponder the significance of such interpretations. Steklis and Lane suggest that the combination of the human medial prefrontal cortex and particular types of developmental experiences result in the hypothesized unique human capacity to reflect on and communicate emotions. If it is the case that advantageous experiences during development are crucial for the emergence of emotional reflection, then it may be possible for other species to acquire some form of similar reflective skills if they are provided with the requisite experiences. Enculturated chimpanzees acquire symbolic reasoning capabilities that their non-enculturated counterparts lack (Kuczaj and Hendry 2003; Premack 1983; Premack and Woodruff 1978), and so providing species such as chimpanzees, dolphins and African grey parrots with empathetic companions that communicate their emotional experiences might result in some capacity for emotional reflection on the animals' part. Of course, the extent to which the human medial prefrontal cortex is essential for such reasoning will determine the extent to which such enculturation attempts are successful.

Another chapter in this volume takes the comparative approach to a new realm by examining the interactions between children and robots. Itakura, Moriguchi and Morita introduce the field of developmental cybernetics, a field they believe will become increasingly important as robots become more commonplace in households. But one of the biggest stumbling blocks in the way of popular acceptance of household robots may concern emotions. The "uncanny valley" phenomenon refers to the point at which robots (or androids) cease to be cute and non-threatening and instead are perceived as "creepy". The uncanny valley experience occurs when a robot becomes very human-like in appearance but lacks some quality (or qualities) that would allow it to pass as "fully" human. Although it is not clear exactly what perceptual factors contribute to the uncanny valley experience, the human-like robots' emotional expressions seem likely to part of this package. Robotics has not progressed to the point that a human-looking android can express emotions in a human-like way. In addition, the ability to perceive others' emotions accurately and respond appropriately seems far beyond any current robot's capabilities. Perhaps developmental cybernetics will provide some of the answers that enable robots to appear human but not frightening or disturbing. However, our evolutionary history has resulted in an impressive ability to recognize emotional states that are "off" in some manner. We can recognize inappropriate or odd emotional responses in other humans, and it will be interesting to see if robots will ever be indistinguishable from humans, even if the uncanny valley phenomenon is overcome.

Finally, Codognet's chapter provides a review of the history of mnemonic systems, including the relationship of such systems to emotional memory. Of course, anyone who has experienced a hard drive crash and belatedly realized that everything on the hard drive was not backed up has first hand knowledge of the emotions involved in the failure of mnemonic systems. Codognet's consideration of the proposal to use cockroach DNA as a means to archive and preserve data might disturb those of us who shudder at the thought of these ubiquitous bugs, which illustrates that emotions influence our perception and acceptance of mnemonic systems, and that this can occur in all phases of the evolution of mnemonic systems, not just when they fail.

The wide range of topics covered by the authors of the chapters in this book illustrates the extent to which emotions are part and parcel of life on Earth, and not just human life. Individuals from a variety of species experience emotions, and one of the challenges we face is to determine the emotional range of individual species. What emotions do individuals from different species experience? How do they react to different emotional experiences? How adept are they at recognizing emotions in others? What do they do when they perceive particular emotional states in others? Is their ability to recognize emotions limited to members of their own species? Could animals recognize imperfect models of themselves and so experience the uncanny valley? Clearly, the study of emotions has a long way to go, and it promises to be an exciting journey.

References

Darwin C (1872) The expression of emotions in man and animals. Murray, London
Kuczaj SA II, Hendry JL (2003) Does language help animals think? In: Gentner D, Goldin-Meadow S (eds) Language in mind: advances in the study of language and thought. MIT Press, Cambridge, pp 237–275
Premack D (1983) Animal cognition. Annu Rev Psychol 34:351–362
Premack D, Woodruff G (1978) Does the chimpanzee have a theory of mind? Behav Brain Sci 4:515–526

Index

A
Advanced emotion, 223–230
Affects, 3, 4, 7, 12
Agents, 208
Aggression, 23
Animal emotions, 278
Animal personality
  coding, 72
  rating, 72, 73
Anthropomorphic, 71
Apes, 113–126
Arginine vasopressin (AVP), 27
Arginine vasotocin (AVT), 28
AVP. *See* Arginine vasopressin (AVP)
AVT. *See* Arginine vasotocin (AVT)
Awareness, 125

B
Beauty, 130–132, 144, 146–147, 151, 156
Binary notation, 266, 267, 272–273, 275
Bonobo, 116
Brain-imaging, 68, 69

C
Category, 132, 137, 139, 146, 147
Chimpanzee, 116
Cichlids, 24
Clade, 125
Co-construction, 125
Code, 271, 272, 274–275
Coding, 123
Cognitive processing, 50
Comparative, 279
Comparative cognition, 129–157
Comprehension, 121
Computers, 266, 267, 271–273, 275
Concept, 130–132, 142, 145–147, 149
Conscious, 278
Consciousness, 245–264
Consonance, 134, 139, 140, 142
Continuities, 114
Cultural scaffolding, 115

D
Darwin, C., 64, 66
Darwinian aesthetics, 131, 132
Declarative, 123
Developmental cybernetics, 208
Discriminative property, 129, 132, 145–146
Dissonance, 134, 139, 140

E
Emotional awareness, 165–198
Emotional phenotype, 3, 4
Emotional reflection, 278
Emotional responses, 50
Emotions, 87–106, 207–220
  behavioral measures (*see also* expressions of dolphin emotion)
    facial expressions, 68–70
    postures, 69, 70
    vocalizations, 69, 70
  definition, 64, 65, 72, 80
  expression, 63–80
  lateralization of, 70, 71
  physiological measures
    brain-imaging, 68, 69
    heart-rate, 68, 69

Emotion symbols, 113–126
Episodic memory, 245–264
Evolution, 245–264
Experimental aesthetics, 130–133
Expressions of dolphin emotion
　facial, 69
　postural, 76
　S-posture, 75, 76
　tactile, 76
　vocalizations, 74, 75

F
False-belief task, 208
Functional autonomy, 133, 152, 154–157
Functional neuroimaging, 223–225

G
Generalization, 138–142, 145–149
Gesture, 118
Global, 147, 148
Goal-directed actions, 211
Grey parrots, 49
Grief
　in dolphins, 71, 77
　in elephants, 77

H
Happy, 114
Hurt, 114

I
Internal state, 114
Isotocin, 28

K
Keyboard, 117

L
Language, 266–269, 271–275
Lateralization of emotion, 70
Lexigrams, 117
Local, 135, 147–150

M
Mad, 114
Medial prefrontal cortex, 166, 172–174, 196, 224–226, 228, 230–232, 235, 237, 238
Memory, 265–275
Mentalizing, 207–220
Mesotocin, 28
Motor skill, 129, 131, 132, 151, 152, 154–157
Music, 130, 132–144, 151, 155, 157

N
Negotiation, 125

O
Oxytocin, 27

P
Personality, 223–238
Phylogenetic contingency, 151
Play, 87–106
Play signals, 95, 98–102
Positive affect, 96, 105, 106
Positive emotion, 88, 93, 95, 96, 103, 106
Primary emotions, 115
Primates, 114, 167, 174, 175, 178–183, 185, 190–193, 195, 196

R
Reinforcing property, 129, 130, 132–137, 143, 144, 155–157
Representation, 152–154
Robot, 279

S
Scared, 114
Secondary emotion, 115
Self-recognition, 218
Self-reinforcement, 130, 133, 143, 152, 155–157
Sensory reinforcement, 130, 132, 133, 144, 157
Social interaction, 49
Social organization, 278
Social play, 88, 90, 91, 93, 95–100, 103, 105
Symbolic capacity, 115

T
Teleost, 23–42
Temperament, 3, 4, 13–14
Theory-of-mind, 174, 178, 209, 224, 225, 228, 230–232, 234–237
Tinbergian levels, 4

U
Uncanny valley, 279

V
Value based words, 114

Printed by Publishers' Graphics LLC
SO20121018.19.24.99